Alkaloid Biology and Metabolism in Plants

Alkaloid Biology and Metabolism in Plants

George R. Waller

Oklahoma State University
Stillwater, Oklahoma

and

Edmund K. Nowacki

Oklahoma State University
Stillwater, Oklahoma
and Institute of Plant, Soil, and Nutrition Science
Pulawy, Poland

PLENUM PRESS • NEW YORK AND LONDON

Library of Congress Cataloging in Publication Data

Waller, George R.
 Alkaloid biology and metabolism in plants.

 Includes bibliographical references and index.
 1. Alkaloids. 2. Botanical chemistry. I. Nowacki, Edmund, 1930- joint
author. II. Title.
QK898.A4W34   581.1'9242   76-30903
ISBN 0-306-30981-5

Published as Article No. P-299
of the Agricultural Experiment Station,
Oklahoma State University, Stillwater, Oklahoma U.S.A.

© 1978 Plenum Press, New York
A Division of Plenum Publishing Corporation
227 West 17th Street, New York, N.Y. 10011

Printed in the United States of America

To

Our Wives
Hilda
and
Danuta

Preface

This book is designed for the use of the advanced student and professional worker interested in the international scientific community, particularly those in the fields of agronomy, agricultural sciences, botany, biological sciences, natural products chemistry, pharmaceutical chemistry and biochemistry. The purpose is to inform the reader about significant advances in the biology and metabolism of alkaloids in plants. Since alkaloids are generally referred to as "secondary metabolites," the reactions discussed are not, for the most part, involved with the main metabolic pathways. The reactions that we are interested in are pathways that have been developed for the formation of these secondary metabolites, using as their starting molecules one of the compounds produced via a main or primary metabolic pathway. The primary metabolic pathways are common to all plants, indeed to most living organisms, whereas the highly specialized branches leading to alkaloid formation are found in only about 10 to 20% of the known plants. The reason for these diversities in plant metabolism is not clear; however, it seems likely that the formation of highly individualized and specialized pathways resulted as a response to the pressure of natural selection. Nevertheless, the genetic peculiarity that controls alkaloid production has provided many extremely interesting problems for scientists and constitutes convincing evidence of nature's superior ability in biochemistry.

This attempt to pull together in one book most pertinent information is a result of a desire by both authors to present an authoritative, up-to-date treatise. Frequent references are made to the outstanding series *The Alkaloids,* edited by R. H. F. Manske, *Biosynthese der Alkaloide* by K. Mothes and H. R. Schuette, *Secondary Metabolism in Plants and Animals* by M. Luckner, and *The Biochemistry of Alkaloids* by T. Robinson. The authors recognize the contributions of the series *The Alkaloids—Specialist Periodical Reports* edited by J. E. Saxton; of *Chemistry of the Alkaloids* by S. W. Pelletier; and those of the early very authoritative volumes

The Plant Alkaloids (4 volumes) by T. E. Henry, *Die Alkaloide* by Georg Trier, and *The Vegetable Alkaloids* by Amé Pictet (translated into English by H. C. Biddle). This list is not all-inclusive. In presenting this material for *Alkaloid Biology and Metabolism in Plants* we have adopted the view that biosynthesis (which is such a large field that it requires a separate book), catabolism, genetic, chemotaxonomic, environmental influences, sites of alkaloid formation, and the role of alkaloids in plant physiology must be considered. Many of the studies presented have been carried out on whole plants, with results understood poorly, i.e., in terms of genetic control, chemotaxonomic relationships, environmental influences, and sites of alkaloid formation. It would be preferable to rely on the results of isotopic labeling studies, such as have developed over the past 25 years, and some are reported in this book. The study of alkaloid metabolism has not reached a level of sophistication as high as that of some other areas of intermediary plant or animal metabolism.

Writing a book of this type presents many difficulties when the coauthors live more than five thousand miles apart. This book was originally drafted in 1968–69 in Stillwater, where we could work on it together. Recently we met and worked on some of the problems; the others have been settled by correspondence. The coverage of the literature has been completed through 1975. We would appreciate that any errors of commission or omission are called to our attention.

We are indebted to those who reviewed the individual chapters: Dr. Robert M. Ahring, Dr. Donald F. Banks, Dr. Eddie Basler, Dr. B. R. Clay, Dr. Margaret Essenberg, Dr. Robert K. Gholson, Dr. Wilfred E. McMurphy, Dr. Jay C. Murray, Dr. E. C. Nelson, Dr. Jimmy F. Stritzke, Dr. Glenn W. Todd, and Dr. Ronald J. Tyrl (all faculty members of Oklahoma State University), Dr. H. J. Floss (faculty member of Purdue University), Dr. W. W. Weeks (faculty member, North Carolina State University), Dr. Ian Forbes, Jr. (Research Agronomist, USDA, Tifton, Georgia) and also Dr. Otis C. Dermer, who read the entire book. We express our sincere appreciation to Mr. Jack I. Fryrear, who did the drawings in this book.

GEORGE R. WALLER
Stillwater, Oklahoma

EDMUND K. NOWACKI
Pulawy, Poland

Contents

Introduction

Twenty years ago the number of known alkaloids was about 800. At present it is estimated that in excess of 6000 compounds wih alkaloid-like properties have been described. Most of them are found in the Angiospermeae and only a few in the Gymnospermeae (including the *Taxus* and *Pinus* pseudoalkaloids). The Lycopodiales and Equisetales do contain some alkaloids, but the ferns and the mosses have none.

A comment on the definition of alkaloids is appropriate. The term "alkaloid" is applied to nitrogen-containing compounds, produced primarily in higher plants but also in lower organisms and in some animals, that have significant pharmacological activity. They are grouped together because of the presence of a basic nitrogen atom in their structures rather than by ring size or type. They represent one of the largest and most diverse families of natural compounds, and they contain some of the most complicated molecular structures.

Alkaloids were considered for a long time as specialized products solely of plant metabolism. Yet in recent times alkaloids have been isolated from both vertebrate and invertebrate animals. Some of the animal alkaloids can clearly be traced to a food plant ingested. As an example, the alkaloid castoramine, isolated from the beaver *(Castor canadense),* resembles the alkaloids of the water lilies, *Nuphar* spp., which serve as food for the beavers. Some caterpillars accumulate alkaloids from the plants on which they feed. Other alkaloids, however, such as the ones found in toads, salamanders, and some fishes, are true products of animal metabolism.

Alkaloids can be recognized as resulting from aberrations of metabolic pathways in both plants and animals. Most alkaloids have a disturbing effect on the animal nervous system; therefore, it is unlikely that a mutation (followed by hybridization) in metabolism causing the synthesis of an alkaloid can be established because it would be self-destructive (lethal). Only in instances where the mutation occurs in a tolerant (preadapted?)

system can the mutant survive. In many laboratories alkaloids are considered merely offshoots of normal metabolism, so-called aberrant forms. Then, what is the reason for a scientist to spend time and resources to investigate these metabolic curiosities—the alkaloids—particularly when our knowledge of the basic metabolism of higher plants is negligible? We shall address ourselves to answering this question. Over 90% of the biological compounds on the earth's surface are produced by higher plants, yet we only have a fragmentary knowledge of the metabolic pathways that operate in these plants. If our knowledge of higher plant basic metabolism is so limited, why bother with alkaloid research?

Alkaloids are the oldest drugs. Since the times of Hippocrates, extracts of plants have been known to serve as medicine for a number of diseases. Later on, many of the respective active principles of those plant extracts have been identified as alkaloids. Even with the dramatic progress of organic chemistry which has resulted in an enormous production of synthetic drugs, some of the most powerful remedies are still of plant origin. Whereas organic synthesis of complicated molecules is often extremely costly, *plants produce them with apparent ease and at little cost; so plants will remain important sources of some.* In the last 40 years, the attention of the pharmaceutical industry has been directed to the lower plants, predominantly to fungi, as sources of antibiotics. There are no reasons to believe that higher plants are entirely void of antibiotics since they do survive in a world infested with bacteria and fungi. The search for alkaloids with anticancer activity has revealed a number of compounds which can retard the development of tumors. A surprise for many scientists was the finding that a common ornamental plant, the periwinkle *(Vinca rosea)*, produces such alkaloidal compounds. Other plants such as *Tylophora crebiflora, Ochrosia elliptica, Acronychia baueri,* and *Camptotheca acuminata* have yielded still other substances of this type. Moreover, some steroid alkaloids are important to the pharmaceutical industry. While the price for some synthetic steroids is prohibitive, certain plants accumulate compounds which after a simple chemical modification become steroid hormones, just like those regulating metabolism in animal and human bodies. With millions of women throughout the world taking "the pill," the demand for steroid alkaloids is increasing. Our knowledge of the distribution of those alkaloids in the plant kingdom has increased considerably. The best sources are in tropical flora, but some countries in temperate zones want to limit import of raw materials. Thus a search for indigenous plants containing steroid alkaloids is going on continuously, and experiments leading to the highest production of these compounds in plants are being performed.

The *Thallophyta* were, for a long time, regarded as alkaloidless, but recently a number of compounds of alkaloid character have been isolated from various lower plants. These new alkaloids may represent new alkaloid families according to their structure, and it is hard at present to estimate their significance both for plant metabolism and for chemotaxonomy. The metabolism of lower plants is still mostly unknown; however, heterotrophic organisms such as the fungi have been helpful in some areas of biological research, e.g., *Neurospora, Aspergillus, Claviceps,* and *Saccharomyces.* The concept of "lower plants" is an artificial one since it refers to a conglomerate of taxonomic units which can have common ancestors as far back in time as the pre-Cambrian era. Extended research on the lower plants would be likely to yield more new compounds that show some alkaloid-like character. Indeed some such compounds have been isolated, and in some instances they (the peptide-like compounds) are of value for their bacteriostatic properties. A number of antibiotics have been shown to be alkaloidal in nature.

In the last 10–15 years the pathways leading to alkaloid biosynthesis have been partly elucidated by the work of numerous laboratories. In spite of the enormous number of alkaloids only a few pathways leading to alkaloid biosynthesis are known. Several pathways lead from essential amino acids to alkaloids, one from a vitamin, nicotinic acid, and/or its coenzyme form NAD, and one from isoprenoids and certain other small molecules that are active intermediates. Usually it is easier to isolate an alkaloid than an amino acid or a vitamin. The purity of the isolated compound is also much higher. Knowing from recent research that a certain alkaloid is a derivative of an amino acid, we can investigate the biosynthesis of this amino acid, using the alkaloid as a model compound. It is a less cumbersome approach, although less direct. Degradation procedures that were elaborated when the structures of the alkaloids were elucidated were helpful in demonstrating how the *precursors were transformed into the amino acid,* which in turn served as the *substrate for alkaloid biosynthesis.* This indirect study of amino acid metabolism is applicable only to the few known amino acids which are intermediates in alkaloid biosynthesis. These very important amino acids are lysine, tryptophan, tyrosine, phenylalanine, and aspartic acid. They are essential for humans and certain animals. We need more knowledge in the near future to be able to manipulate plant metabolism in such a way that the plants will produce more and higher quality proteins. The biosynthesis pathway for nicotinic acid, which proved to be entirely different in plants and in animals, was probably learned only with the help of alkaloid research; specifically, the route of biosynthesis of ricinine and nicotine provided proof of the pyridine nucleotide metabolic cycle in higher plants. It is extremely difficult to examine the biosynthesis of a compound such as nicotinic acid in plants where the concentration is less than 1 mg/g of dry weight. Yet, in several examples, an alkaloid can accumulate in concentrations up to 10% of the dry weight of the substance. An isolation procedure for a substance that occurs in low concentration is usually cumbersome and expensive; in contrast, the isolation of an alkaloid, once the proper procedure is found, is relatively simple and quick.

Interdependence of individuals is a fundamental characteristic of societies, be they plant, animal, microorganism, or human. Humans today depend more on intellect for survival and an understanding of food problems is essential; some food habits and attitudes toward eating may become quite strong. Some civilizations have died out while others have survived; some alkaloid-containing plants may have been responsible. It is well-known that certain animals have the capability to smell and/or taste the plant which has a high alkaloid content. Thus if given a choice they will avoid the high-alkaloid-containing plant and consume the low-alkaloid-containing plant; but if they have no choice then they will consume the high-alkaloid-containing plant, sometimes to the animals' own demises.

We are living in a world where most plants produce secondary metabolites that are not without significance for our health. Properly used, they can be medicines, but they can also be poisons. Because of progress in processing food, mass poisoning by alkaloids accidentally introduced is much less common today than it was in the past, e.g., with bread made from flour contaminated with the sclerotia of *Claviceps*. Yet new dangers arise. Increased application of synthetic fertilizers can upset the precarious balance of metabolism in some cultivated plants with the result that the plant can start to produce compounds of alkaloid character. Until recently there were only a few known examples, the victims being herbivorous domesticated animals. Edible varieties of cultivated plants bred under certain conditions proved to be poisonous when grown under other conditions. Our knowledge of the processes leading to changes in alkaloid metabolism is still limited, but some clues are already available. Some alkaloids which at first glance seem to be harmless have, after further investigation, been recognized as mutagenic or tumorigenic, e.g., lysergic acid diethylamide (LSD) and some pyrrolizidines. Minute amounts of alkaloids are taken daily by everybody in food. People may be fully concious of their intake in the case of beverages like coffee, tea, and cocoa (all of which contain caffeine), but they may ingest some unaware, since increased alkaloids may be present as a result of the changes in the environment of the plants.

With all the above-mentioned data in mind it is understandable that research on alkaloids should and will proceed. In some plants they have acquired the name "a metabolic curiosity," important to study both on its own merit and because of its contribution to understanding of basic nutrition processes. In this manner even alkaloids can be nutrition research tools. Alkaloids in medicine remain important sources of drugs both for physical illnesses and for increasing numbers of mental diseases. So, there is need for research on alkaloids produced in plants that serve as food and feed; for example, scientists breed a disease-resistant crop variety and then a new destructive strain of the disease develops and the scientists have to start over again. When the superior varieties bred in one country are grown in other countries under new environmental conditions, studies on alkaloids as well as on the nutritionally valuable components should be made.

As the world turns to alternative energy technologies the role of agricultural output will become more important. Of particular interest are those plants which will be able to provide the world with huge amounts of energy (via photosynthesis and nitrogen fixation, the two energy processes of plants and microorganisms) that humans will need during the next century to maintain and/or improve economic growth and the standard of living. We know very little about how the alkaloids and their production interface with the increasing emphasis on energy. *It will be vitally important that the agriculture, biology, medicine and biochemistry of the alkaloids be included in the study of all of the living plants during the next 100 years.*

The future of research in alkaloids will be inextricably tied in with the new scientific effort to boost the food output of the world. The biological and chemical techniques involved in advances in farming may result in producing new types of alkaloids. Scientists have recently succeeded in imparting new qualities to a whole tobacco plant which was generated from fused single cells of two tobacco species. Cells of more than a dozen plants have already been fused in laboratories in several countries. We can expect cell fusion and cell culture to become commercially useful after another several decades of experimental work. This is the most promising way to tap the unused biological potential of plants and animals. The effort is just beginning. Scientists by the hundreds from university, government, and agricultural-industrial laboratories the world over are conducting imaginative experiments to improve the mechanisms of plant growth so that the plants will yield more food. Other researchers are seeking to get more offspring from farm animals and to propagate animals of superior quality. *The production of "secondary metabolites" of these natural species may be both desirable and undesirable; thus the future of alkaloid research will be exceedingly exciting, difficult, and challenging.*

1 | Alkaloids in Chemotaxonomic Relationships

1.1. Introduction

The idea of utilizing results obtained by chemical analyses for the revision of plant systematics based on morphology is relatively new. James Petiver in the 17th century endeavored to prove that plants with similar morphological aspects "of the same make of class" possess similar forces, which he described as "the like virtue." Subsequently, considerable evidence has been accumulated, and at present chemotaxonomy is well established. Some of the first studies were made on seeds of Leguminosae, subfamily Papilionoideae, which contain much protein, and on cereal grains (wheat, rye, and corn), which are high in starch content. At the beginning of the present century, Gohlke (1913), Mez and Gohlke (1913), and Mez and Ziengenspek (1926) tried to apply serological reactions in an attempt to establish a systematic relationship among these plants (Figure 1.1, Table 1.1). Although only rather primitive procedures were available, they managed from their crude results to construct a phylogenetic tree, the content of which was partially in agreement with modern views on the relationship between classes and families of plants. For example, Mez and Gohlke obtained a precipitation reaction with anti-*Lens* serum (from *Lens esculenta,* the common lentil) using the following dilutions: *Lens* serum and *Vicia faba* proteins 1:25,600; *Pisum* 1:12,800; *Phaseolus* and *Trifolium* 1:6400; while proteins from *Astragalus, Lotus, Melilotus, Laburnum, Lupinus, Acacia,* and *Mimosa* gave the precipitation reaction in a dilution of 1:3200. Proteins of nonleguminous plants showed no reaction; however, some of the Rosaceae plants showed traces of positive reaction after a very long time. The idea of combining chemical and morphological character was first developed into a systematic approach in

1

Table 1.1. List of Names Corresponding to the Numbers in Figure 1.1

1. Chroococcus	53. Chiloscyphus	92. (omitted)	142. Lythraceae
2. Nostocaceae	54. Scapania	93. Blechnum	143. Lecythidoideae
3. Scytonemataceae	55. Ptilidium	94. Aspidium	144. Punicaceae
4. Tetrasporaceae	56. Marsupella	95. Cystopteris	145. Araliaceae
5. Mougeotia	57. Plagiochila	96. (omitted)	146. Nyctaginaceae
6. Protococcee	58. Madotheca	97. (omitted)	147. Lentibulariaceae
7. Hydrodictyaceae	59. Georgia	98. Trichomanes	148. Chenopodiaceae
8. Chrysomonadales	60. (omitted)	99. Aneimia	149. Basellaceae
9. Peridinales	61. Anthoceros	100. Pilularia	150. Proteaceae
10. Siphonocladiales	62. (omitted)	101. Araucaria	151. Julianiaceae
11. Cladophora	63. Hostimella	102. Selaginella	152. Salicaceae
12. Saprolegnia	(fossil group)	103. Walchia	153. Moroideae
13. Saccharomycetes	64. Asteroxylon	(fossil group)	154. Betulaceae
14. Aspergillaceae	(fossil group)	104. (omitted)	155. Berberis
15. Exoascus	65. Hyenia	105. (omitted)	156. Capparidaceae
16. Hypocreaceae	(fossil group)	106. Picea	157. Dilleniaceae
17. Exobasidium	66. Sphenophylla	107. Sciadopitys	158. Hydrastis
18. Stereum	(fossil group)	108. Glyptostrobus	159. Lardizabalaceae
19. Vuilleminia	67. Calamitaceae	109. Pseudolarix	160. Mercuriales
20. Tulostoma	(fossil group)	110. Nilssonniaceae	161. Euphorbia
21. Geaster	68. Equisetum	(fossil group)	162. Aceraceae
22. Melanogaster	69. Pseudoborneales	111. Caytoniales	163. Rutaceae
23. Hymenogaster	(fossil group)	(fossil group)	164. Simarubaceae
24. Secotium	70. Aneurophytum	112. Cyclanthaceae	165. Burseraceae
25. Hysterangium	(fossil group)	113. Nymphaeaceae	166. Empetraceae
26. Dacrymyces	71. Eofilices	114. Trochodendraceae	167. Staphyleaceae

27. Uredineae
28. Craterellus
29. Peniophora
30. Tremellaceae
31. Pilacre
32. Clavaria
33. Hydnum
34. Fistulina
35. Boletus
36. Cantharellus
37. Schizophyllum
38. Lentinus
39. Limacium
40. Paxillus
41. Russula
42. Clitocybe
43. Tricholoma
44. Chaetophora
45. Coleochaete
46. Fegatella
47. Pellia
48. Blasia
49. Sphagnaceae
50. Archidiaceae
51. (omitted)
52. Fossombronia

 (fossil group)
72. Callixylon
 (fossil group)
73. Mesoxylon
 (fossil group)
74. Cordaitales
 (fossil group)
75. Lepidospermae
 (fossil group)
76. Kaulfussia
77. Helminthostachys
78. Botrychium
79. Baiera (fossil group)
80. Stangeria
81. Ceratozamia
82. (omitted)
83. Cycadedoidea
 (fossil group)
84. Cycadofilices
 (fossil group)
85. Marattia
86. Angiopteris
87. Todea
88. (omitted)
89. Cyathea
90. Alsophila
91. Ceratopteris

115. Potamogetonaceae
116. Lauraceae
117. Ranunculaceae
118. Ceratophyllaceae
119. Menispermaceae
120. Cephalotaxus
121. Podocarpus
122. Chamaecyparis
123. Butomaceae
124. Pontederiaceae
125. Dioscoreaceae
126. Iridaceae
127. Burmanniaceae
128. Orchidaceae
129. Zingiberaceae
130. Scirpoideae
131. Caricoideae
132. Restionaceae
133. Eriocaulaceae
134. Connaraceae
135. Platanaceae
136. Pittosporaceae
137. Crassulaceae
138. Thymeleaeaceae
139. Elaeagnaceae
140. Halorrhagaceae
141. Saxifragaceae

168. Hippocrateaceae
169. Linaceae
170. Erythroxylaceae
171. Tiliaceae
172. Caricaceae
173. Caryocaraceae
174. Ochnaceae
175. Oleaceae
176. Gentianaceae
177. Buddleia
178. Apocynaceae
179. Myoporaceae
180. Selaginaceae
181. Acanthaceae
182. Labiatae
183. Plantaginaceae
184. Dipsacaceae
185. Caprifoliaceae
186. Boraginaceae
187. Turneraceae
188. Droseraceae
189. Frankeniaceae
190. Styracaceae
191. Passifloraceae
192. Cucurbitaceae
193. Campanulaceae

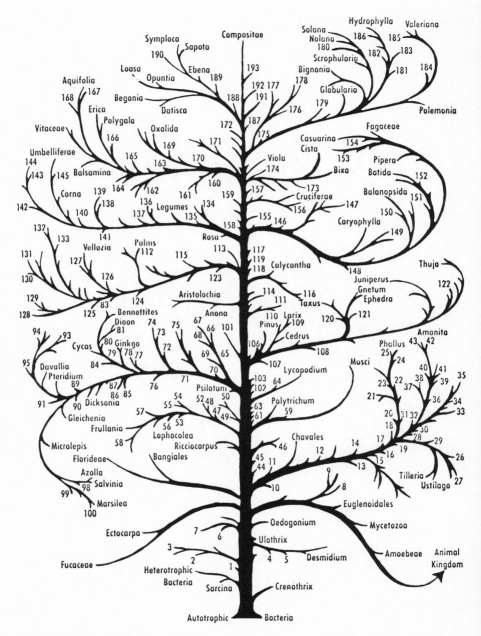

Figure 1.1. The "Stammbaum" of Mez, a phylogenetic tree purportedly constructed, in part, from serological data. The ending "aceae" is omitted from many of the families. For the names of taxa that correspond to the numbers indicated in the figure, see Table 1. Adapted from R. A. Gortner, 1929, *Outlines of Biochemistry,* courtesy of John Wiley & Sons.

the 1930s by Molisch and McNair. Molisch and McNair encountered considerable criticism because their classifications were not satisfying to the botanist, organic chemist, and pharmacist, each of whom held differing views about the subject (Molisch, 1933; McNair, 1930, 1965; Hennig, 1966; Swain, 1966, 1974; Gottlieb, 1972; Florkin, 1974; Hughes and Genest, 1973).

1.2. Taxonomic Character of Alkaloids

Probably the differences among judgments made on the taxonomic value of alkaloids were caused partly by insufficient recognition of alkaloid-accumulating taxa[1] as well as by general lack of data concerning other secondary natural products, i.e., terpenes, flavones, phenolic compounds, etc. (Nowacki and Prus-Glowacki, 1972). A systematic taxonomic classification of higher taxa such as orders and classes of plants must be based not on a single compound but on a number of compounds. Such an attempt was made by Hegnauer (1962). Hegnauer was convinced that alkaloids and other natural products could be of use in plant taxonomy; his view is probably best expressed by the motto cited from Linné at the beginning of his six-volume work *Chemotaxonomie der Pflanzen:*

> Plantae, quae *Genere* conveniunt, etiam virtute conveniunt; quae Ordine Naturali continentur, etiam virtute propius accedunt; quaeque *Classe* naturali congruunt, etiam viribus quodammodo congruunt. C. Linné, *Philosophia Botanica,* 1751.
>
> [Plants which are in the same genus have also the same quality; those which are in the same natural orders have similar properties; those which are collected in natural classes have the powers in accordance.]
>
> In short this can be translated as: plants which are related in form also have similar properties.

A different viewpoint was expressed by Mothes (1966a,b): "Secondary plant substances are not essential to the existence of plants. Therefore, they can (may) be missing from individual members of a related group (of plants). They are not an inherent taxonomic characteristic."

The idea that alkaloids are without value for taxonomic purposes was firmly upheld until about a decade or two ago because organic chemists and plant taxonomists had difficulty communicating with each other. Around the turn of the century, the chemical classification of alkaloids was based on readily distinguishable degradation products, and the following classes

---

[1] Further evidence is provided by the establishment of a Commission on Chemical Plant Taxonomy by IUPAC at the 4th International Symposium on Natural Products held in Stockholm, Sweden, June 26–July 2, 1966.

of alkaloids were known: pyridine, pyrrolidine, indole, quinoline, isoquinoline, tropane, quinolizidine, and pyrrolizidine derivatives. This classification proved to be not useful for taxonomic purposes. To construct a
phylogenetic tree by use of such chemical data would be meaningless to the
botanist; plant families would be grouped in most unexpected orders. For
instance, *Equisetum* and some species of Solanaceae would be grouped in a
"nicotineferous" family, while the poppies and some of the Liliaceae
would be "isoquinolineferous," and the Loganiaceae (Bisset, 1974a,b;
1975), some grasses, and a parasitic fungus, *Claviceps purpurea,* would be
included in "indoleferous." Thus the chemical classification of alkaloids was
unsatisfactory when directly transposed to botanical classification.

For a long time it was believed that the indole alkaloids of the ergoline
type derived from tryptophan were restricted to the fungus *Claviceps* and
some other fungi. In 1960 Hofmann and Tscherter reported the isolation of
ergoline alkaloids from higher plants, especially in the Convolvulaceae.
Recently, these alkaloids have been found in 28 species representing 6
genera of the morning glory family (Chao and DerMarderosian, 1973).

In Figure 1.2, quinolizidine alkaloids found in completely unrelated
taxonomic units are shown. All quinolizidine alkaloids have the same basic
ring structure, but the compounds are formed by different biosynthetic
pathways. There is no reason to suppose that a certain well-defined chemical compound found in two unrelated species is necessarily formed in both
by the same pathway. Kasprzykowna (1957), for example, writes, "One

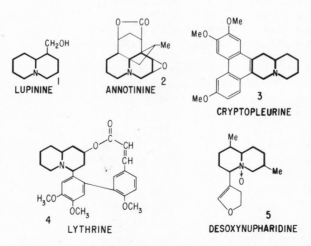

Figure 1.2. Some taxonomically unrelated quinolizidine alkaloids.

does not find any dependence (connection) between the alkaloid structure and the relationship in systematic families within which they occur." Canadine (berberidine) is found in six families according to Hutchinson's (1959) systematics. Three of these families, Berberidaceae, Papaveraceae, and Ranunculaceae, belong to the order Ranales, while three others, Rutaceae, Menispermaceae, and Annonaceae, are placed in the order Magnoliales. Nicotine occurs not only in Ranales and Magnoliales but also in Lycopodiales and Equisetales, which are sporophytes. On the other hand, alkaloids varying in structure are often encountered within closely related species.

Since Hutchinson's work in 1959, a considerable amount of information regarding phylogeny has been accumulated. These data have been synthesized into excellent classifications giving precise subclass, order, and family relationships. The currently accepted classifications[2] of Cronquist (1968) and Takhtajan (1959) are essentially the same and present a better framework for discussion of alkaloid distribution patterns than the work of Hutchinson. According to Cronquist and Takhtajan systematics the distribution of some alkaloids coincides with cytological and anatomical patterns. For example, the occurrence of betalains, alkaloidlike plant pigments, trinucleate pollen, bitegmic crassinucleate ovules, and free central placentation delimit the Caryophyllideae with some 11,000 species. However, there is still much controversy when some taxonomical works are compared. Cronquist's and Takhtajan's classification is, at the moment, one of the possible arrangements of plant families, but not the exclusive one. This situation leads to some conflicts since in the same publication a plant order is named Caryophyllales, in others Centrospermae or Phytolaccales. The phytochemist who is trying to use the alkaloid distribution data for chemotaxonomical purposes wishes usually to accept the plant system which corresponds to his work. But even with this bias in some cases it is impossible to recognize that plants with the same groups of alkaloids are related to the plants which produce alkaloids. For example, derivatives of tyrosine, converted into alkaloids on a pathway related to norlaudanosoline, belong to such different species as *Papaver, Annona, Croton, Sinomenium, Stephania,* and *Erythrina* (Romeo and Bell, 1974; Hargreaves *et al.,* 1974; James and Butt, 1957; Bhakuni *et al.,* 1972a,b). And there is no taxonomy which would recognize all the species as related. When in one systematics some species are included in one family or order and recognized

[2]Cronquist (1968)–Takhtajan (1959)–Mowszowicz (1974) is the classification that is the most appropriate with the evolution of angiosperms; however, most botanists and a fair number of biochemists, natural products chemists, pharmaceutical chemists, and agronomists are still familiar with using the Wettstein and Hutchinson classification.

relatives are included in another order, the affiliations will be very different. It is acceptable that an alkaloid has originated in plants several times and that only in some genera or families it was established as a permanent character (Cronquist, 1960; Engler, 1954; Hutchinson, 1959; Takhtajan, 1959; Wettstein, 1935; Zukowski, 1951; Mowszewicz, 1974; Szafer and Kostyniuk, 1962; Strasburger et al., 1960).

An attempt to divide plants into "alkaloid-bearing" and "alkaloid-free" types would be as unjustified as an effort to divide plants into "pilose" and "glabrous." A chemical character is like a single morphological character and is an insufficient basis for defining a large taxonomic unit. But when an additional character which originates by a separate pathway is available, as in the case of the alkaloids, it becomes possible to devise an artificial classification system. This system is based on the presence or absence of the particular pathway as represented by end-product analysis. If a compound can arise by different pathways, this type of chemical classification is meaningless. Thus an attempt to form a natural classification of alkaloids is necessary. During the first part of the 20th century, numerous hypotheses of alkaloid biogenesis were proposed (e.g., Winterstein and Trier, 1910; Schöpf, 1949; Wocker, 1953; Goutarel et al., 1950; Robinson, 1917, 1955; Wenkert, 1959, 1968; Marion, 1958; Battersby, 1961, 1965; and Leete, 1969). All authors, except Wocker and Wenkert, considered some amino acids as the key precursors of alkaloids. Starting in the early 1950s Marion, Mothes, Battersby, Byerrum, Arigoni, and later Waller, Schuette, Nowacki, Coscia, and a number of other natural-products chemists and biochemists provided experimental data, using labeled precursors, on biosynthesis of alkaloids. Only five amino acids, phenylalanine, tyrosine, tryptophan, lysine, ornithine, and in exceptional cases a sixth, aspartic acid, were found to be the precursors of the majority of alkaloids. An appreciable number of alkaloids are derived from nicotinic acid or from the "pyridine nucleotide pool," which suggests the validity of the Wocker original "codehydrogenase"[3] hypothesis, subsequently validated by Waller et al. (1966). Some alkaloids are derivatives of monoterpenes, sesquiterpenes, diterpenes, and steroids; and a smaller number are derived from anthranilic acid, purines, and pyrimidine bases. Part of the nicotinic acid family of alkaloids, all of the isoprenoid alkaloids, and the purine derivatives are sometimes referred to as "pseudoalkaloids" or "alcaloida imperfecta" (Hegnauer, 1958, 1959, 1963). Hegnauer bases the restriction of the term "true alkaloids" on the strict and literal application of Winterstein and Trier's (1910) definition of alkaloids (Figure 1.3):

---

[3]Old name for NAD^+.

Alkaloids are more or less toxic substances which act primarily on the central nervous system. They have a basic character, contain heterocyclic nitrogen, and are synthesized in plants from amino acids or their immediate derivatives. In most cases they are of limited distribution in the plant kingdom.

This definition excludes an enormous number of plant constituents which are generally acceptable as alkaloids. Some such compounds can, for example, be nontoxic, as are the betalains, and have either a neutral or acid character, as do oxolupanine and aristolochic acid and colchinine, which have no ring nitrogen. The substrate can be a compound other than an amino acid, as with the hemlock alkaloids, where it is acetic acid (Leete and Adityachaudhury, 1967). The *Conium* (hemlock) alkaloids are structurally related to the tropane *(Datura stramonium)* and pelletierine *(Punica granatum)* alkaloids, which are biosynthesized from diamines. We will therefore apply the Winterstein and Trier (1910) definition of alkaloids only within certain limits. Deviations from this definition will be made when its strict application would exclude a compound that is generally accepted as an alkaloid. This liberal approach towards describing the alkaloids is slightly dangerous. As an example, consider mimosine, a compound produced by some *Mimosa* and *Leuceana* species, which was defined as an alkaloid following its isolation by Mascré (1937). Its structure was in accord with the alkaloid definition: it was toxic, and it is now known that the pyridine ring is derived from lysine (Hylin, 1964; Notation and Spenser, 1964). Recently some related species of Mimoseae and Acaciae have yielded a number of compounds which were acceptable as alkaloids, as is mimosine; but Gmelin (1959) and Gmelin *et al.* (1958, 1959), who isolated those compounds, were more impressed by the presence of a carboxyl group and an α-amino group than by that of the ring and thus described the compounds as the amino acids willardine and albizzine (Figure 1.4). A similar compound, lathyrine (tingitanine), was found in *Lathyrus* by Bell (1961) and Nowacki and Przybylska (1961), and again the amino acid part of the structure caused it to be described an an amino acid. Thus it is readily apparent that broadening the definition of alkaloids to include certain amino acids such as tryptophan and proline becomes possible. Perhaps certain vitamins such as nicotinic acid could be included. To restrict sensibly the term "alkaloid," Nowacki and Nowacka (1965) have proposed the following definition:

An alkaloid is a substance with nitrogen in the molecule, connected to at least two carbon atoms. The molecule must have at least one ring, but it is not necessarily heterocyclic. The compound cannot be a structural unit of macromolecular cellular substances and cannot serve as a vitamin or hormone.

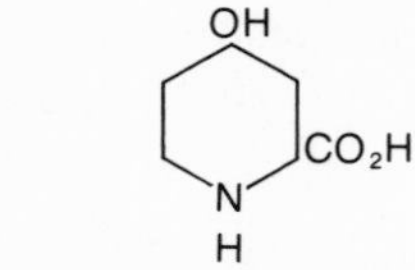

4-HYDROXYPIPECOLIC ACID

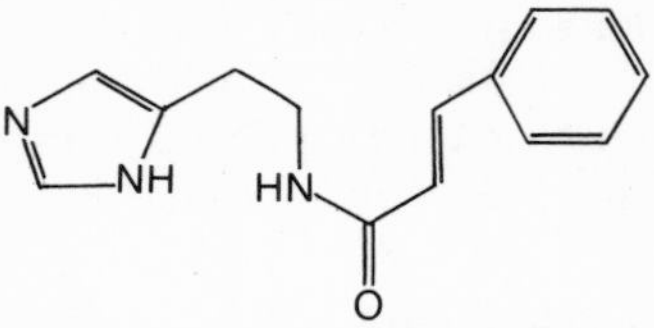

N-CINNAMOYLHISTAMINE

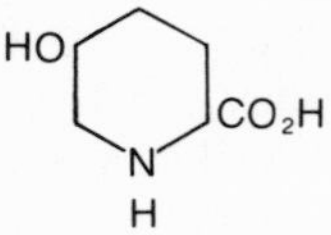

5-HYDROXYPIPECOLIC ACID

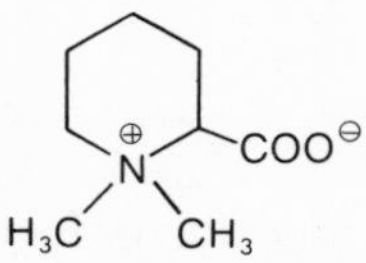

HOMOSTACHYDRINE

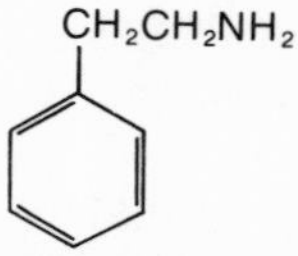

β-PHENETHYLAMINE

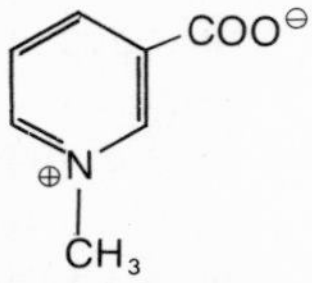

TRIGONELLINE

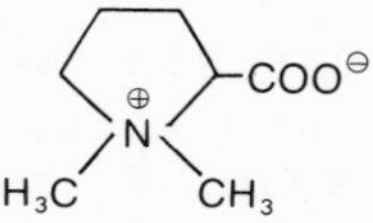

STACHYDRINE

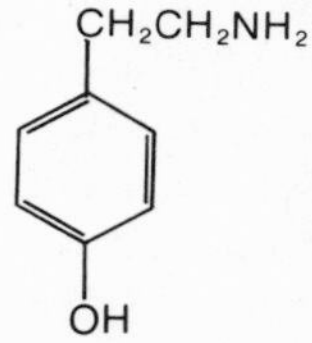

TYRAMINE

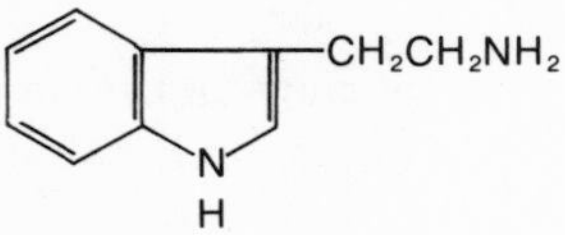

TRYPTAMINE

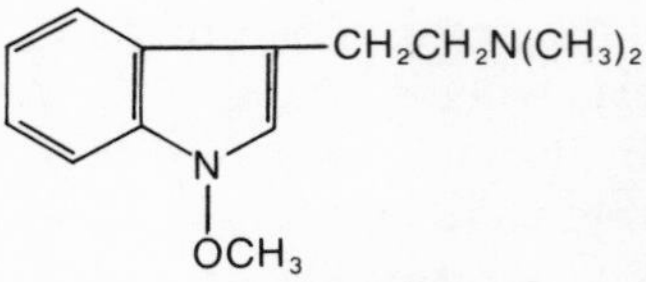

LESPEDAMINE

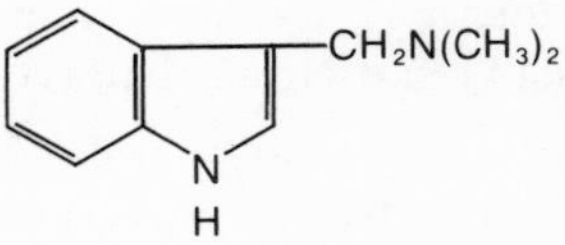

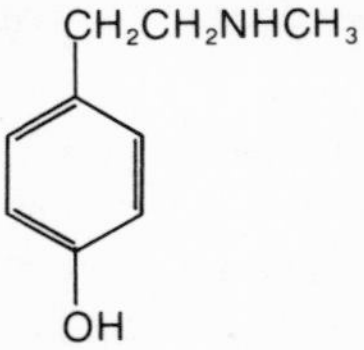

N-METHYLTYRAMINE

GRAMINE

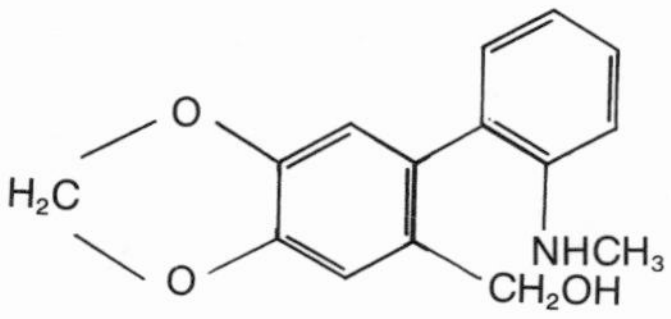

ISMENINE

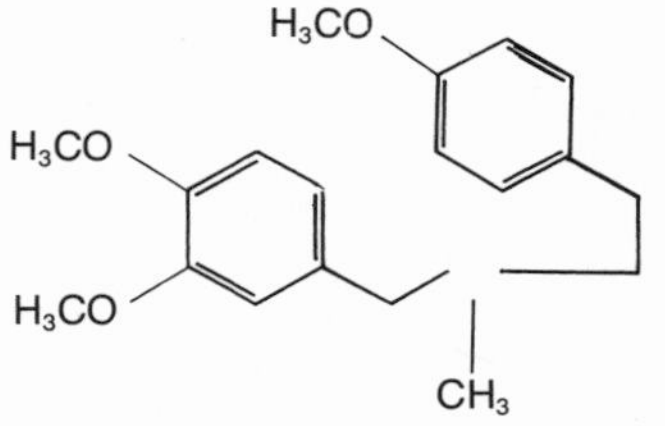

BELLADINE

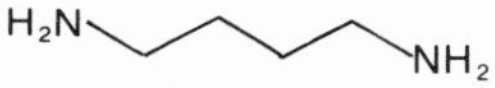

PUTRESCINE

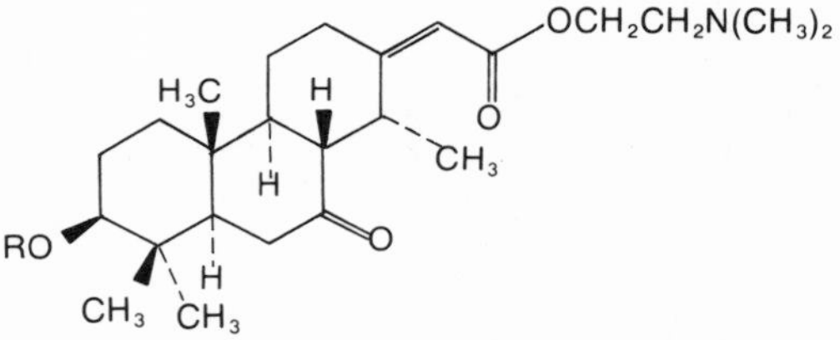

COUMINGIDINE

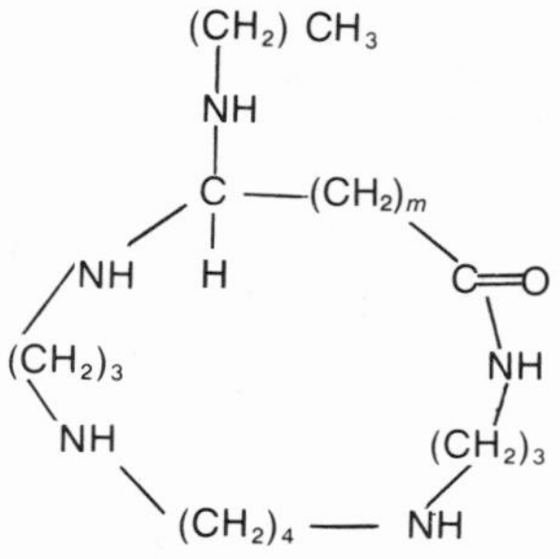

PITHECOLOBINE

Continued

Figure 1.3. Examples of compounds not generally accepted as alkaloids. In pithecolobine ($m + n = 9, 10, 11$). a ($m = 3, n = 6$), b ($m = 1, n = 8$), c ($m = 1, n = 10$). (Weisner, K., Valenta, Z., Orr, D. E., Liede, V., and Kohan, G. (1968), *Can. J. Chem. 46,* 3617.)

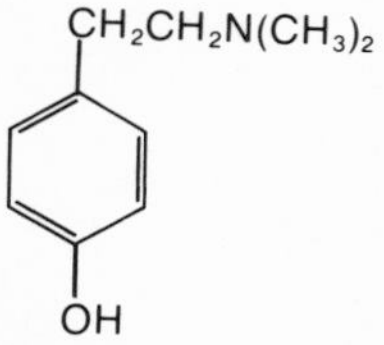

CH$_2$CH$_2$N(CH$_3$)$_2$

OH

HORDENINE

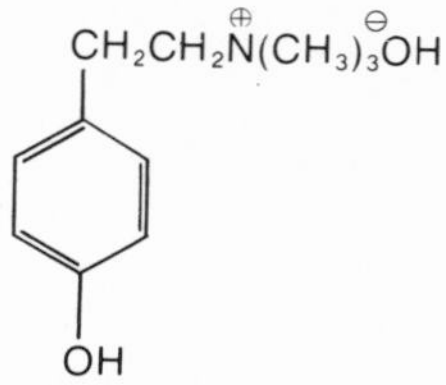

CH$_2$CH$_2$$\overset{\oplus}{N}$(CH$_3$)$_3$$\overset{\ominus}{OH}$

OH

CANDICINE

CH$_2$CH$_2$NH—C

O

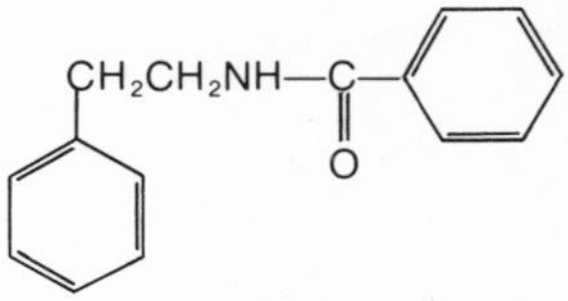

N-BENZOYL-β-PHENETHYLAMINE

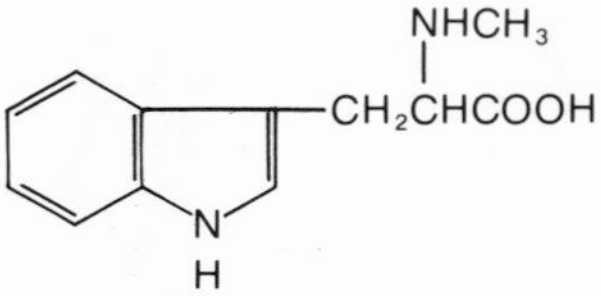

NHCH$_3$

CH$_2$CHCOOH

N
H

ABRINE

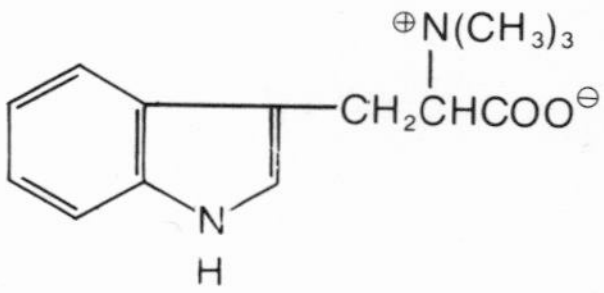

$\overset{\oplus}{N}$(CH$_3$)$_3$

CH$_2$CHCOO$^{\ominus}$

N
H

HYPAPHORINE

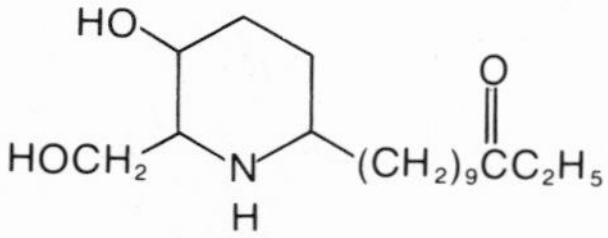

HO

HOCH$_2$

N
H

O
(CH$_2$)$_9$CC$_2$H$_5$

PROSOPININE

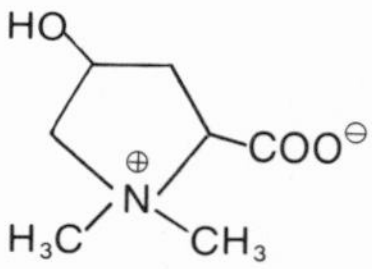

HO

N$^{\oplus}$
H$_3$C   CH$_3$

COO$^{\ominus}$

BETONICINE

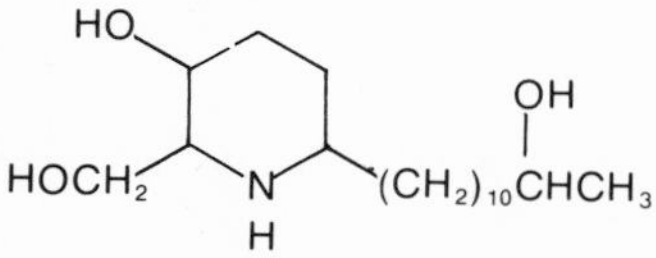

HO

HOCH$_2$

N
H

OH
(CH$_2$)$_{10}$CHCH$_3$

PROSOPINE

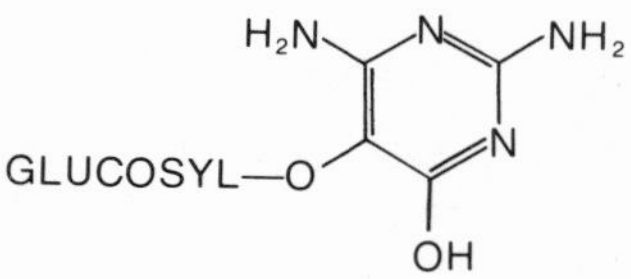

H$_2$N   N   NH$_2$

N

GLUCOSYL—O

OH

VICINE

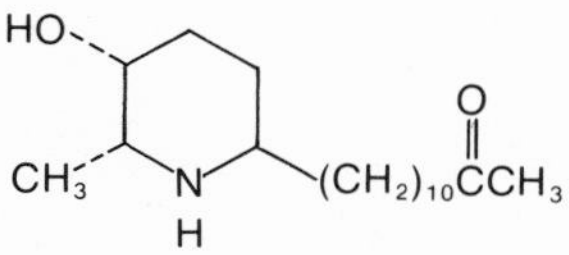

HO

CH$_3$ N
H

O
(CH$_2$)$_{10}$CCH$_3$

CASSINE

H

HN   N   CH$_3$

CH$_3$

NH$_2$

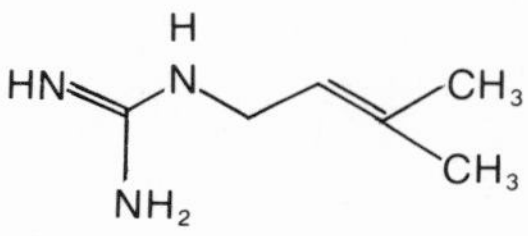

GALEGINE

Figure 1.3. Continued

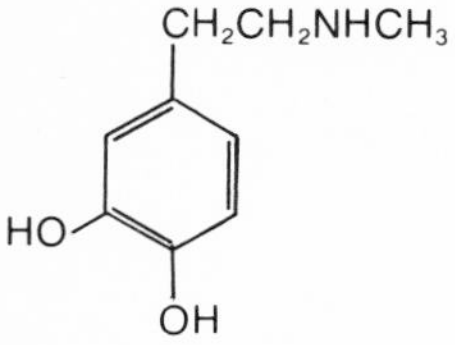

EPININE

$$^{\ominus}OH \quad HOCH_2CH_2—\overset{\oplus}{N}(CH_3)_3$$

CHOLINE

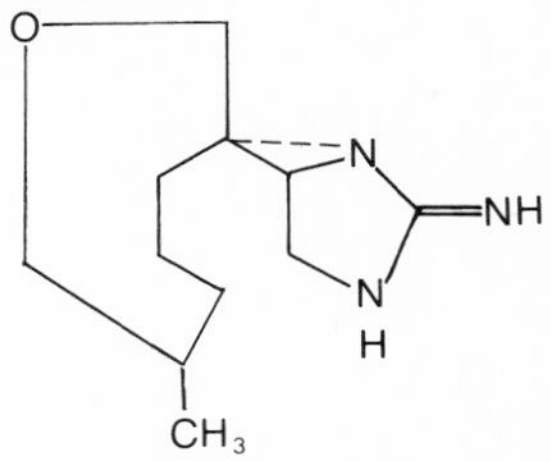

CHAKSINE

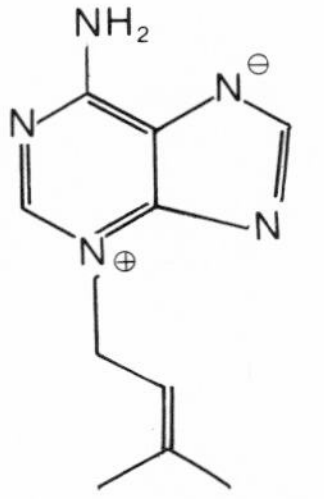

TRIACANTHINE

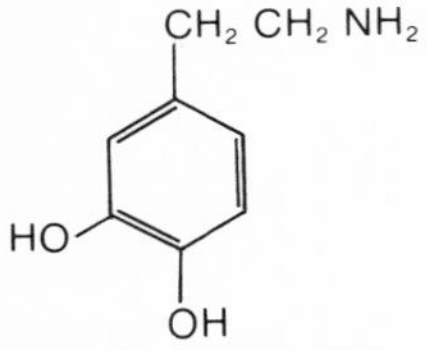

HYDROXYTYRAMINE

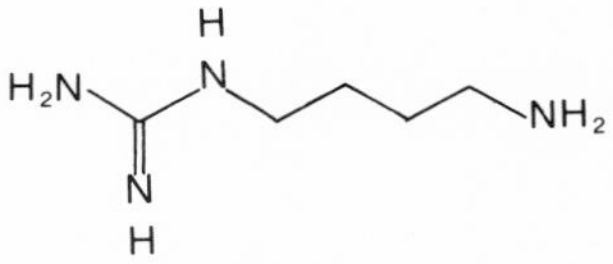

AGMATINE

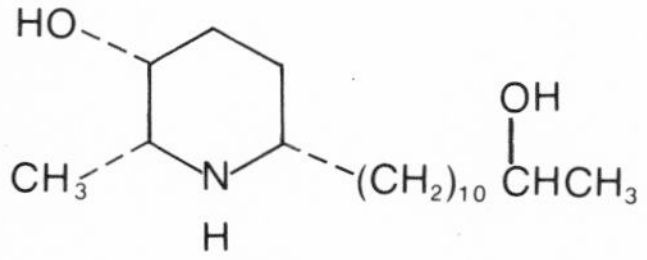

CARNAVOLINE

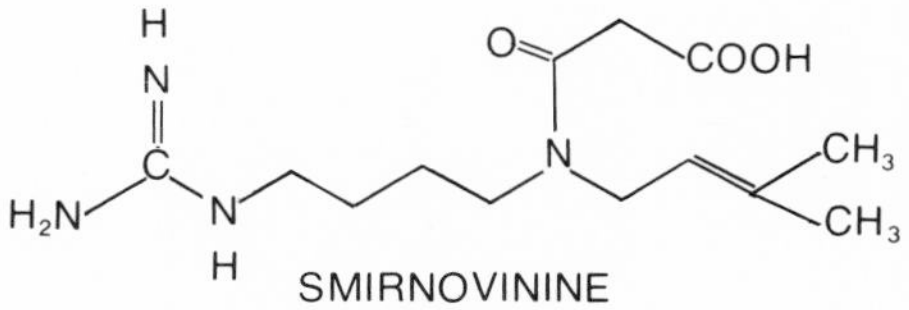

SMIRNOVININE

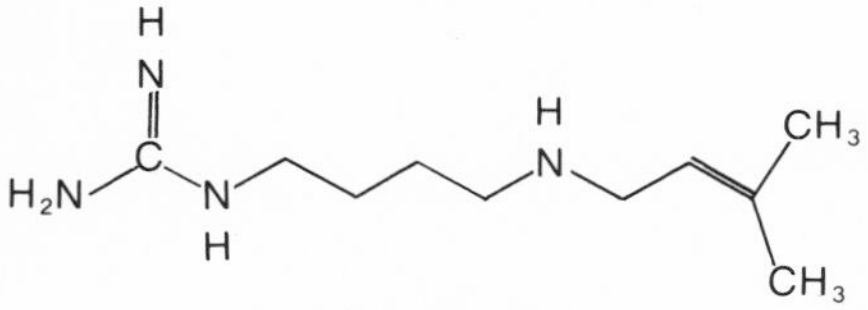

SPHAEROPHYSINE

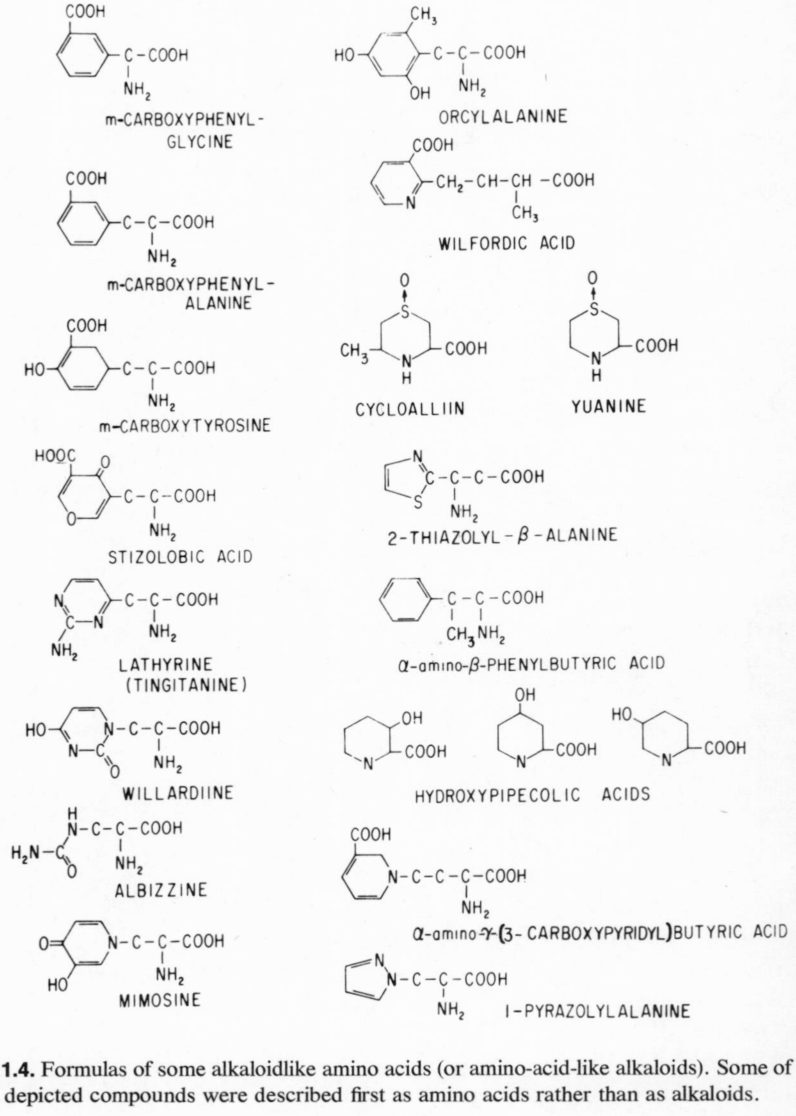

Figure 1.4. Formulas of some alkaloidlike amino acids (or amino-acid-like alkaloids). Some of the depicted compounds were described first as amino acids rather than as alkaloids.

This definition does include some betaines and atypical amino acids, but, on the other hand, a similar exception for these types of compounds was made by Hegnauer (1966):

> The large number of compounds which have no heterocyclic ring, such as the biological amines (e.g., ephedrine, hordenine, mescaline, narceine, and galegine); betaines such as betaine itself (glycylbetaine); and aliphatic quaternary bases such as choline, acetylcholine, muscarine, sinapin, stachydrine (proline betaine), and tryptamine are also excluded, although in this case they are clearly derived from amino acids and contain heterocyclic nitrogen. All of the compounds mentioned above are usually referred to as biological amines (Guggenheim, 1951) or "protoalkaloids" (Ackermann and List, 1958). When such "protoalkaloids" occur in the same genus or family as "true alkaloids" to which they are biogenetically related (e.g., hordenine, candicine, and mescaline in Cactaceae; narceine in Papaveraceae), then it is usual to classify them as alkaloids also.

Actually there are no good definitions of alkaloids (Bate-Smith and Swain, 1966) since each one is either too narrow or too broad. Even in the restricted Winterstein and Trier definition, at least five alkaloid families exist that can be derived from different amino acids; consequently, *there is a need to establish the proper biosynthetic pathways to permit the application of the alkaloid character to chemotaxonomy.* It has been mentioned above that canadine (berberidine) may be found in plants of six partially unrelated botanical families. This fact is not surprising when considered in relation to the biochemical investigations of canadine biosynthesis. Many reactions are necessary to convert tyrosine into canadine; consequently, one might even wonder why the distribution of this alkaloid is so limited. In contrast, other plants (and even some that produce canadine) can produce many alkaloids that are derived from tyrosine but have a marked difference in structure. Tyrosine serves as the key precursor of alkaloids of the isoquinoline type, but other types of alkaloids, such as colchicine and the Amaryllidaceae and the Erythrina alkaloids, may be synthesized from this amino acid. The nucleus of an alkaloid molecule can arise from different precursors; thus the indole nucleus in Erythrina alkaloids arises from tyrosine, while in brucine it comes from tryptophan (Figure 1.5). The alkaloids cinchonamine and cinchonine differ in that cinchonamine has an indole nucleus, while cinchonine (like quinine) has a quinoline nucleus; however, they exist in a precursor–product relationship (that is, the quinoline type is derived from the indole type in a one-step reaction).

Some plants produce two or more types of alkaloids. In some cases, all types are actually derivatives of the same precursor, while in others it is not so (Bu'Lock, 1965). For example, some Rutaceae contain five different types

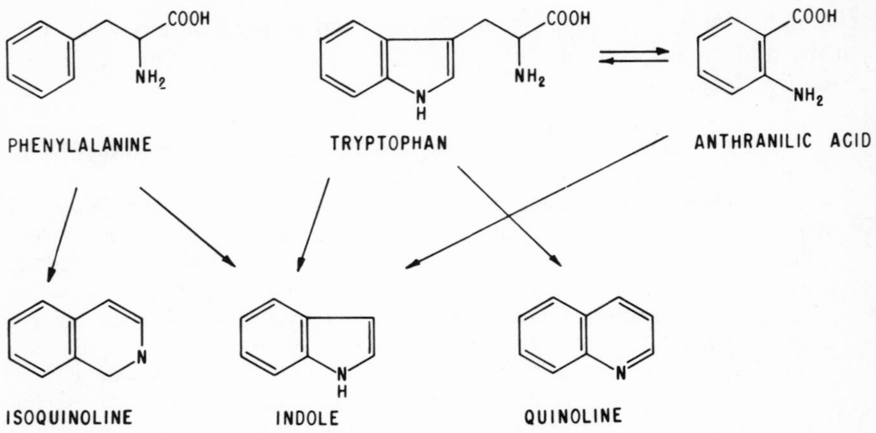

Figure 1.5. The origin of isoquinoline, quinoline, and indole from two different sources.

óf alkaloids, as in the genus *Casimiroa* (Figure 1.6a). A similar case was found in *Euphorbiales,* Figure 1.6b. In Leguminosae two related genera, *Cytisus* and *Crotalaria,* of the tribe Genisteae contain different alkaloids. *Cytisus,* like the remainder of the Genisteae (except *Crotalaria*) (Steinegger and co-workers, 1968), accumulates quinolizine derivatives (lupine alkaloids), whereas *Crotalaria* accumulates necine (senecio) alkaloids (Figure 1.7). A single species, *Laburnum anagyroides,* was found which had both types of alkaloids. Biosynthetic experiments of Schuette (1960), Nowacki (1961), Nowacki and Byerrum (1962), Warren (1966), and others, show that there are similar but nonidentical enzyme systems in these groups of plants and that the only difference is that the quinolizidine alkaloids are derived from lysine, whereas the pyrrolizidine alkaloids arise from ornithine. There is still more proof that lysine and/or ornithine can be converted into alkaloids by similar enzyme systems; this is the case with nornicotine and anabasine in tobacco, stachydrine and homostachydrine in alfalfa, cuscohygrine and anaferine in cocoa plants, etc. These biosynthetic relationships make classifying alkaloids in higher taxa for chemotaxonomical purposes more difficult.

The already mentioned occurrence of certain alkaloids or alkaloid types in unrelated plants creates some problems for the chemotaxonomist (Matveyev, 1959; Sharapov, 1962). The pyrrolizidine alkaloids, for example, are produced by plants like *Crotalaria* in Fabaceae, *Senecio* in Compositae, *Lolium* in Glumiflorae, *Cynoglossum* in Boraginaceae, and *Securinega* in Euphorbiaceae (Crowley and Culvenor, 1962; Culvenor and Smith, 1967; Culvenor *et al.,* 1968a,b). The alkaloids derived from the pyridine nucleotide cycle are distributed in unrelated plants. The most common nicotinic acid derivative, trigonelline, is encountered in a large number of plants, whereas the distribution of nicotine has already been mentioned. Other alkaloids which are biosynthesized from nicotinic acid are ricinine and nudiflorine, both in Euphorbiaceae. Wilforine and related alkaloids are accumulated in *Triptergium wilfordii,* which together with the genus *Euonymus* belongs to Celestraceae. The species *Euonymus europea* is noted for the alkaloid evoninic acid, which has strong insecticidal properties and is used in folk medicine for extermination of lice.

In recent years, however, sufficient data have been accumulated to make possible a new attempt to classify plants according to their alkaloid distribution. There is a need for the proper experimental evidence for the biosynthesis of alkaloids, since many of the pathways are still hypothetical. While a plant that contains alkaloids is simple to detect, an alkaloid-free plant is much more difficult to identify; it can be a plant which gives no reaction with the commonly used alkaloidal detection reagents, but it might contain traces of alkaloid-like compounds, thereby giving a false negative reaction. The alkaloid-free plants can be divided into two groups: those which never produce alkaloids, and those which actually produce alkaloids but degrade them at a high level of efficiency so that a false negative reaction is observed. These two groups may be classified as "alkaloid-free" and "alkaloid-trace," respectively. Excluding some exceptional cases, there is no proof that alkaloid-free plants have enzyme systems capable of metabolizing alkaloids. In some instances, it may be that the acid which often forms an ester such as the tropane, lupine, and necine hydroxyamine esters with hydroxylated alkaloids is of greater taxonomical importance that the alkaloid itself. Since over 70% of plants flourish without producing alkaloids, it must be concluded that alkaloid metabolism is of minor importance in the plant kingdom. Although alkaloid-trace induced mutant lupines grow reasonably well, in contrast mutants with wild types have a much higher yield of seeds (Zachow, 1967; Aniol *et al.,* 1972). On the other hand, in some desert floras over 60% of plants accumulate alkaloids, whereas the others accumulate additional structures which are also "secondary metabolites" and provide additional structures for the chemotaxonomic division of plants. It is necessary to construct a hypothesis describing alkaloid origins, but, since there are no paleobotanical data (and there is little hope that someone will furnish them), the classification is only speculative and based on observations of contemporary flora. Since the majority of alkaloids are derived from amino acids, the amino acid composition of some leaf proteins is depicted in Figure 1.8. This figure depicts an average of data from different plants. It is striking in that of the eleven amino acids least abundant in plant proteins, five can readily produce alkaloids and two (isoleucine and valine) produce sulfur-containing glucosides. Since the

Laurifoline

Chelerythrine

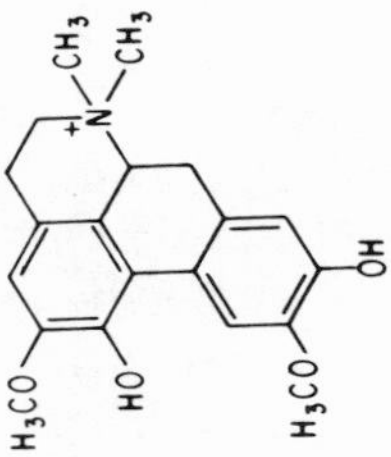

N-Methylcanadine

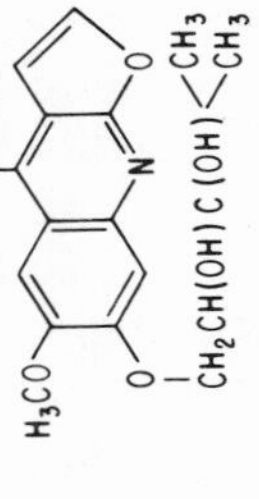

Hortiacine

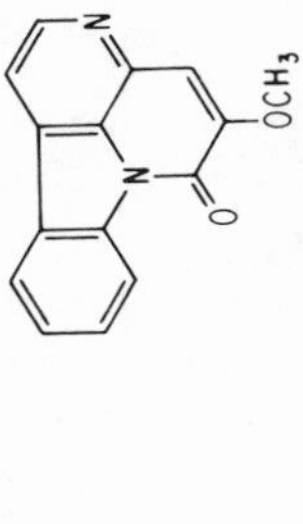

5-Methoxycanthin-6-one

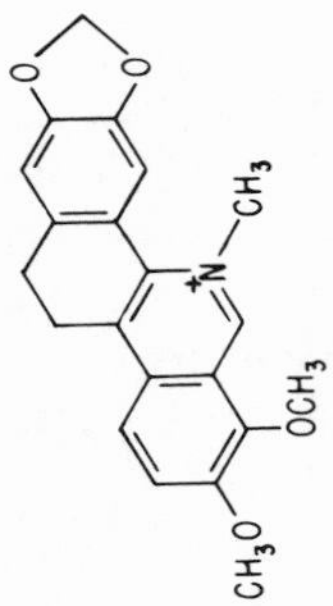

Arborine

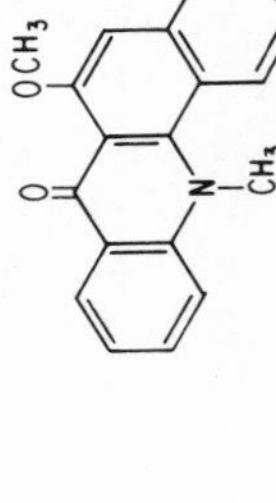

Balfourodine

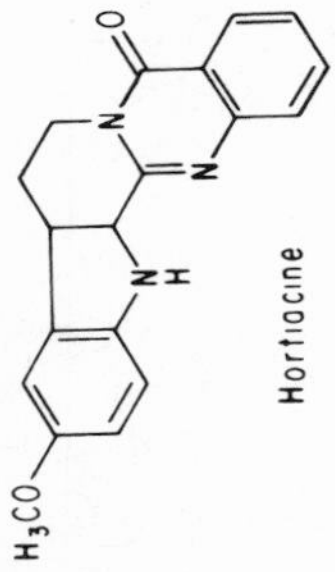

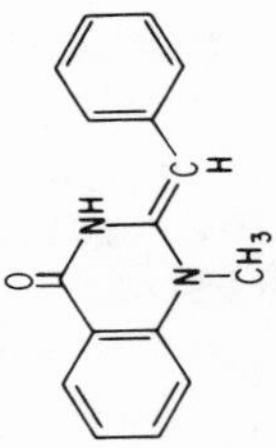

Evolatine

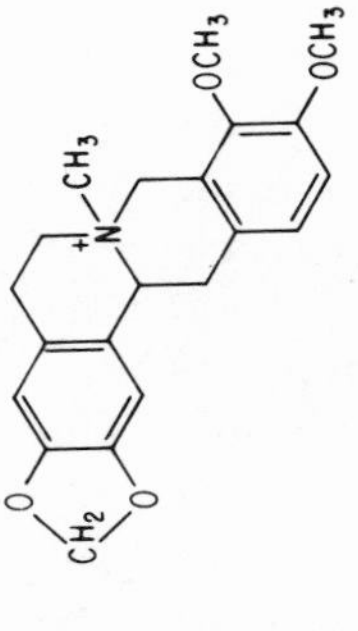

Acronycine

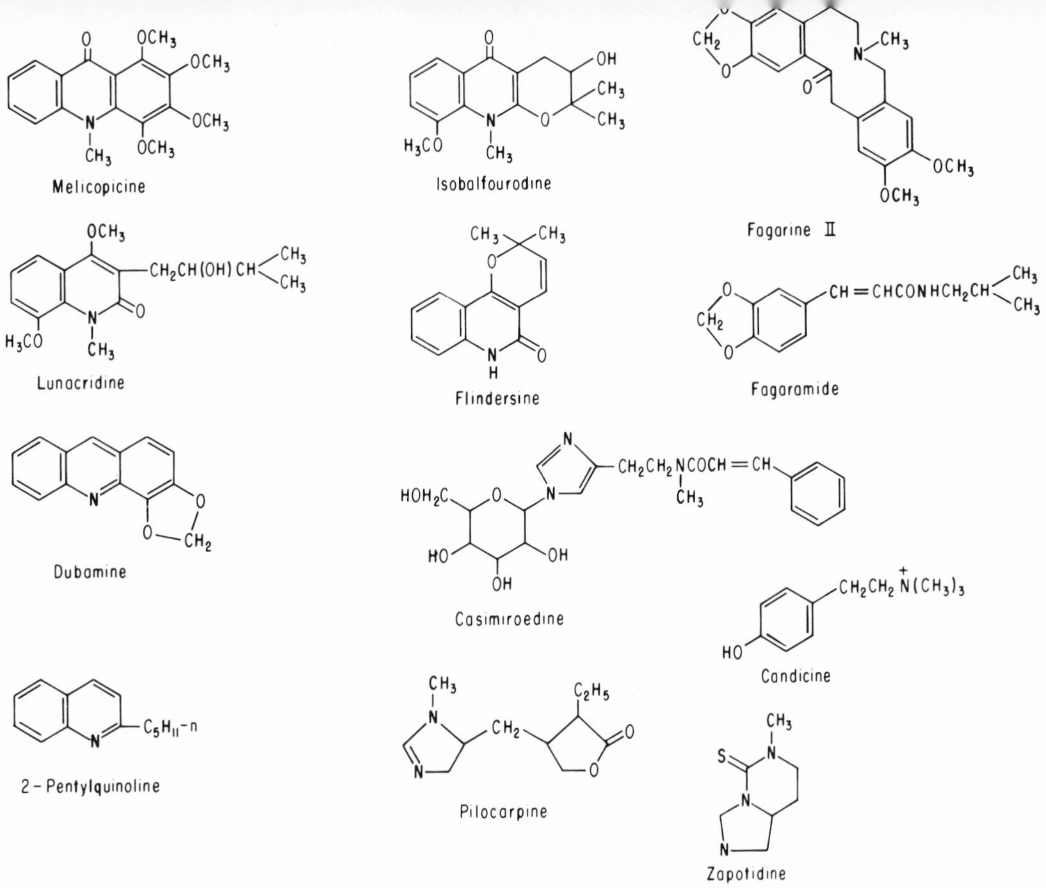

Figure 1.6a. Structures of some Rutaceae alkaloids. This order is recognized for accumulating alkaloids derived from different biosynthetic pathways, e.g., from anthranilic acid, tyrosine, tryptophan, and histidine.

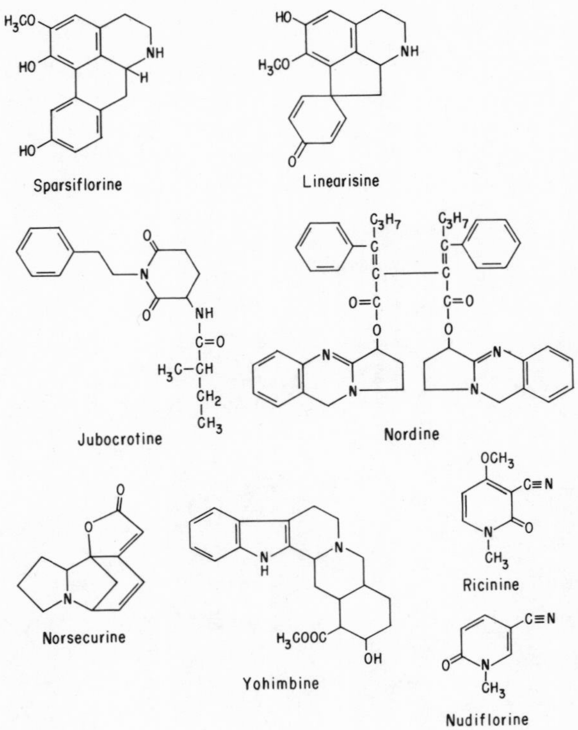

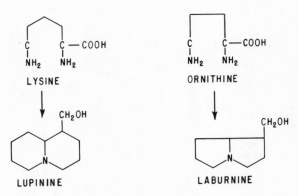

Figure 1.6b. Euphorbiaceae alkaloids.

Figure 1.7. Lupinine and the necine (e.g., laburnine) alkaloids and their precursors.

number of molecules of any given amino acid built into a protein molecule is governed by the content and structure of RNA and DNA, a plant in which a repressor or inhibitor system cannot be switched off, owing to a mutation, will produce more of a given amino acid than required for protein synthesis. When the level of this amino acid passes a certain threshold value, the plant will either perish or the accumulated amino acid will induce enzyme synthesis so that further metabolism can occur. *Alkaloids are reasonable products of this metabolism;* hence the name "secondary metabolites" is quite appropriate. The mutation may have no effect otherwise on the ability of the plant to survive, or it might decrease or increase its viability. In either case, the mutant plant will be eradicated from the population by means of selection or genetic drift, but it can be of some advantage to the plant to accumulate compounds with toxic properties, bitter taste, or other characteristics unpleasant to herbivorous animals. This accumulation will increase the survival capacity of the genotype, and the mutant will rapidly displace the normal type. How rapid such a displacement can be is shown in the example of the moth *Biston betularia.* The black mutant was found during the middle of the 19th century in an industrial area of England, where it was better adapted to live on the bark of the soot-covered trees than the grey form which predominated before the industrial revolution. This mutant *(carbonaria)* is at present the only known type in northwestern industrial areas of Europe; only single specimens were found in 1900 in Upper Silesia industrial areas, but at present they are the prevalent species. An example from the plant kingdom is the European *Hypericum perforatum,* which was introduced into California (Holloway, 1957). *Hypericum* outgrew the other drought-tolerant grasses

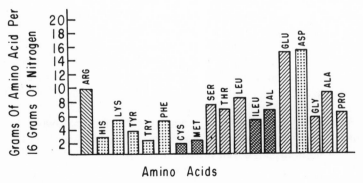

Figure 1.8. The average composition of the amino acids of plant proteins. Amino acids marked with dots are alkaloid precursors; the amino acids that are crosshatched are precursors for the synthesis of mustard glucosides. The other amino acids are less common precursors for alkaloids.

and virtually supplanted them wherever it spread. It was inedible to indigenous animals because of its bitter taste. To control the weed, a specialized leaf-eating beetle *(Chrysolina gemellata)* from Europe was introduced. In nature, a species of plant that turns to alkaloid production probably escapes certain predators and some parasites (Nowacki and Waller, 1973) for a while since it takes time for herbivorous animals to adapt to the new (and already common) mutated plant.

The alkaloid character most probably arises as a result of a mutation in the regulatory system governing the intermediary metabolism. The mutation can be established in a taxon by the stress of environmental selection forces (Mettler and Gregg, 1969). Yet it can happen that a mutation has no selective value by itself and that its propagation in the population can be due to a genetic association with another factor biogenetically unrelated but of great selective value. In this case, the mutation without any selective value will be propagated as long as the association with the selective plausible factor is maintained. When this association is broken, by chromosome translocations, the mutant will be protected no more. Then a new complex will be established, with part of the surviving population possessing the protecting factor still associated with the mutated gene. Another part of the population will possess the protective factor, but in dissociated form, so the populations will evolve into separate taxa. But they can also undergo evolution without regard to the mutation; thus we will encounter in related units plants with and without a certain chemical character. Since most of the alkaloids are produced by enzymes already present in some alkaloid-free plants, and since some of the enzymes can belong to the class of compounds without selective value, they can be distributed at random throughout the entire taxonomical unit. The enzymes of secondary importance to the plant can transform the product of mutation in many different ways. For example, the Leguminosae tend to produce an excess of the diamino acids lysine, ornithine, or diaminobutyric acid (Figure 1.9). Their accumulation is probably without any importance to the plant involved. The compounds are enzymatically transformed through oxidation by diamine oxidase, decarboxylation, transformation to their carbonyl derivatives, methylation, etc. All possible pathways were used by the legumes, and in 1960 Nowacki constructed a scheme for the main pathways of lysine and ornithine metabolism in legumes. There was a pathway leading to the production of homoarginine, but it was purely theoretical, built only on the basis of exact similarity to the metabolism of ornithine. Three years later homoarginine was indeed discovered in the genus *Lathyrus,* which belongs to the same tribe Vicieae (Bell, 1961, 1971). Arginine accumulates in the genus *Vicia* (Figure 1.10). Similar pathways can be constructed for some other amino acids as well as for nicotinic acid and isoprenoids.

Once a mutant type is established, further mutations may occur in the

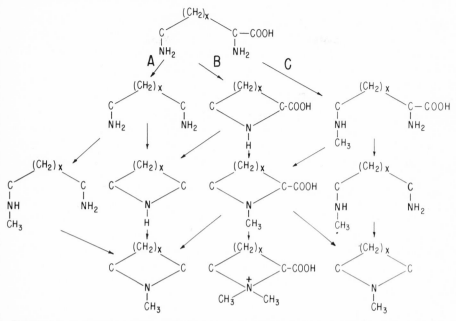

Figure 1.9. Proposed general metabolic grid of the biosynthesis of diamino acids. *A*. decarboxylation; *B*. cyclization; *C*. methylation. Pathway *A* leads to symmetrically labeled compounds; in pathways *B* and *C* the original asymmetry is retained.

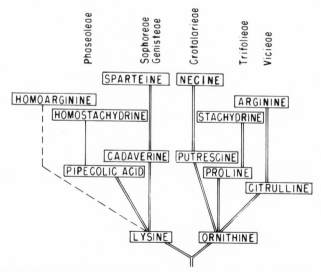

Figure 1.10. Lysine and ornithine conversion pathways in Fabaceae (Nowacki, 1964). The lysine–homoarginine conversion was discovered in the genus *Lathyrus* (Vicieae). Courtesy of the journal.

alkaloid-bearing plant; these usually involve oxidation reactions. In the Fabaceae, three closely related tribes contain "lupine alkaloids." While the more primitive shrubby and tree species contain racemic sparteine, matrine, and dipiperidines, those lupines that are more progressive in evolution accumulate only derivatives of (−)-sparteine and sometimes other compounds elaborated by specific enzymes (Figure 1.11). It seems conceivable that an alkaloid that has become established in a given taxonomic unit will be exposed to new enzymes produced in the course of evolution of this taxonomic unit; the more specialized compounds will be restricted to the smallest and most progressive units (Haensel, 1956), Figure 1.12. The ultimate possibility is a complete degradation of the alkaloid to nonalkaloidal products. Such catabolic reactions must be

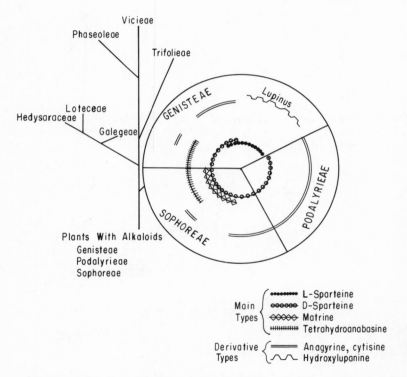

Figure 1.11. The three tribes of Fabaceae with quinolizidine alkaloids. The Sophoreae are close to the original stock of all Fabaceae. The more advanced tribes (Phaseoleae, Vicieae, and Trifolieae) are either entirely alkaloid-free or accumulate alkaloids of different origin, as in some species of Phaseoleae, Loteceae, Hedysaraceae, and Galegeae, although more primitive, are without alkaloids of the quinolizidine type. Probably an independent evolutionary line of the Fabaceae evolved which was not directly related to Sophoreae, Genisteae, and Podalyrieae tribes (Nowacki, 1960). Courtesy of the journal.

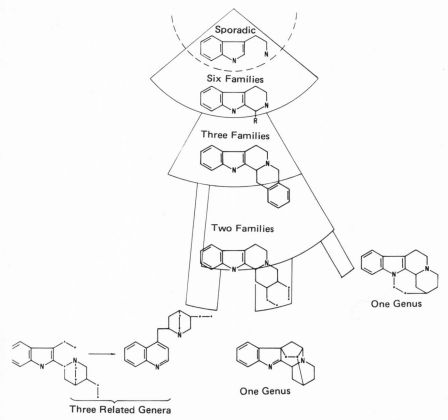

Figure 1.12. Percentage of frequency rule as demonstrated by occurrence of a group of indole alkaloids in plant groups derived from tryptophan. The more complex members of the series have the more limited distributions (Haensel, 1956). Courtesy of the author and the publisher of *Arch. Pharm.*

achieved without a setback to the plant, since the alkaloid which has accomplished the adaptation of the plant with its predators is no longer playing a protective role. The degradation products, for the most part unknown, are suspected to be unusual amino acids, phenolic compounds (Figure 1.13), etc. The conversion of the sparteine type of alkaloids was investigated using different methods:

1. Conversion of labeled alkaloids introduced into the plants
2. Conversion of alkaloids in an allogenic graft of different genotypes of lupine
3. Genetic experiments.

In all examples, a pathway leading from a tetracyclic quinolizidine on a

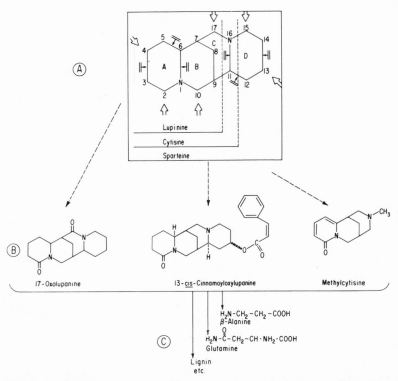

Figure 1.13. Degradation of quinolizidine alkaloids: *A*—general structure. *B*—end product alkaloids, and *C*—products derived from alkaloids. The structure of lupine alkaloids: four ring—sparteine type; three ring—cytisine type; two ring—lupinine type; large hollow arrow—most common oxidation and hydroxylation; arrows from parallel lines—most common dehydrogenation site.

hydrogenation level of sparteine to derivatives with dehydrogenated ring A, that has at least one carbon atom oxygenated, could be demonstrated. Further oxidation leads to compounds like 17-hydroxylupanine and methylcytisine. The compounds are detected with conventional alkaloid detection reagents and no more are detected. Another possibility is a degradation of ring D, which gives 13-*cis*-cinnamoyloxylupanine. The alkaloids are ultimately degraded to small molecular compounds such as β-alanine, glutamine, and lignin, or they may be incorporated into the cell wall. Both the tetrahydroanabasine and the quinolizidines commonly form derivatives with some acids, like acetic, tiglic, benzoic, cinnamic, ferulic, and other acidic compounds. The acids can form two kinds of derivatives, *N*-substituted or esters with hydroxy groups. This topic is further examined in Chapter 6.

That an enzyme system of an alkaloid-free plant is capable of synthesizing alkaloids was proven in experiments performed by Hasse and Maisack (1955), Skursky and Novotny (1966), Macholan (1965), Nowacki and Waller (1973), and others. The ultimate degradation of alkaloids needs some experimental proof in plants that can be suspected of being alkaloid-free or alkaloid-trace. The capability to transform foreign alkaloids into new alkaloids was demonstrated by Piechowski and Nowacki (1959), Nalborczyk (1968), Neumann and Tschoepe (1966), and Taylor and Shough (1967), using different groups of alkaloids as well as different plants. Hadorn (1956, 1961), working on a mutant in *Drosophila,* established that mutation of a single gene caused the disappearance of at least three compounds. It was established that all three compounds resulted from the activity of a single enzyme, xanthine oxidase, in the wild-type organism (Figure 1.14). This is described as a functional gene unit, as are those involved in the control of glycolysis and gluconeogenesis.

A widespread alkaloid family is made up of phenylalanine–tyrosine derivatives. These amino acids are precursors of a variety of compounds: alkaloids, flower pigments, phenylpropane acids, lignins, etc. In most cases, deamination is the first reaction, but sometimes decarboxylation, *O*-methylation, or *N*-methylation occurs. The resultant "protoalkaloids" can be transformed into an impressive number of alkaloids. In closely related orders, e.g., Ranales, Berberidales, Aristolochiales, and Rhoeadales, the commonly synthesized compound is norlaudanosoline, and it acts as a precursor for all alkaloids in these plants. In some orders such as Aristolochiales, the alkaloids can be converted into compounds that do not give an alkaloid-positive reaction. Another group of plants with tyrosine- and phenylalanine-derived alkaloids are the Amaryllidaceae and related taxa.

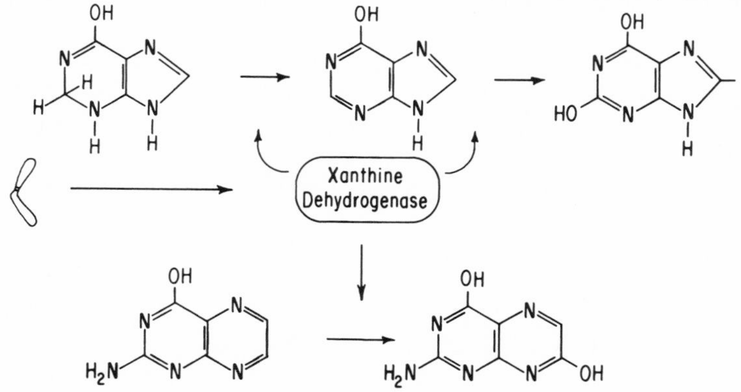

Figure 1.14. Examples of the effects of a mutation. The mutation "rosy" in *Drosophila* is lacking the enzyme xanthine dehydrogenase. Therefore, at least three compounds are not synthesized. Arrows show the actions catalyzed by the enzyme.

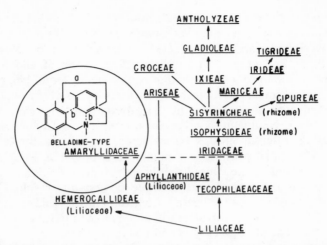

Figure 1.15. Relationships in the Liliaceae. Only the Amaryllidaceae have alkaloids of the norbelladine type. Other families are either alkaloid-free or contain only the colchicine alkaloids or unusual amino acids.

Here the first reaction is different from that in Ranales, since the product is norbelladine. This taxonomic unit manages to form alkaloid degradation products that are no longer recognizable as alkaloids (e.g., ismenine and belladine, see Figure 1.3) (Figure 1.15). The closely related Iridaceae do not accumulate alkaloids, but they store some unusual amino acids like *m*-carboxyphenylglycine and similar compounds. Perhaps these can be recognized as degradation products of alkaloids.

Information is needed regarding the biosynthesis and catabolism of alkaloids in each taxon if scientists are going to gather some meaningful information on the chemotaxonomy of the distribution of alkaloids. An example is provided by the closely related iridoid compounds, which include methylcyclopentane monoterpenoid alkaloids and indole alkaloids (Hegnauer, 1967; Waller, 1969; Gross, 1969; Bisset, 1975). They present a rather complicated picture of alkaloid (and terpenoid) distribution among many families.[4]

It is impossible to give all instances of good agreement between classical taxonomy and that based on distribution of alkaloids; it is also too early to do so, since data remain insufficient.

The distribution of nicotine alkaloids and tropane alkaloids in Solana-

---

[4]Tétényi (1973) says that "neither is the frequency of occurrence of the biosynthesis of a certain substance [Figure 1.12] to be taken as decisive; rarity is rather an indication of the aptitude to serve as a taxonomic marker."

ceae is worth mentioning. The genus *Nicotiana* accumulates nicotine and anabasine. The pyridine ring is derived from nicotinic acid, the piperidine ring in anabasine from lysine, and the pyrrolidine ring in nicotine from ornithine. As was mentioned for the Fabaceae alkaloids, both α-amino acids can be converted to the ring precursors by the same or similar enzyme systems. The genera *Hyoscyamus, Datura,* and *Atropa* accumulate tropane alkaloids; the genus *Duboisia* generates both the *Nicotiana* and tropane types. Moreover, there is a great diversity in the proportion of both types of alkaloids from species to species and still more diversity among genotypes.

The genera *Nicotiana, Datura,* and *Duboisia,* according to Bentham and Hooker (1876), Wettstein (1895, 1935), and Engler (1954), belong to three different tribes of the Solanaceae. The tribes are Nicotianinae, Datureae, and Salpiglossideae (Evans and Treagust, 1973). According to serological investigations of Hawkes and Tucker (1968) the tribes are closely related; the serological relationship also indicates a genetic or evolutionary relationship (Nowacki and Waller, 1973) (Figure 1.16). Other genera of the Solanaceae

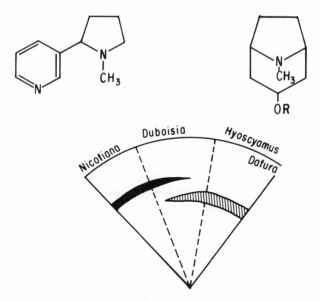

Figure 1.16. Alkaloids of the genera *Nicotiana, Duboisia,* and *Hyoscyamus,* family Solanaceae. At the top left is nicotine, and on the right is the tropane alkaloid type. Below there is a schematic chromatogram. The genus *Duboisia* includes species containing both alkaloids, and sometimes a difference exists between genotypes of a single species (Nowacki and Waller, 1973). Courtesy of the publishers of *Flora,* V. E. B. Gustav Fisher Verlag.

accumulate lupine alkaloids, such as the species *Solanum lycocarpum*. All of the above-mentioned Solanaceae alkaloids are produced by only minor modifications of one basic reaction pathway, which involves methylation of the amino acid nitrogen, decarboxylation, and ring closure. It is conceivable that these alkaloids had become established in plants as substances that offer some protection to the plant (at least in the early stages of its evolution) from herbivorous animals and insects and probably from some parasitic fungi.

A newly discovered but significant group of alkaloids, betalains, surprisingly not toxic, are also plant pigments (Piattelli and Minale, 1964, 1965; Minale *et al.*, 1966). An important group of plants lacks the usual anthocyanine or carotene pigments; the corolla is colored by betalains derived from tyrosine (Mabry and Dreiding, 1968; Chang *et al.*, 1974). The taxonomic significance of any given character is perhaps best ascertained by the constant presence of that character within any taxon and its absence from any other taxon which might be under consideration. The betalains are encountered in all but two families of the order Centrospermae (Mabry, 1966). On the other hand, plants which readily biosynthesize a compound will not establish themselves all over a taxon under conditions that give no selectional advantage to such plants. There will always be competition between plants that produce the particular compound and those that do not. Thus a poisonous but tasteless amino acid, canavanine, can be synthesized by some Fabaceae (Reuter, 1957; Tschiersch, 1959; Tschiersch and Hanolt, 1967; Birdsong *et al.*, 1960). It is actually produced only in some species which belong to different genera throughout the family. Sometimes in a species there are genotypes capable of producing this compound and others that are not able to produce it. Since it is tasteless and the poisoning effect is delayed, there is no way animals can learn to avoid the plant; so even the mutation which makes possible the synthesis of canavanine is just not characteristic of any genus of Fabaceae. As it has been maintained earlier, the significance of alkaloids for distinguishing genera or higher taxonomic units is small; the alkaloids can be of some value in establishing relationships only within genera or tribes.

There is another application of chemotaxonomy, actually chemotaxonomy *a rebours* (in reverse). A chemist, pharmacologist, or biochemist can be guided by the botanist in searching for plants which produce compounds of special interest. The castor bean plant, *Ricinus communis*, produces a peculiar alkaloid, ricinine. Studies by Waller *et al.* (1966), Essery *et al.* (1962), and Jackanicz and Byerrum (1966), showed some of the pathway of its biosynthesis. The genus *Ricinus* is monotypic, represented by the single species *R. communis*. No other plant among the Euphorbiaceae is known to accumulate ricinine. In 1964, a new alkaloid, nudiflorine, was isolated by

Mukherjee and Chatterjee from *Trewia nudiflora,* the closest relative of the genus *Ricinus.* Again, this genus is monotypic. Nudiflorine is a stereoisomer related to ricinine. Sastry and Waller (1972) isolated from *R. communis* and *T. nudiflora* some natural products recognized as derivatives or precursors of the main alkaloids (Figure 1.17).[5] Both genera *Ricinus* and *Trewia* belong to the same subfamily of Euphorbiales—to the Crotonoideae. Botanists recognize monotypic genera and higher taxa as units on the verge of extinction (Szafer and Kostyniuk, 1962). Therefore, it is conceivable that both *Ricinus* and *Trewia* are the only survivors of a previously more crowded taxonomic unit. Both alkaloids are synthesized from nicotinic acid or quinolinic acid via the pyridine nucleotide cycle[6] (Figure 1.17) (Waller *et al.,* 1966, 1975: Johnson and Waller, 1974). Their primitive ancestors most probably synthesized both isomers of α-cyanopyridone, but in the course of evolution the branch leading to *Trewia* lost the ability to accumulate the 2-pyridone derivatives and the *Ricinus* ancestors lost the ability to accumulate the 6-pyridone derivatives. The finding of Robinson (1965) and Robinson and Cepurneek (1965)—that a cell-free preparation of castor seedlings could catalyze the oxidation of *N*-methylnicotinonitrile to the corresponding 6-, 4-, and 2-pyridones in a 70:20:10 ratio—is difficult to explain since ricinine is not oxidized in the 6 position. The situation is comparable to a hypothetical situation in Fabaceae; suppose that all the thousands of alkaloid-bearing species of Sophoreae, Genisteae, and Podalyrieae were extinct except for one species of *Sophora* that could accumulate matrine and one species of *Lupinus* that could accumulate 13-methoxylupanine. One plant is a tree, the other an herb, and they have different flowers; but most botanists recognize them as the surviving numbers of tribes that were once widespread. The biochemist would furnish confirmatory data to the effect that the two isomeric bisquinolizidines are synthesized from lysine. In Fabaceae, the alkaloids are restricted to the above-mentioned three tribes. Some more primitive species of the genus *Cytisus* store both quinolizidine and dipiperidine alkaloids. The more primitive *Sophora* accumulate, in addition, three basic isomers of bisquinolizidines, the matrine type and (+)- and (−)-sparteines. The matrine type is absent in all more advanced genera, but some of these still accumulate racemic sparteines. Only the most advanced lupines store predominantly the (−)-sparteine derivatives, while members of

[5]The pyridine nucleotide cycle is a series of reactions that are ubiquitous in nature, differing only in the biosynthesis of quinolinic acid. It (quinolinic acid) occurs from tryptophan in animals, fowl, molds, and in certain microorganisms. It may come from either aspartate and glyceraldehyde-3-phosphate in higher plants and bacteria. In other plant systems highly specialized examples exist, such as mimosine, fusaric acid, and actinidine, where other precursors are used.

[6]For more complete references see Chapter 6.

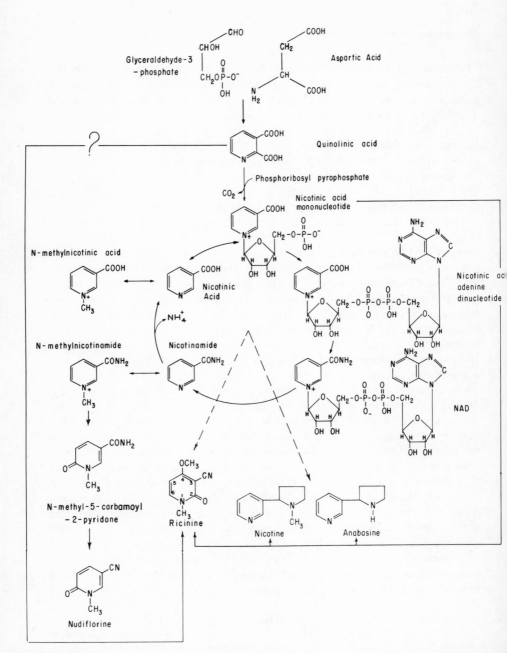

Figure 1.17. Biosynthesis of pyridine alkaloids from compounds in the pyridine nucleotide cycle (Waller *et al.,* 1966, 1975). Courtesy of the *J. of Biological Chemistry,* edited by the American Society of Biological Chemists, Inc.

the genus *Baptisia* contain only (+)-sparteine derivatives (Nowacki and Drzewiecka-Roznowicz, 1961).[7] The species with (−)-sparteine derivatives usually contain a higher level of alkaloids, and only in exceptional cases are there dehydrogenated derivatives. In contrast, the plants with (−)-sparteine usually have a lower level of alkaloids, and most of them are dehydrogenated. Sometimes the four-ring system is severed biologically and the D-ring is completely lost. It seems that, when starting with the first reaction from lysine, there are always alternative pathways, but these are used predominantly in the formation of the more oxygenated and/or dehydrogenated products (White, 1943, 1946, 1964; Mears and Mabry, 1971).

By combining the biosynthetic data obtained in our earlier studies (Nowacki and Waller, 1975a,b) with published evidence from the work of others and the distribution of alkaloids in the Fabaceae (Papilionaceae), a computer program to outline the possible pathways for the biosynthesis, conversion, and ultimate degradation of quinolizidine alkaloids in plants was designed (Nowacki and Waller, 1975c). The results are presented in the form of a "metabolic grid" (Figure 1.18a and Table 1.2). The "metabolic grid," together with the available data, permits some speculation on the origin of the alkaloids in species of Sophoreae, Podalyrieae, and Genisteae. As mentioned earlier, these three tribes are closely related and all but a few species have lupine alkaloids. The absence of lupine alkaloids in *Podalyria* and *Crotalaria* may be secondary in character; both genera actually produce alkaloids but of different structure. While *Podalyria* has tyrosine-derived amines, *Crotalaria* accumulates pyrrolizidine alkaloids which can be synthesized from ornithine on a pathway homologous to the synthesis of lupinine from lysine (Nowacki and Byerrum, 1962). With these exceptions, all other lupine alkaloids can be derived from lysine (Figure 1.18b). The next step in the biosynthesis must be a symmetrical compound since the distribution of radioactivity in the rings is such that it is explainable only when a symmetrical intermediate is assumed (Schuette, 1960, 1965; Schuette *et al.,* 1961). The incorporation of cadaverine (B11) with a lower dilution factor than lysine and the decrease of lysine incorporation when fed simultaneously with a diamine makes the assumption highly probable that cadaverine is the proper intermediate. Cadaverine can be oxidized to form a piperidine ring both by a diamine oxidase and an amine transferase. The diamine oxidase can be excluded as the enzyme actively synthesizing the alkaloids, since no relation could be found between the actual alkaloid synthesis and the level of diamine oxidase in the plants. Some sweet lupines have higher levels of this enzyme than the bitter lupines (Nowacki, 1964).

---

[7] There seems to be some enzymatic predisposition in both groups of species.

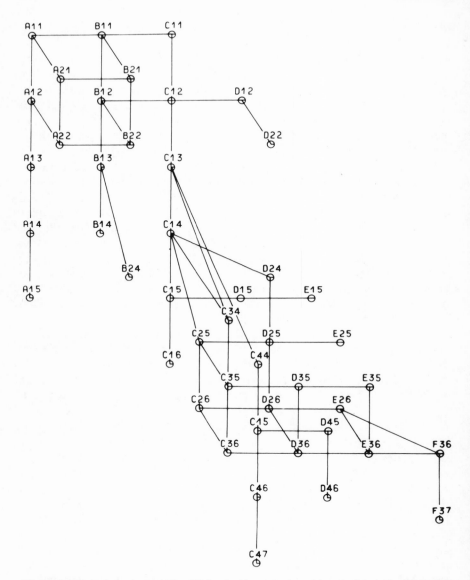

Figure 1.18a. Proposed metabolic grid for the biosynthesis and interconversions of quinolizidine alkaloids and related products in Fabaceae (Papilionaceae). For definition of symbols, see Table 1.2. Construction of the metabolic grid of lupine alkaloids: The data on the various conversions of the lupine alkaloids were used to construct short metabolic pathways. Data concerning the distribution of lupine alkaloids in Leguminosae were drawn from Boit (1961); Cranmer and Turner (1967); Mears and Mabry (1971); Aslanov *et al.* (1972); Wicky and Steinegger (1965); Faugeras and Paris (1968); Bratek and Wiewiórowski (1959); Balcar-

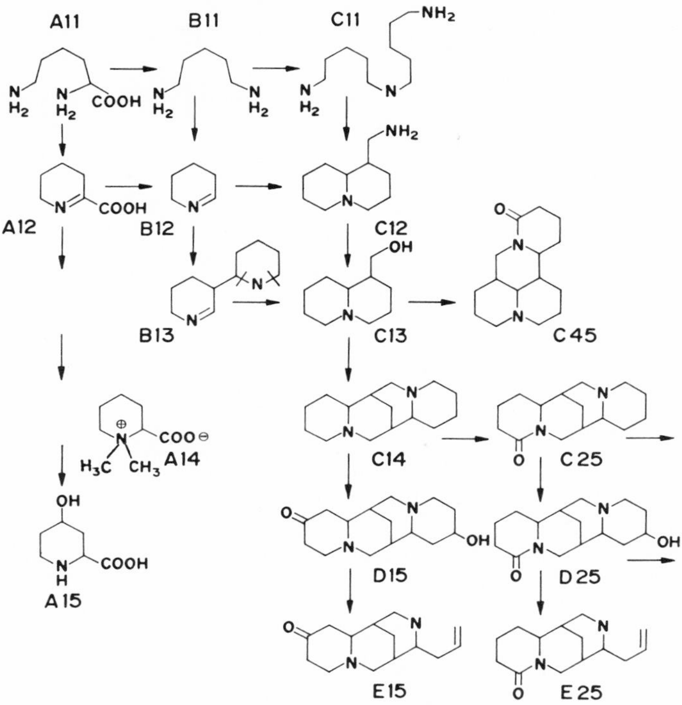

Figure 1.18b. Pathways for the conversion of lysine into alkaloids (see Table 1.2 for names of compounds).

Skrzydlewska and Barkowski (1966, 1972); Gill and Steinegger (1963, 1964a,b); Kustrak and Steinegger (1968); Steinegger and Moser (1967); Steinegger and Wicky (1965); Steinegger *et al.* (1963, 1968); Shalaby and Steinegger (1964); Bernasconi *et al.* (1965a,b); Goldberg and Moates (1961a,b,c). The pathways were connected when they overlapped. Tentative connections were drawn where no experimental data were available, but the structural formulas of the alkaloids enabled us to propose a conversion. In cases where the conversion of one substance to another was known to exist, but there were two possible intermediates and no proof was available that would permit us to choose one of the intermediates and disregard the other, both presumable intermediates were assigned the same probability. Each reported entry of an alkaloid was assigned a separate symbol (Table 1.2) in the computer program. Tracing backward from the substance to the common precursor, lysine, each substance on the pathway was assigned a separate symbol. In case there were two or three similar possibilities, e.g., thermopsine, which can arise theoretically from both anagyrine and isolupanine, and no data were available that indicated which of these alkaloids was the real substrate, both were assigned an equal chance for a symbol. To simplify the program, D and L isomers were treated together and the *Ormosia* alkaloids were disregarded.

Table 1.2. Metabolic Grid Symbols and Extent of Occurrence of Fabaceae (Papilionaceae) Alkaloids and Precursors[a]

Grid symbol	Compound name	Number of times produced	Number of times occurred as end product
A11	Lysine	1002	0
A12	Pipecolic acid	28	3
A13	Methylpipecolic acid	24	20
A14	Homostachydrine	4	2
A15	Hydroxypipecolic acid	2	2
B11	Cadaverine	972	0
B12	Δ^1-Piperideine	498	1
B13	Tetrahydroanabasine	23	0
B14	Ammodendrine	7	7
C11	Dicadaverine	473	0
C12	Aminolupinine	947	1
C13	Lupinine	944	32
C14	Sparteine and dehydrosparteine	847	168
C15	Multiflorine	23	14
C16	Dehydromultiflorine	1	1
D12	Lusitanine	2	1
D15	Hydroxymultiflorine	8	2
E15	Albine	6	6
A21	N-Methyllysine	2	0
A22	Homostachydrine	3	2
B21	N-Methylcadaverine	1	0
B22	N-Methylpiperidine	2	2
B24	Orensine	16	16
C25	Lupanine	478	123
C26	Anagyrine	243	119
D22	Hystrine	1	1
D24	Hydroxysparteine	153	83
D25	Hydroxylupanine	141	26
D26	Baptifoline	217	9
E25	Angustifoline	6	6
E26	Rhombifoline	193	9
C34	Isosparteine	50	9
C35	Isolupanine	82	4
C36	Thermopsine	31	16
D35	Isohydroxylupanine	62	1
D36	Isobaptifoline	46	0
E35	Tetrahydrorhombifoline	46	0
E36	Rhombifoline	138	0
F36	Cytisine	276	154
F37	Methylcytisine	122	122
C44	Desoxymatrine	40	4
C45	Matrine	36	9
C46	Dehydromatrine	23	3
C47	Leontine	20	20
D45	D-Leontine derivative	4	0
D46	D-Leontine derivative	4	4

[a]Nowacki and Waller, 1975c (courtesy of Pergamon Press).

The cyclic derivative of cadaverine can be dimerized or polymerized enzymatically in the cell. Thus, the origin of the dipiperidine alkaloids such as ammodendrine (B14), sanguinarine, and hystrine (D22) is explainable without considering special enzyme systems being required to convert the piperidine nuclei. The dipiperidine alkaloids are encountered in the most primitive species of the tribes Sophoreae and Genisteae. A conversion of the dipiperidine into lupinine is possible only by means of a rearrangement in the molecule coupled with the loss of a nitrogen atom; unfortunately, the experimental data do not confirm this assumption. Actually, the dipiperidines are poorer substrates than cadaverine; but lupinine is easily converted into sparteine in *L. luteus* (Schuette, 1961) and should be the key compound leading to the diverse sparteine, matrine, and ormosianine type of alkaloids. Ultimately the branching of the pathways leads to the production of more than 40 compounds. Some of these may be changed structurally so that they do not react with the usual alkaloid detection reagents. The ultimate result would probably be a total degradation of the alkaloid. The catabolic steps remain, for the most part, unknown. A posible degradation product of the bisquinolizidine type of alkaloid may be 3-methoxypyridine, found in some species of *Thermopsis*.

A similar pathway, yet one more complicated, is found in the Ranales. Genera which belong to Berberidaceae, Papaveraceae, and some Ranunculaceae all accumulate a great number of alkaloids which are derivatives of norlaudanosoline (Figure 1.19). The biosynthetic pathways, like those in Papilionaceae, are multibranched (Sárkány *et al.*, 1959a,b). Some pathways are restricted to smaller units, that is, to sections in a genus. Such is the case for the morphine type of alkaloids, which are only found in the section *Mecones* in the two species *Papaver somniferum* and *P. setigerum*. The closely related *P. orientale* and *P. bracteatum* belong to the section *Macrantha,* and they accumulate either thebaine, a substrate for morphine, or isothebaine, which can be demethylated to give oripavine, an isomer of codeine. Both species of *Macrantha* do hybridize with *P. somniferum,* giving an infertile F_1 generation. The rhoeadines are found only within the genus Papaveraceae. The alkaloid trait in Polycarpicae was established probably very early in the phylogenetic development; therefore, botanists have different arrangements of the families and tribes in their taxonomies. Generally they agree that all the above-mentioned families are related (Figure 1.20). A different but more natural and younger unit is the family Amaryllidaceae. It belongs to the class Monocotyledonae, subclass Corolliferae, order Liliales. In this family, as in the Ranales, the amino acid precursor of alkaloids is tyrosine, but the first reaction is different. The most common intermediate en route to a majority of Amaryllidaceae alkaloids is norbelladine. Modification of the pathway leads to a large number of derivatives, some of which can be utilized as taxonomic characters for the determination of plant relationships (Figure 1.20).

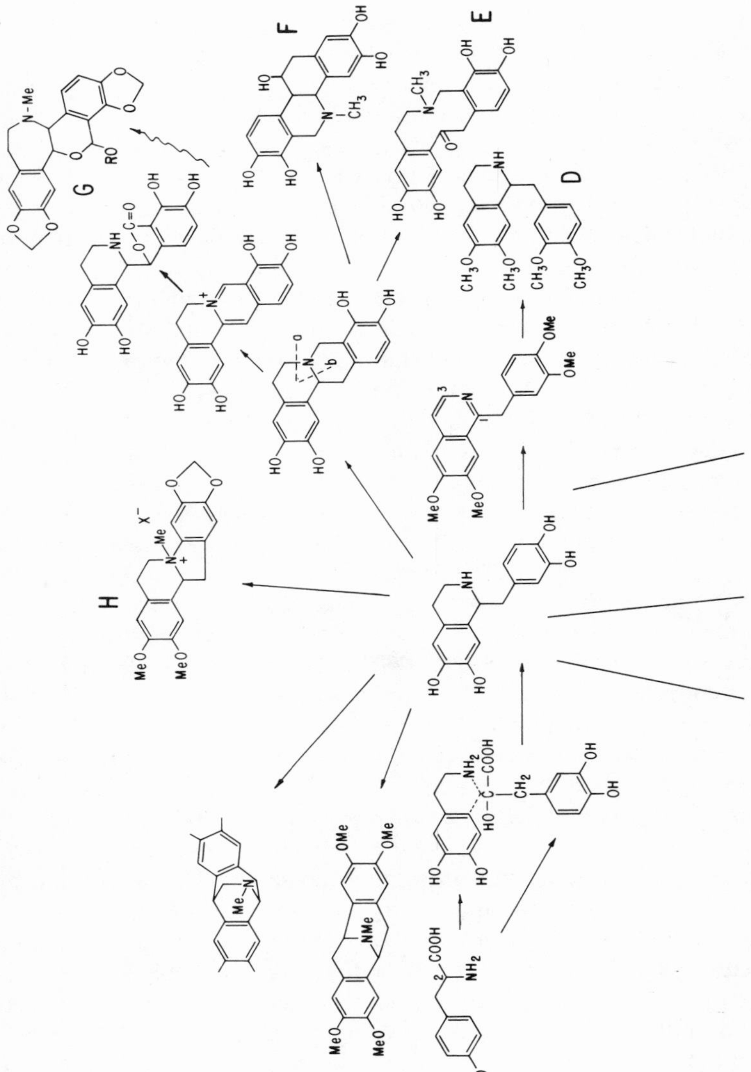
D
E
F
G
H

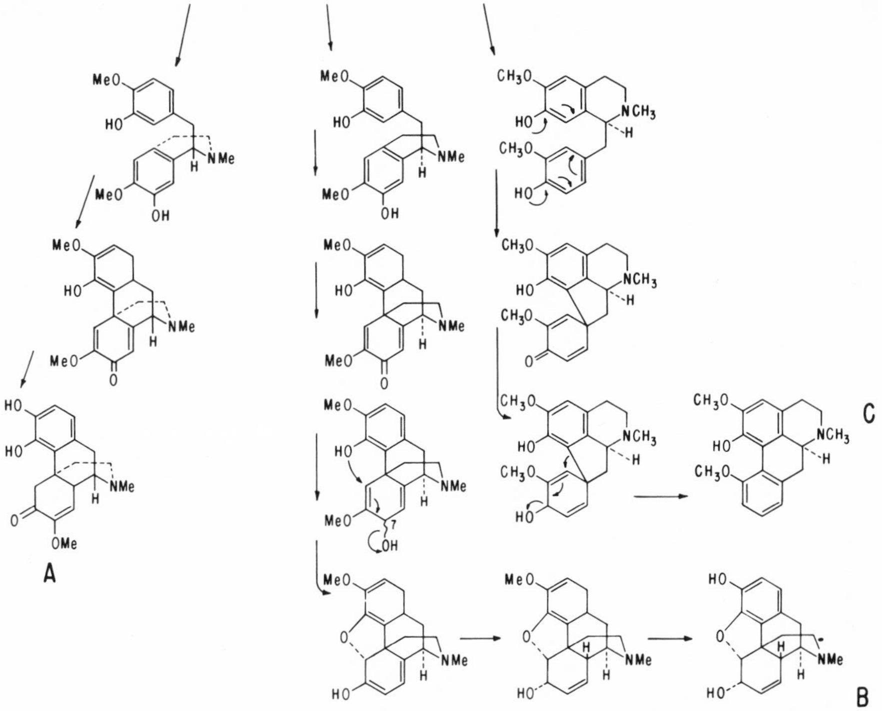

Figure 1.19. Tyrosine-derived alkaloids in the Rhoeadales, Berberidales, and Ranales. The first compound synthesized is norlaudanosoline, and nine distinct types of alkaloids are encountered either in restricted taxa or widely distributed: *A*—salutaridine; *B*—morphine pathway restricted to *Papaver* section *Mecones; C*—isothebaine pathway, characteristic of *Papaver* section *Macrantha; D*—*Papaver* section *Miltantha; E*—*Papaver* section *Scapiflora; F*—encountered in *Chelidonium* (Papaveraceae–Rhoeadales) and *Berberis; G*—characteristic for *Papaver* section *Orthoroedes*, the *Corydalis dicentra* and *Chelidonium* spp.; *H*—cryptowoline; *I*—*Papaver* section *Scapiflora*, genus *Escholtzia*, and *Argemone; J*—*Papaver* sections *Orthoroedes* and *Pilosa.* Data are compiled from publications by Stermitz (1968, 1974), Santávy (1966), Santávy *et al.* (1965, 1966) Slavík (1955), Slavík *et al.* (1963), Pfeifer (1962), and Tétényi and co-workers (1961, 1965, 1967, 1968), Bandoni, *et al.* (1972, 1975). The reaction pathways are according to Barton and Widdowson (1972). Only pathways *A*, *B*, and *C* are shown in detail; the others are only suggestions. The pathways leading to the dimeric (bis) alkaloids are not shown.

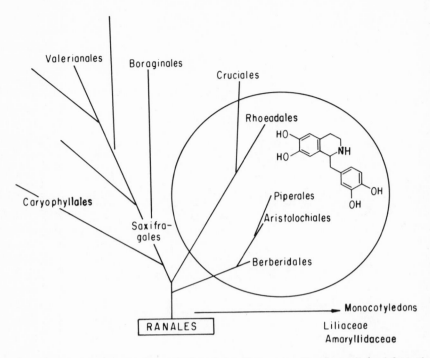

Figure 1.20. Some possible phylogenetic relationships among the plants derived from the Ranales (Hutchinson, 1959). Courtesy of the author and Academic Press, Inc., London.

Besides the alkaloids that are restricted to phylogenetically related taxa, there are a number that are so widely and randomly distributed in the plant kingdom that it is impossible to use them for taxonomic purposes. It is notable that alkaloids of quite simple structure are the most randomly distributed. Some methylated derivatives of tyrosine and tryptophan, as well as condensation products of their compounds with simple aldehydes like formaldehyde and acetaldehyde, are encountered in small units distributed all over the plant families. Since the reactions are very simple and relatively inexpensive in energy requirements, it is not surprising that their products are so widespread. Some of these compounds are either very weak poisons or not toxic at all.

Any attempt to apply biochemical data to taxa with very uncertain affiliations is difficult. Also, to form a taxonomical classification without regard to the morphological, histological, and cytological data is unjustified. On the other hand, botanists sometimes are so divided about where to place a certain unit that they welcome any additional evidence regarding its proper placement. Such tedious subjects for study are the Loganiaceae, Rubiaceae, Euphorbiaceae, Apocynaceae, Valerianaceae, and Buxaceae (see Figure 1.21). Some of these families are included by Hutchinson (1959) in the order Magnoliales, and others in Ranales, while Engler (1954), Cronquist (1968), and Takhtajan (1959) arrange the groups in a very different way. As far as the alkaloids are concerned, the group is also very diversified. Some of the plants accumulate alkaloids derived from tryptophan, others only pseudoalkaloids of the mono-, sesqui-, and diterpene types, while some store alkaloids derived from tyrosine or diamines. Yet this seemingly artificial agglomeration of families has some common characteristics. The first is the wide distribution of irididiol-like compounds within it, and the second is the lack of sufficient morphological data to permit arranging the families to satisfy most scientists (botanists). Compounds such as irididiol and related lactones are quite easily converted into the corresponding aldehydes. An aldehyde can react not only in plants but also sometimes even more easily *in vitro* with an amine to form an "alkaloid-like" compound. Experiments *unter Zell-möglichen Bedingungen* (under cell-feasible conditions) have usually been disregarded in recent years as too artificial, yet a hypothesis by Wenkert (1968) concerning the biosyn-

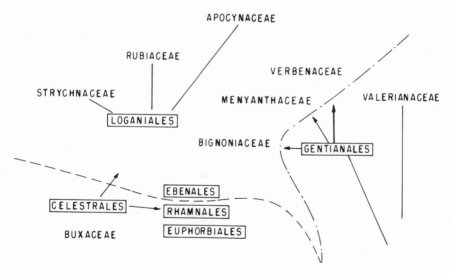

Figure 1.21. Phylogenetic relationship in plants of the Celestrales, Loganiales, and Gentianales orders. The dash–dot line is according to Engler's (1954) systematics. The dashed line is according to Hutchinson (1959). In each case, plants on different sides of a dividing line should be unrelated, but it is important to recognize that they may have some characteristics in common.

thesis of the C_9-C_{10} unit of the indole alkaloids belonging to the disputed group of plant families was confirmed in the last few years. According to findings of Barton (1964), Battersby (1961, 1966), Battersby et al. (1965a,b, 1966, 1968a,b, 1969), Battersby and Gregory (1968), Guarnaccia et al. (1970, 1971), Arigoni (1972), Brechbuhler-Bader et al. (1968), Loew and Arigoni (1968), Loew et al. (1968), Inouye et al. (1968), Bu'Lock (1965), and Scott et al. (1973), such indole alkaloids as catharanthine, serpentine, ajmalicine, and vindoline, as well as the isoquinoline alkaloids emetine and proemetine, are biosynthesized from the corresponding amines and loganin, geraniol, or a related monoterpenoid unit. It is presumed that for the differently arranged taxa the amine precursor is less important from the taxonomic point of view than the monoterpenoid unit. This group of plants needs reconsideration by the taxonomist, since it seems to be more homogeneous than was once thought. The division of Dicotyledoneae into only two subdivisions seems to be unjustified, and it is probably more natural to separate them into a greater number of parallel developing units (Sokolov, 1952).

It was pointed out at the beginning of this chapter that there is a lot of disagreement between the natural products chemist and the taxonomist. Errors have been made on both sides. Because the chemist is mostly concerned with the chemical and physical properties of isolated compounds, she or he pays no attention to the botanical source from which the compound was isolated. Remarks like the following are frequent: "An alkaloid was isolated from the bark of a Congo tree not unquestionably identified botanically" and the plant taxonomist usually is not sufficiently trained to apply the best knowledge of alkaloid distribution to systematic efforts (Mirov, 1963). Hence, the discipline of biochemistry, which has come on the scene recently, offers an alternative solution; indeed it may be the best solution, since the biochemist, by training, is both a biologist and a chemist.

The evolution of Angiospermae is 150 million years old according to the fossil data. In this long time, a countless number of mutations have occurred. Some mutations have no plus or minus selective values that are preserved in all the taxa in which they are scattered, and sometimes they are scattered without regard to their taxonomy. At other times, they are restricted to some higher or lower units. When by chance a plant already has the capability of performing a reaction (resulting from a mutation) leading to the accumulation of a substrate, this plant starts to synthesize alkaloids. Only when this type of mutation occurred in a newly developing taxonomical unit and had a high selective value would it have been distributed all over the new unit (Figure 1.22). Aberrations from the main pathway may already have occurred. Thus the Rhoedales, the taxonomically remotely located Berberidaceae, and some of Ranunculaceae all have

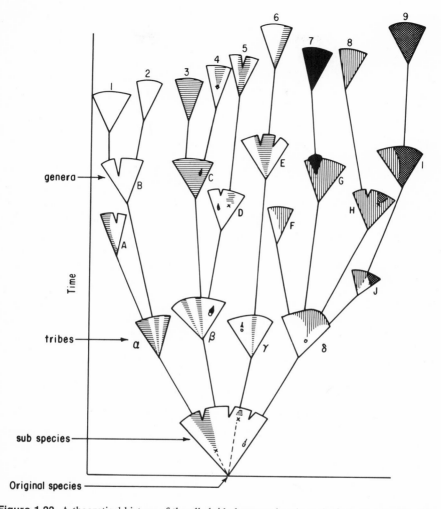

Figure 1.22. A theoretical history of the alkaloid character in a hypothetical plant family. At the beginning, the family was represented by only one species divided into four subspecies. Mutations leading to alkaloid synthesis occurred three times (marked by *x* and *o*). The mutations marked with *x* were selectively favorable or neutral, and those marked with *o* were inferior. In the course of evolution, the original form gave four new species, which later developed into four tribes (marked with Greek letters). Three of the tribes, α, β, and γ, inherited by drift the alkaloid character of the fourth, δ, and a new and superior mutation took place. The character spread over the entire species. The four tribes gave rise to nine genera (marked with capital letters). The alkaloid marked with horizontal lines was randomly distributed by drift into the derivative genera. The alkaloid marked with the vertical lines, since it was selectively superior, was distributed to all descent forms. New mutations may occur (*x* selective, neutral, or superior; or *o* negative), rapidly weeding out the negative. The cross-hatched field represents plants capable of further conversion of alkaloids.

alkaloids of the protopine group, and the modifications in their structure seem to be similar throughout. All plants that belong to the group of Rhoedales have alkaloids, but a certain alkaloid, canadine, which occurs in many of the species, can be present in distantly related units (Vágújfalvi and Tétényi, 1967; Lörincz and Tétényi, 1966). This alkaloid may be absent either in some plants or in the progeny of a single plant. For the Rhoedales group, it seems important to produce tyramine-derived alkaloids, because no single plant in this group is known to be alkaloid-free; but the particular alkaloid that is produced seems to be without any importance (what is important is that some alkaloids are produced), so the genes for the conversion of norlaudanosine into the hundred-plus derivatives are scattered all over the group. A similar case (on a much smaller scale) was found in the genus *Baptisia* by Turner (1967), Cranmer and Turner (1967), and others at Austin (Cranmer and Mabry, 1966; Mears and Mabry, 1971). The genus *Baptisia* accumulates alkaloid derivatives of (+)-sparteine. The derivatives are anagyrine, lupanine, baptifoline, cytisine, *N*-methylcytisine, rhombifoline, and 13-hydroxysparteine. The distribution of the mentioned alkaloids was found unrelated to the taxonomic division of the genus. On the other hand, this same group of lupine alkaloids in another genus *(Lupinus)* are good markers of the species; consequently, it is possible to place a plant in its proper place using only alkaloid distribution (by TLC or paper chromatography) as a guide, without visually observing the plant. In Papaveraceae, a similar tendency is found but only in a few sections of *Papaver* and *Argemone* (Santávy, 1962, 1963, 1966; Pfeifer and Mann, 1966; Pfeifer and Heydenreich, 1961; Slavík, 1955, 1965; Slavík and Slavikova, 1963; Stermitz, 1968; Stermitz and Coomes, 1969; and Stermitz *et al.,* 1969, 1973a,b). It is remarkable that in most cases the plants with a close relationship of the alkaloid spectrum to the morphology of the various species are regarded by a taxonomist as either "old" or "good" species. This means that a given species does not cross with relatives and that in the past the species was limited to a small population from which some genes were eliminated by accidental loss (drift). The section *Mecones* in *Papaver* has only five surviving species, and the section *Macrantha* only four. These sections are closely related, yet they are difficult to cross, and the cross usually produces only sterile hybrids. The difference in alkaloids is small but remarkable. The *Mecones* accumulate codeine and morphine, while the *Macrantha* transformation stops with the formation of thebaine or its homologue, isothebaine. The *Mecones* differ from *Macrantha* by an additional character: the main alkaloids are all derivatives of (+)-reticuline, and at least two pathways of alkaloid conversion from norlaudanosine in *P. orientale* operate. One proceeds through reticuline, the other through orientaline. In lupines, the now "good" and "old" species operate similarly, with only limited pathways for alkaloid conversion. In *L. luteus*, the only pathway leads to (−)-sparteine and stops there. A gene for the conversion of this compound was lost, but the species still has the capability of converting hydroxylupanine to a compound that does not react with alkaloid detection reagents (Reifer *et al.*, 1962). In *L. angustifolius*, only a single pathway converts (−)-sparteine to the derivatives lupanine, hydroxylupanine, and angustifoline. In sections *Albus* and *Pilosus*, two pathways lead from (−)-sparteine to both oxosparteines (lupanine and multiflorine) and involve a reaction catalyzed by alcohol dehydrogenase. The American species of lupines are much more difficult to classify by the alkaloids they produce, but this oddity is explainable. The Mediterranean species are the only survivors of the harsh climatic conditions of the glacial period, and then only in a few isolated places. The American species gradually became less common in the northern part of the continent and were subsequently found primarily in Mexico and countries south, where they emerged as young species, and much introgression took place. As they began to repopulate Mexico, they were still in the hybridizing stage, and some of the genetic carriers were insufficient to prohibit introgression of genes. Lupines as a unit are far advanced in evolution from the common stock and have lost the ability to accumulate (+)-sparteine and the dipiperidine types of alkaloids; they contain only (−)-sparteine derivatives (Nowacki, 1968).

In summary, an alkaloid metabolism pattern can be established in a taxon when it proves of high selective value; thus if a mutation conferred unusual adaptive benefits on an individual bearing it, that individual would leave more offspring than others in the population, and all of its progeny bearing the mutant allele would do likewise. If the whole population were large, the individuals bearing this mutated character would at first be only a small proportion of the total. With the difference in survival rates, however, the new type would tend to increase by geometric proportions. The mutant, while increasing in number, will cross with the nonmutated members of its species. In some combinations it can encounter varieties of genotypes with preadapted enzyme systems capable of transforming the product of the mutant gene. Thus, while the mutated allele is increasing in number, it will be combined with an enormous number of combinations of other alleles capable of transforming the first product so that the alkaloid spectrum produced by the species will tend to become diversified. The superior type will now tend to form a new species, and the character will be transmitted to the derived units with all or part of the inherited diversity. As long as no genetic barriers are imposed, a gene flow can tend to maintain the broad spectrum of alkaloids. But when in the course of evolution the species is restricted in number, good working genetic barriers in the meantime are formed, and the species can lose alkaloids. This loss may be due to further mutations or to selection of specific alkaloid conversion (biosynthetic) pathways. There will be a tendency to use a uniform but specialized pathway common to all members of the species. If such a species starts again to diversify morphologically into new units, all of these units will have the same or similar alkaloids.

There are only a few steps necessary to transform a product of intermediary metabolism into an alkaloid, and, therefore, the same product can be metabolized independently in unrelated taxa. In some rare cases, such as in the Rutaceae, plants produce as many as six different types of alkaloids (Price, 1963) (see Figure 1.6). This can be rationalized with a model similar to that used for explaining the origin of the single type of alkaloids. The assumption is made that in a certain species of different populations two independent mutations occurred and that both were favorable (Figure 1.23). The mutants took over the population, and, when the

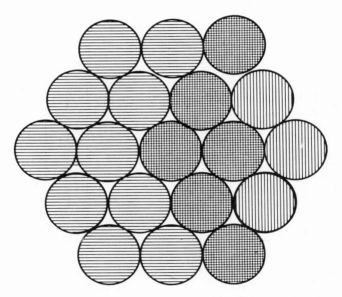

Figure 1.23. Explanation of possible ways of the origin of plants containing two types of alkaloids, in this case furoquinolines (derivatives of anthranilic acid) and benzylisoquinolines (derivatives of tyrosine), in the closely related genera *Fagara* and *Zanthoxylum*. According to the hypotheses the alkaloid character originated independently in two separate postulations. By means of introgression it spread over the whole unit. In some cases the two alkaloid-bearing types met and both characters were retained in the offspring (crosshatched circles). The horizontal lines in circles are species with benzylisoquinoline alkaloids, and the vertical lines are species with furoquinoline alkaloids. Data on alkaloids are from Boit (1961) and Vaquette *et al.* (1974).

populations merged thereafter by means of introgression, the different compounds (alkaloids) were established throughout the newly formed uniform population. A reverse of this situation, the loss of alkaloids due to mutation, is also possible; but the plant metabolism is usually so well adapted to the production of these compounds that this rarely occurs. Equilibrium mechanisms are involved, and an alkaloid-free mutant can be rendered nonviable or become extinct by an overproduction of organic acids which can no longer be buffered by the alkaloid (Nowacki, 1969).

1.3. Conclusion

The alkaloid character is now established in some taxa due to environmental pressure (this subject will be dealt with in more detail in Chapter 4). Alkaloids are of little value in the larger taxonomical units, but they can prove to be of great value in smaller ones. Knowledge of biosynthetic pathways to them, as well as of their distribution in taxa, aids in the search for new compounds that have predictable structures. On the other hand, it is helpful to construct, from short pathways established in biosynthetic experiments, larger ones involving the entire unit, as has been done for the lupine alkaloids (see, for example, Mankinen *et al.*, 1975).

2 | Genetic Control of Alkaloid Production

2.1. One Gene, One Enzyme, One Reaction Hypothesis

In 1865, J. G. Mendel, an abbot in an Augustinian monastery in Brno, now in Czechoslovakia, published a paper on the inheritance of certain morphological characters in the pea, *Pisum sativum* (Peters, 1959). One of these was a chemical character; while crossing red and white flowering peas, Mendel actually crossed varieties with and without the ability to synthesize the red flower pigment known as anthocyanidin. The ability to synthesize the pigment was a dominant trait, thus presenting the first proved example of the inheritance of a chemical character. It was significant that the trait was dominant. A lapse of eighty years occurred before the rules governing the inheritance of chemical characters were properly understood. In the meantime, hundreds of new examples were added, and in most cases the genetic transmission of the chemical character was a dominant trait (Nowacki, 1964a,b, 1968).

The explanation of this dominance is that a gene is necessary to produce a compound and that a recessive character is actually due to the absence of a gene (Srb and Owen, 1952). Crossing two organisms which both lack a certain compound usually results in the production of offspring that also lack the particular substance. An exception is complementation, when defects are in different enzymes of the same pathway. In the opposite case—the crossing of individuals possessing a particular substance—the offspring all have the character, or, in cases where the parents were heterozygous, a majority of the next generation have the character and the minority are without it. The segregation ratio is 3:1, where only one gene is involved, but there can be many different segregation types when more than one gene is involved.

49

The concept that an organism's metabolic characteristics are controlled by its enzyme makeup, which in turn is controlled by genes, raises the next question: How do the genes function? In 1941 Beadle and Tatum began a study with a simple organism which they found admirably suited for this purpose. It was *Neurospora crassa,* the common pink bread mold. It was an easily cultivated organism since it needed only a source of carbohydrate, inorganic nitrogen, other inorganic salts, and a trace of biotin. Beadle and Tatum irradiated the mold with X-rays and produced mutants which lacked the ability to synthesize certain amino acids. In all examples, the strains of molds which could not synthesize a substance were recessive to those able to perform the synthesis. The clearest way to account for such a situation was to suppose that the X-rays had damaged the gene responsible for the formation of an enzyme necessary for manufacturing the substance. In the absence of a normal gene, *Neurospora* could no longer make the specific enzyme required for the synthesis of a specific amino acid.

There is a certain analogy between experiments involving *Neurospora*—or other molds—and those involving alkaloids. The substances investigated in *Neurospora* were essential amino acids, vitamins, etc. It was possible to breed those organisms on media supplemented with substances which the organisms were unable to produce. A higher plant is an extremely difficult organism to cultivate on media supplemented with organic substances. Since a plant lacking the ability to synthesize an essential product is unable to survive, genetic research on essential substances is difficult.

It has, however, become possible to investigate the inheritance of the ability to accumulate compounds that are of secondary importance to the plant. Alkaloids are commonly regarded as substances of this type, *secondary metabolites.* Plants which use different metabolic processes in alkaloid synthesis have survived, in both pot and field cultivation conditions. The basic life processes of these mutants remained *ipso facto* unchanged, so it becomes feasible to study the biosynthesis of secondary metabolites, i.e., the alkaloids produced. Of course, some of the mutants are less viable than the original wild type, but usually several exceptional examples can be cultivated in the greenhouse (Böhm, 1969; Mothes, 1960).

From the very beginning of alkaloid research a common trait became apparent: usually the alkaloid-bearing plant produced not a single compound but a series of chemically related substances. For example, the poppy produces not only morphine but also thebaine, papaverine, norlaudanosine, codeine, as well as other compounds having a similar structure (Battersby *et al.,* 1964; Manske and Holmes, 1950–1975). Tobacco plants were found that produced not only nicotine, but also nornicotine, anabasine, and a series of minor, related, alkaloids. The structural similarities of alkaloids from a given plant suggest that they are intermediates and products of the same biosynthetic pathway. On the other hand, some plants belong to alkaloid-bearing species or genera that are depleted of alkaloids and are considered very alkaloid-poor; hence, the isolation of alkaloids is difficult if not impossible. Examples are the "bitter" and "sweet"[1] lupine varieties and the alkaloid-rich *Nicotiana rustica* and alkaloid-poor *Nicotiana alata*.

Knowledge obtained from experiments performed on *Neurospora* and other organisms showing that the dominant substance is the product and that the recessive one is the substrate permits the scientist to deduce the biosynthetic pathway. It becomes possible not only to investigate the dominance of a single substance but also to outline a complete reaction pathway. As this will be shown later, the pathways elucidated by means of genetic experiments are generally in agreement with results obtained from biochemical investigations employing the use of isotopic tracers.

With all the previous statements in mind, it is possible to propose a research program of alkaloid biosynthesis and metabolism employing mutants that specifically involve alkaloids and studying the reactions involved in their anabolism and catabolism.

Genetic research is usually possible only by analyzing hundreds and thousands of individuals. Without a simple analytical method it would be a very tedious project. Relatively simple methods were developed by Tswett (1903), who is considered the father of paper chromatography. Now advanced techniques—paper chromatography, thin-layer chromatography, gas–liquid chromatography, and high- and low-pressure liquid chromatography—are being used by scientists conducting research in this field. For purposes of genetic experiments, any method is satisfactory which allows one to perform a great number of analyses while sacrificing only a small part of the plant material, leaving the rest to be used in breeding the next generation (Kraft, 1953; Feigl, 1966; Becker, 1956; Nowacki, 1959, 1963a,b, 1966; Buzzati-Traverso, 1960; Schwarze, 1963).

2.2. Alkaloid-Rich and Alkaloid-Poor Plants within a Species: The Inheritance Pattern

There are two well-documented series of inheritance. The alkaloids involved are of distinctly different metabolic origin. The first example is the *lupine alkaloids*. In 1929 a German chemist, Sengbusch, with encour-

---

[1]"Bitter" and "sweet" lupines refer to "alkaloid-rich" and "alkaloid-poor," respectively.

agement from Baur, a geneticist, developed a simple chemical method to distinguish alkaloid-bearing plants from nonalkaloid plants. Using this method, he was able to find in millions of bitter lupines three "sweet" plants that gave a negative alkaloid test (Sengbusch, 1930; Seehofer, 1957). In the following years, other alkaloid-poor plants in other lupine species were found (Sengbusch, 1931). The analytical method was patented, and since no other publication resulted, the techniques used were not made available to other scientists. Two years later similar results were obtained by the Soviet scientists Ivanov and co-workers (1931a,b, 1932) and Fedotov (1932), who published their results. Some time later the Germans decided to publish their results (Sengbusch, 1938, 1942; Hackbarth and Sengbusch, 1939; Hackbarth and Troll, 1941; Nowacki, 1959; Hackbarth, 1961). From the first publications it was clear that the genes for alkaloid-poor character were recessive, which indicated that the plants that produced only traces of alkaloids were actually *defective in this character*.

In a cross of a bitter lupine with any of the sweet mutants, the F_1 was bitter, and in the F_2 generation a segregation of 3:1 (bitter:sweet) resulted. In crosses of sweet plants from different mutants, the F_1 generation was bitter. This finding indicated that the first three isolated mutant plants from the F_2 generation had the alkaloid biosynthesis pathway interrupted by a different mutation. The three first sweet plants gave rise to three breeding lines of sweet yellow lupines: Stamm 8, Stamm 80, and Stamm 102, in which the genes were, respectively, named *dulcis, amoenus, and liber* (Schwartze and Hackbarth, 1957). In a cross of a plant of one line with a plant of another line, the F_2 generation segregated with a ratio of 9:7; out of the nine bitter plants only one was found to breed true in the F_3 generation, while the rest were segregated in a 9:7 or a 3:1 ratio. The seven sweet plants when backcrossed with the parental strains revealed that three were homozygous in the defect from one parent, another three from the other parent, and one was a homozygous double recessive. This means that $\frac{1}{16}$ of the F_2 plants had accumulated both of the defective genes (Figures 2.1a and 2.1b).

A pattern corresponding to that described above was found in the inheritance of flower pigments in sweet peas by Scott-Moncrieff (1936) and Harborne (1963). Scott-Moncrieff crossed two strains of white flowering plants, one lacking in leucoanthocyanin, the other in an enzyme capable of transforming the leucoanthocyanidin to the corresponding colored compound. The F_1 product was pink "wild" colored, and the next generation segregated in exactly the same manner as the lupine did (Nowacki, 1966) (Figure 2.1b). To date, over 18 genes for alkaloid-poor character in lupines have been reported, but the number will certainly be reduced after additional genetic experiments are performed. For example, the genes a_1, a_2, and a_3 in *Lupinus angustifolius,* and the genes *tertius* and *primus* isolated by Miko-

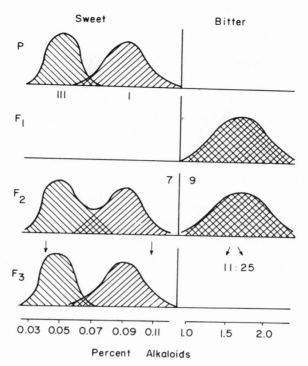

Figure 2.1a. The inheritance pattern observed from crossing two varieties of "sweet" lupines with different genes. P—parent plants; I—plants with the recessive gene *primus;* III—plants with the recessive gene *tertius* (also called *pauper*). In the F_1 generation all plants are "bitter" and their alkaloid content is well over 1% dry weight. The F_2 generation had a segregation ratio of 9 bitter: 7 sweet. The sweet plants formed two distinct and partially overlapping groups. Plants with the lowest alkaloid content of the F_3 generation gave only progeny with the gene *tertius,* while plants with the higher alkaloid level gave progeny with the gene *primus.*

Figure 2.1b. The segregation pattern observed in the F_2 generation for the dominant alleles T and P responsible for alkaloid production. The recessive alleles are p and t. The symbols along the margins denote genes from two F_1 parents, and each box shows an F_2 progeny type. Symbols in the corners give genotype, and hatchings give phenotype. Crosshatching indicates bitter plants, horizontal lines represent sweet plants of the homozygous *tertius* gene, vertical lines indicate sweet plants of the homozygous primus gene, and a clear field stands for the double recessive homozygote (Nowacki, 1968). Courtesy of the journal.

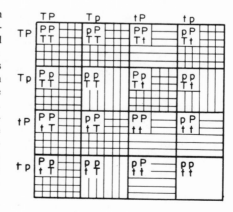

lajczyk and Nowacki (1961) and Mikolajczyk (1961) in Polish white lupine plants, are identical to three genes given different names published by Hackbarth and Troll (1955) and Porsche (1964) (see Table 2.1).

The lupine alkaloid inheritance pattern is simple and easily explainable by means of genetics (Figure 2.2). The alkaloid-poor plant either produces no enzymes for alkaloid synthesis at all or the enzyme is altered in such a way that its efficiency is decreased and the alkaloid production is much less than in the wild type. The pathway leading from a nonalkaloidal substrate to the alkaloid may be sufficiently long to be readily interrupted by mutations at several loci which control the reaction sequences leading to the substrates. Even if it turns out that there are 18 genes for alkaloid character, it must be pointed out that they are in four different and uncrossable species; consequently, it may be that some of those mutations interrupt the same reactions in different species. This ambiguity arises since there is no possibility of a genetic experiment with noncrossable species. It thus becomes necessary to identify the interrupted reactions by performing isotope experiments using labeled substrates. A careful search to locate the point where an interruption in the biosynthetic pathway has occurred must be made. No such experiments have been performed to our knowledge.

In lupines there are genes retarding the biosynthesis of alkaloids at different levels, e.g., the genes *primus* and *tertius* in *L. albus*. The gene *primus* decreases the alkaloid level, which may be as high as 10% of the wild type; *tertius* is responsible for a reduction of alkaloids to only trace levels. When a cross between those types is performed, an F_2 generation segregates according to the previously described pattern in a ratio of 9:7. By quantitative analysis of alkaloid production by the sweet plants, it is possible to distinguish between the two genes using only chemical methods,

Table 2.1. Genes That Govern Alkaloid Production in Lupines[a]

L. luteus	Percent alkaloids	*L. angustifolius*	Percent alkaloids	*L. albus*	Percent alkaloids	*L. polyphyllus*	Percent alkaloids
Wild type	0.9–1.2	Wild type	1.5–1.7	Wild type	1.7–2.2	Wild type	1.5–2.0
dulcis	0.04	a_1	0.06	*primus*	0.1	M	0.01
amoenus	0.01	a_2	0.01	*tertius*	0.04	T	0.05
liber	0.01	a_3	—	*pauper*	—	P	0.03
V345	—	*iucundus*	—	*nutricus*	—	B	0.01
		esculentus	—	*mitis*	—		
		depressus	—	*suavis*	—		
		tantalus	—	*reductus*	—		
				exiguus	—		
				minimus	—		

[a]From Nowacki (1964c) and related references.

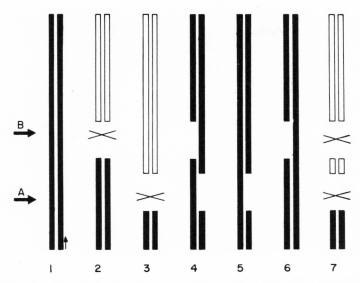

Figure 2.2. Schematic illustration of the pathways for alkaloid synthesis in a bitter (wild type) form (1) of lupines and in two different sweet mutants (2 and 3). The pathways are depicted as two parallel lines since in a plant usually two sets of genes are operating. The × indicates the blocked reaction in the sweet form. In the F_1 generation of the cross between 2 and 3, i.e., 4, the biosynthetic pathway is regained, and the resulting plant is bitter. In crosses involving a bitter form and a sweet form the F_1 generation is always bitter (5 and 6). In 7 there is a double recessive homozygote, and when these plants are crossed with either 2 or 3 only sweet plants are produced (Nowacki, 1966). Courtesy of the journal.

without the help of test crossing (backcrossing) to the parents (Mikolajczyk and Nowacki, 1961).

Another example of an alkaloid-rich:alkaloid-poor inheritance pattern is the genus *Nicotiana*. For a long time it was known that certain *Nicotiana* species, i.e., *N. tabacum, N. rustica,* and *N. glauca,* are alkaloid-rich, while on the other hand, such species as *N. alata* are extremely poor in alkaloids. If it were not for the botanical relationship with other *Nicotiana* species, it would probably not have been noticed that *N. alata* was capable of producing alkaloids. A careful examination of *N. alata* revealed that this species does indeed produce alkaloids, but, contrary to the situation in other *Nicotiana* species, the alkaloids are quickly degraded to nonalkaloid substances. In a cross of *N. alata* with an alkaloid-rich plant the offspring in the F_1 generation is like the *N. alata* parent—alkaloid-poor (Goodspeed, 1959).

In the past few years some *N. tabacum* plants were found with a lower nicotine content than usually noted (Valleau, 1949; Wegner, 1956; Burk and Jeffrey, 1958; Jeffrey and Tso, 1964; Koelle, 1961, 1965a,b, 1966a; Blaim and Berbéc, 1968; Mothes *et al.*, 1955). These plants, when crossed with nicotine-rich *N. tabacum* plants, behaved in a manner like those of *N. alata*. The hybrid generation resulting from the cross of a plant rich in nicotine and a plant poor in nicotine usually resembles the alkaloid-poor parent; since the level of alkaloids is variable and is influenced in some way by vegetation period, it sometimes approaches the higher content of the alkaloid-rich parent. Another feature is the alkaloid composition of the hybrid; without regard to the distribution of alkaloids in the alkaloid-rich parent the hybrid always produces nicotine and nornicotine in a 1:2 ratio. Nornicotine is a derivative of nicotine and is probably an intermediate in the nicotine catabolic pathway.

The alkaloid accumulation pattern in *Nicotiana* is different from that described previously in *Lupinus*. In *Lupinus* the alkaloid-rich plants are dominant, in *Nicotiana,* recessive. The low level of alkaloids in lupines is due to an interruption of the biosynthetic pathway in *Nicotiana,* whereas the high level of nicotine is caused by a defect in nicotine degradation.

2.3. The Inheritance Patterns of Individual Alkaloids

Using genetics it is possible to investigate not only such characters as the presence of smaller or larger amounts of single alkaloids, but also the relation between two alkaloids. Three distinct possibilities are considered:

1. Two compounds A and B have no biosynthetic relation to each other.
2. Compound B is produced from compound A in a one-step reaction.
3. Both compounds A and B are produced from a common precursor.

In the following discussion the simplifying assumptions are made that all genes involved in the experiments are located on different chromosomes and that no disturbances in chromosome behavior occur.

In the first case, suppose that both substances are dominantly inherited. In a cross of a plant producing A with a plant producing B, the F_1 generation must produce both substances. The F_2 generation will segregate in a ratio of 9:3:3:1 (Figures 2.3a, 2.3b, 2.3c, and 2.3d), i.e., nine out of sixteen plants will produce both alkaloids, three only alkaloid A, another three only B, and one in every sixteen will be alkaloid-free. There are no good published examples of this inheritance pattern. A cross between a *Lupinus luteus* plant which produces only gramine, an indole-type alkaloid, and *L. luteus* which produces only quinolizidine alkaloids

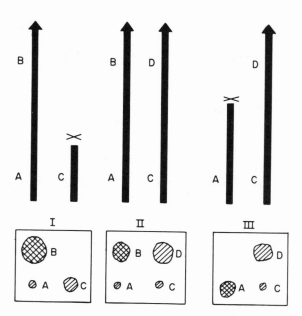

Figure 2.3a. Schematic illustration of a situation when two alkaloids originate from separate pathways. I—a parent form producing only compound B; III—a parent producing only compound D, II—an F₁ hybrid producing both compounds.

would probably give this inheritance pattern, but to the authors' knowledge no one has performed this experiment. A related experiment was attempted with *Galega officinalis,* which is known to produce two types of alkaloidlike compounds: galegine and peganine (Becker, 1956; Schröck, 1941; Sengbusch, 1941; Pufahl and Schreiber, 1961, 1963; Schreiber *et al.,* 1961, 1962); however, no alkaloid-free mutants were obtained.

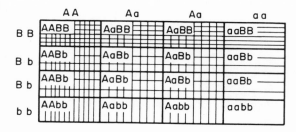

Figure 2.3b. Segregation pattern of the ability to produce the alkaloids. Each alkaloid segregates independently in a 3:1 ratio. The 3 × 3 plants have both alkaloids (crosshatched fields); three have only one compound, three have only the second compound, and one is completely alkaloid-free.

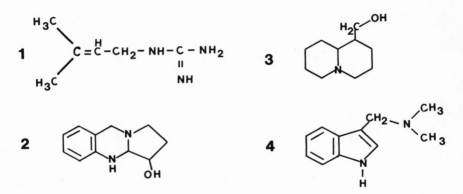

Figure 2.3c. Examples of unrelated compounds with alkaloid character encountered in the following plants: 1—galegine, 2—peganine (both found in *Galega officinalis*), 3—lupinine, 4—gramine, both found in *Lupinus luteus*.

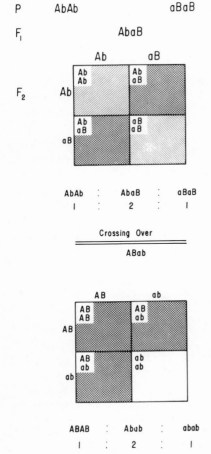

Figure 2.3d. Segregation pattern in a set of hybrid plants with genes for conversion of substrate X into two different derivatives A and B. The genes A and B are on homologous chromosomes; the different dominant alleles are found together only in heterozygotes. A homozygous plant with both dominant or recessive genes can arise by a crossing over.

The second type of segregation pattern is identical to the one just described for the inheritance of alkaloid-rich:alkaloid-poor characters. Details will be discussed later in this section.

The third type can occur in two distinctly different ways: First, if the genes involved in the conversion of an unknown substrate to A or B are located on nonhomologous chromosomes, the F_1 generation will produce both substances, but, if there is competition for the precursor, neither alkaloid will be produced at the normal level. The F_2 generation will segregate in a ratio so that nine plants will contain both alkaloids, while three plants will contain only A, three only B, and one plant out of every sixteen will have no alkaloids at all. In the second case (and this happens in some experiments with flower pigments such as blue:yellow) the genes are on homologous chromosomes which have no or only a slight crossing-over ratio. There will be only three types of plants produced in the F_2 generation: both parent types (having the same as one locus with no dominance) and the F_1 type. Both types of inheritance patterns are expected from crosses of plants bearing alkaloids formed from a common precursor. An example is the inheritance of nicotine and anabasine in the interspecific crosses in the genus *Nicotiana*. Similar results are expected in interspecific crosses in the genus *Papaver* involving *P. somniferum* and *P. orientale* (Kawatani and Asahina, 1959; Tétényi *et al.,* 1961). Some crosses in American lupine species between species with lupanine and species with α-isolupanine will probably yield this type of result.

The best example for a (branched) divergent inheritance pattern when two compounds originate from a common precursor is the inheritance of anabasine and nicotine and/or nornicotine (Goodspeed, 1959; Jeffrey, 1959; Smith and co-workers, 1942, 1963). For these alkaloids, nicotinic acid is the common substrate (see Figures 2.6 or 6.32); the other part of the molecule is synthesized from lysine for anabasine and from ornithine for nicotine. The two amino acids are converted to heterocyclic compound by enzymes with similar properties. The common substrate, nicotinic acid, is used for the synthesis of both types of alkaloids and therefore in some cases can be the limiting factor. In crosses between closely related parents, as with F_2, a segregation is observed. The ratio of segregants suggests a distribution of the alkaloid synthesis genes on homologous chromosomes.

Crosses between *Nicotiana glauca,* a plant containing anabasine, and other species of *Nicotiana* containing nicotine and/or nornicotine were performed some time ago; however, only recently has such work been done in sufficient detail to enable a thorough genetic and biochemical evaluation to be made. Even so, most of the experiments were terminated at the F_1 generation. Weybrew and Mann (1963) crossed these species and obtained an F_1 group of plants which accumulated both nicotine and anabasine. Surprisingly, the plants also produced nornicotine; there was more in these than the trace amounts found in the parent plants (see Figures 2.4a, 2.4b,

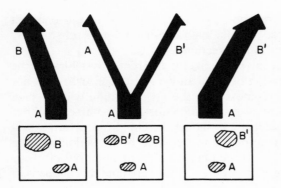

Figure 2.4a. Schematic illustration of a branched pathway. Compound *A* is a substrate for both *B* and *B'*; the hybrid will produce *B* and *B'* in approximately a 50:50 ratio; however, there is usually competition for substrate. Underneath are schematic chromatograms.

and 2.4c). In 1963 Smith and Abashian described combinations of 35 species. While the crosses involving nicotine–nornicotine inheritance will be discussed next, some attention must be given to the anabasine–nicotine inheritance pattern. In all crosses involving *N. glauca* and/or *N. debneyi* the authors (Smith and Abashian, 1963) also obtained hybrids accumulating anabasine as the main alkaloid (Figure 2.5). Anabasine was always found in the F_1 generation, but it was accompanied by the same amount of nicotine or nornicotine. A feature confirming the findings of Weybrew and earlier authors was the appearance of nornicotine, even in cases where the parents did not contain this particular compound. Some multiple crosses were performed combining amphidiploids (alloploids) with a third species. In such cases there was the possibility of investigating the influence of known sets of chromosomes on the alkaloid distribution. In cases where, for example, a multiple hybrid contained one set of chromosomes from *Nicotiana glauca* and two sets from a species containing no anabasine, the level of anabasine produced in the F_1 generation usually dropped to one-third or less of the total amount. Investigation of the second generations was

Figure 2.4b. Paper chromatograms of the alkaloids of *N. glauca* (left), *N. tabacum* (right), and their F_1 hybrid (center) by the method of Griffith *et al.* (1955). Copyright, 1955 by the American Association for the Advancement of Science.

NICOTINE –

Figure 2.4c. Alkaloid chromatograms of greenhouse leaves of *N. glauca* that had been infused through their petioles with water (left) or nicotine (right), demonstrating the capability of this species to convert nicotine into nornicotine (Weybrew and Mann, 1960). Courtesy of the authors and *Tobacco Science and Tobacco International.*

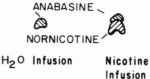

ANABASINE

NORNICOTINE

H_2O Infusion   Nicotine Infusion

seldom performed, and only recently (Blaim and Berbéc, 1968) has there been reported an exhaustive investigation of alkaloids in the generation resulting from a cross of *Nicotiana glauca* and *Nicotiana tabacum* (Blaim and Berbéc, 1968). Exactly as reported in other publications concerning these hybrids, the F_1 plants accumulated nicotine, nornicotine, and anabasine. Since *Nicotiana tabacum* is a natural alloploid and appears to have arisen from the hybridization of two wild South American species (*N. sylvestris* with 24 chromosomes and, in all probability, *Nicotiana otophora,* also with 24 chromosomes), it has 48 chromosomes, and so all sets of genes occur doubly (Allard, 1960) (Figure 2.6a). The *Nicotiana glauca* is a diploid species with 24 chromosomes. The F_1 plants were triploid, as expected, with 36 chromosomes. As triploids they produced only a

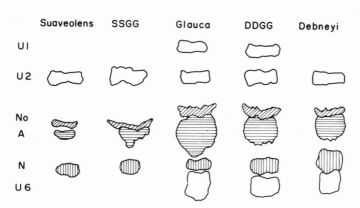

Figure 2.5. Chromatographic patterns of *N. debneyi, N. glauca,* and *N. suaveolens* together with the amphiploids *N. debneyi-glauca* (DDGG) and *N. suaveolens-glauca.* No—nornicotine; A—anabasine; N—nicotine; and U1, U2, U6—unknowns (Smith and Abashian, 1963). Courtesy of the *American Journal of Botany.*

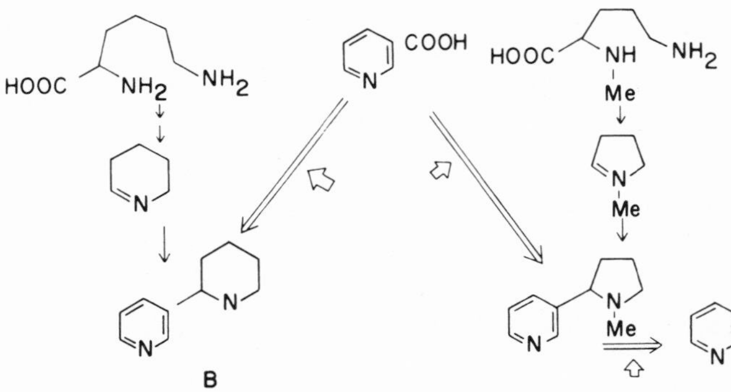

Figure 2.6a. Biosynthesis of anabasine, nicotine, and nornicotine from a common substrate, nicotinic acid, in *Nicotiana glauca* (pathway *B*) and in *N. tabacum* (pathway *B'*).

limited amount of pollen and a limited number of seeds and ovules. The genetic composition of the gametes, as expected, was very different from Mendelian expectation, so a reliable genetic analysis was not possible. Even so, the segregation pattern of the F_2 plants was not so very different from that expected, particularly if it is assumed that the *Nicotiana tabacum* parent has a double set of genes governing the biosynthesis of the nicotine type of alkaloid, in contrast with *Nicotiana glauca,* which has only a single set of genes for anabasine production (see Figure 2.6b).

In a cross of plants with different genes on homologous or heterologous chromosomes, F_1 depends upon the location of genes and will be heterozygous. The inheritance pattern in the F_2 generation will be different. Since in a diploid only two homologous chromosomes can meet in a zygote, there are only three possibilities: homozygous dominant, heterozygous, and homozygous recessive (see Figure 2.6b).

The *Nicotiana* alkaloids, nicotine and anabasine, are products of a fusion of nicotinic acid with a heterocyclic compound originating from ornithine or lysine. In case ornithine is the precursor, the second ring in the *Nicotiana* alkaloid will have four carbon atoms but when originating from lysine it will contain five carbons (see Figure 2.6b).

There is a need for a discussion of the nornicotine appearance phenomenon. Nornicotine, according to recent biochemical as well as genetical investigations, is a demethylation product of nicotine (Schroeter, 1958; Neumann and Schroeter, 1966). Anabasine is, according to its methylation status, a homologue of nornicotine. Correspondingly, a homologue of nicotine would be an *N*-methylanabasine. Since nicotine is formed from *N*-methylornithine, the anabasine probably would be synthesized from *N*-methyllysine, and *N*-methylanabasine would likely be involved as an intermediate. No present biosynthetic evidence for *N*-methylanabasine as an

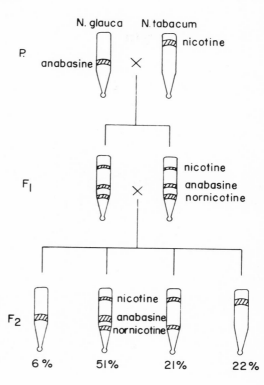

Figure 2.6b. Diagrams of chromatograms of alkaloids occurring in the F_1 and F_2 of the cross *N. glauca* × *N. tabacum* and their quantitative distribution (Blaim and Berbéc, 1968). Courtesy of the authors and the journal.

intermediate is available. The presence of an active N-demethylating enzyme is proven without doubt in *Nicotiana glauca* (see Figure 2.4b), so there remains an open question: Is the first step in anabasine synthesis different from that in nicotine synthesis or is the N-methylanabasine so rapidly demethylated that it never accumulates in more than trace quantities? The authors have devised an approximate genetical analysis of the hybrids resulting from the above *Nicotiana glauca* and *Nicotiana tabacum* hybrid studies (Figure 2.7). The only plausible explanation for the observed mode of inheritance is that enzymes leading to anabasine and nicotine (also nornicotine) compete for a common precursor, thus decreasing the level of both compounds. The observed segregation ratio is quite similar to that predicted by this model. After taking into account the obvious irregularities in the gameto- and embryogenesis, the idea that N-methylanabasine is an intermediate in anabasine biosynthesis seems plausible; however, final proof will rest on isolating the substance from the proper plants and/or elucidation of the biosynthetic pathways using isotopic tracers. The common substrate necessary for synthesis of both kinds of alkaloids is nicotinic acid.

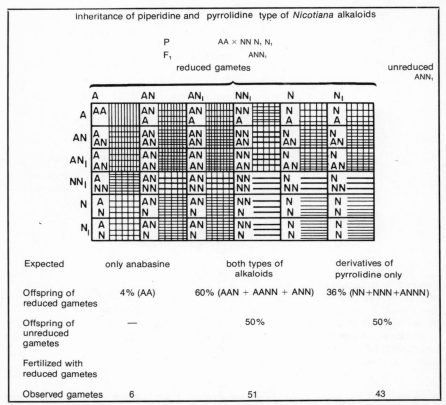

Figure 2.7. The inheritance pattern of nicotine alkaloids in the F_2 generation of *N. glauca* × *N. tabacum*. A—alleles for anabasine production; N—alleles for pyrrolidine derivatives. This diagram is only an approximation, since the parent plants had a different ploidy level. The number of unreduced gametes is not known, but the results fit the theoretically expected numbers surprisingly well. It is assumed that a plant with the genetic formula ANNN produces only traces of anabasine, which are not detectable on paper chromatography.

2.4. Intraspecific and Interspecific Hybridization

Most alkaloid genetics research has been performed on interspecific crosses. There are some disadvantages of such studies. While the parent plants are fertile and viable, the hybrid is usually sterile, and the development of most plants is abnormal; however, occasional hybrids are viable and vigorous. Seeds are produced in some instances, but the ratio of developed seeds to the number of ovules produced is very low. Also, only a small percentage of the pollen grains is viable. To secure the large number of F_2 plants required for studying the inheritance of a chemical character, the F_1 generation must be of significantly greater numbers—a difficult task to achieve! Since elimination of chromosomes is taking place, the observed segregation ratios in the offspring are not always according to Mendel's laws. In some instances, all the F_1 plants may be sterile. To secure seeds, a duplication of the chromosome number is necessary, but increasing the chromosome number will cause an alloploid plant to be produced. Alloploid plants do not segregate at all, so there is no possibility of investigating them further. Thus the F_1 generation remains the only generation available on which experiments can be made. Fortunately, in some cases, the F_1 progeny are sufficient for determination of a relationship between two compounds. As was previously stated, in a cross involving a plant with compound A, which is a precursor, and another plant with compound B, which is the product, the product is in most cases dominant over the substrate. The concentrations of substrate (A) and product (B) can be quite different in different crosses. In the simplest case—when the product is fully dominant—the amount of enzyme produced by the dominant gene is sufficiently high to transform all the precursor available into the alkaloid that is accumulated. Since in a heterozygous condition only one allele of a certain gene is producing the correct enzyme, it is conceivable that the concentration of this enzyme will be insufficient to transform all the precursor which is produced by genes from both parents. Alternatively, inhibitors may also be present in these plants. All kinds of intermediate inheritance can be observed in genetic crossing experiments. The simplest, of course, is seen when the concentration of product is reduced to 50% of that of the parent species able to perform the reactions. Also possible are intermediate concentration changes of the product between 50 and 100% of the amount produced by the homozygous parent of the species. The last possibility is called overdominance. Overdominance seldom occurs, and the cause is thought to be that the amount of precursor produced by the parent species required for subsequent reactions leading to alkaloid production is unusually low. The second species produces a much higher level of precursor. In the dominant parent species the level of product was limited by the level of precursor and not by the lack of an enzyme. In the hybrid individual it can happen that even in the heterozygous condition the plant will be able to produce much more product than the homozygous parent. An example of this phenomenon was found by Weybrew *et al.* (1960) and Weybrew and Mann (1963) using *Nicotiana* (Figure 2.8). The authors performed a cross between *N. tabacum* and *N. sylvestris*. *N. tabacum* has nicotine as its major alkaloid, and the concentration is rather high (3.85%). *N. sylvestris* contains two alkaloids, nicotine and nornicotine; the level of nicotine is low (approximately 0.5%), whereas the level of nornicotine is rela-

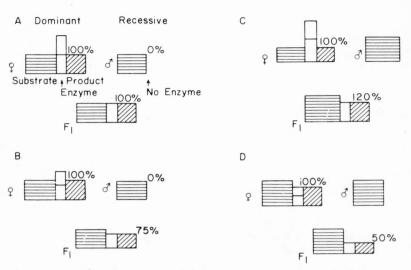

Figure 2.8. Explanation of dominance, partial dominance, overdominance (transgression), and intermediate dominance in the inheritance of alkaloid content. *A*—Full dominance pedigree. The female parent produces the precursor, enzyme, and the product, while the male produces the precursor but not the enzyme, and consequently contains no product. The hybrid has an intermediate level of enzyme, but it still has sufficient to convert all of the precursor, and thus the amount of product is exactly the same as in the dominant parent. *B*—Partial dominance. The enzyme level in the hybrid heterozygous plant is not sufficient to convert all the precursor; thus the hybrid contains an intermediate amount of product. *C*—overdominance. The recessive parent, while not capable of producing the end product, produces much more precursor than the dominant parent. The F_1 plant produces more precursor; the enzyme level in *D* despite the heterozygous state enables the conversion of more substrate into product than in the parent. *D*—Special case. In intermediate dominance the level of the product in the heterozygous plant is exactly 50% of that in the dominant homozygous parent (female above). The explanation is simplified since the assumption is made that an enzyme in a heterozygous plant will perform exactly half as much conversion of precursor to product as the one in the homozygous plant capable of performing the reaction (Nowacki, 1966). Courtesy of the journal.

tively high (1.86%). The F_1 hybrid had the same amount of nicotine as *N. sylvestris,* but the amount of nornicotine was increased to 2.62% (an increase of 40%) (Table 2.2).

Examination of the F_1 plants is sufficient for the determination of the metabolic relationship between two compounds only in cases where dominance is about 100% or where an overdominance exists. In other cases the result cannot be conclusively explained as a simple substrate–product relationship, since it can also be accounted for otherwise. For example, no biogenetic relationship exists when both compounds are synthesized by separate pathways. In the heterozygous condition of the F_1 generation the

Table 2.2. Inheritance of Ability to Store *Nicotiana* Alkaloids[a]

Parent species or hybrid	Nicotine	Nornicotine	Anabasine
From Manske and Holmes (1950–1975)[b]			
N. tabacum	++	+	T
N. sylvestris	+	++	T
sylvestris × *tabacum* F₁	++	+	T
N. tomentosa	+++	−	−
tomentosa × *tabacum* F₁	++	+	−
N. glutinosa	+++	−	−
glutinosa × *tabacum* F₁	++	+	−
glutinosa × *sylvestris* F₁	++	+	−
N. glauca	T	T	+++
glauca × *tabacum* F₁	+	−	++
From Weybrew and Mann (1963)[c]			
N. tabacum	3.85	0.28	−
N. sylvestris	0.50	1.86	−
sylvestris × *tabacum* F₁	0.57	2.62	−
N. tomentosiformis	0.32	0.21	−
tomentosiformis × *tabacum* F₁	0.22	0.76	−
N. otophora	0.35	0.16	−
otophora × *tabacum* F₁	0.13	0.47	−
From Smith and Abashian (1963)[d,e]			
N. bigelovii	++	T	+
N. glauca	T	+	++
bigelovii × *glauca* F₁	+	++	++
N. glutinosa	+	++	T
glutinosa × *bigelovii* F₁	+	++	+
glauca × *glutinosa* F₁	T	+	++
N. langsdorffi	++	T	+
N. sanderae	++	+	−
langsdorffi × *sanderae* F₁	T	++	+
N. suaveolens	++	+	++
suaveolens × *sanderae* F₁	++	+	T
N. debneyi	+	T	++
N. clevelandi	+	T	++
debneyi–clevelandi allop.	++	+	++
debneyi–glauca allop.	T	+	++
glauca–langsdorffi allop.	T	++	++
N. plumbaginifolia	?	++	?
glauca–plumbaginifolia allop.	T	+	++
tabacum–glauca allop.	+	++	++
N. longiflora	++	+	+
tabacum–longiflora allop.	+	++	+
N. tabacum	++	T	+

[a] Key: +++ and ++ = major alkaloids; + = secondary major alkaloids; T = minor alkaloids.
[b] Courtesy of Academic Press.
[c] Courtesy of the authors and Tobacco Science and Tobacco International.
[d] The evaluation made by the authors was such that 1 = ++, 2 = +, and 3, 4, 5 = T in our description.
[e] Courtesy of the American Journal of Botany.

alkaloid level is reduced to one-half. Also the compounds may originate from a common precursor, so the 50% level of both the final alkaloids produced is caused by competition for a single substrate. Since the F_1 plants from this example give no conclusive data, investigation of subsequent generations is necessary. But, as it was stated earlier, the interspecific hybrids are either sterile or semisterile and have a high probability or chromosome elimination and other disturbances which interfere with the production of a clear F_2 segregation. Analyses requiring more than one generation should therefore be specific. The trouble with intraspecific crosses is in obtaining the two desired parental types. The presence of different genotypes was overlooked for a long time. Since the plants containing different alkaloids are morphologically indistinguishable from one another, a time-consuming search for mutants is the prerequisite for the beginning of genetic experiments using intraspecific hybrids. As an example, the authors searched for, but were unable to find a single *Lupinus angustifolius* plant that did not have the ability to transform sparteine into lupanine. In the course of this study, over 10,000 plants were analyzed. Still, it seems likely that such a plant must exist in related species also (according to Vavilov's rules of homologous diversity in related taxa). In other lupines, we found within a species a number of genotypes suitable to chemogenetical research. Recently Harding and Mankinen (1968) investigated the natural diversity of Californian lupines and found three mutants lacking a single alkaloid, *L. nanus*, in each case (Figure 2.9). Further mutants have been found in *L. succulentus*. In *Datura* and *Nicotiana* species a great number of genotypes with different alkaloid distribution patterns were recognized. Sometimes plants belonging to a single species but collected in different geographical areas have quite dissimilar alkaloid

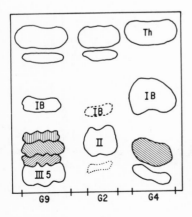

Figure 2.9. Chromatogram alkaloids of three different forms of *Lupinus nanus* found by Harding and Mankinen (1968). *G9*—wild type that produces seven alkaloids; G2 and G4—mutant types that lack the ability to transform compounds II and IB into derivatives.

compositions. *Lupinus luteus* plants from Italy as well as the cultivated varieties in central Europe have an alkaloid ratio of lupinine to sparteine of 3:1, while plants from the Iberian peninsula contain little sparteine. Plants from the eastern Mediterranean countries, Egypt and Palestine, have, in contrast, only sparteine. Successful crosses of such genotypes with different alkaloid distribution patterns will yield, except in rare cases of different ploidy level, a fertile F_1 generation. Ordinarily the hybrids F_2 produce a great number of seeds, so that the number required for proper statistical evaluation of the results can be readily obtained. Since the segregation mode can be predicted for all kinds of relationships between two substances (and in each case it is different), a properly planned genetic experiment can solve many of the problems relating to alkaloid metabolism. A decided advantage is that no exogenous compounds are introduced, as is done in the use of the tracer techniques to follow a metabolic pathway. Such genetic intraspecific studies have been performed (see Section 2.8).

2.5. The Inheritance of Ability to Store Alkaloids of Different Oxidation Levels

For a long time it was known that plants can accumulate alkaloids of different levels of oxidation. Examples are (a) the quinolizidine alkaloids—sparteine, $C_{15}H_{26}N_2$, and lupanine, $C_{15}H_{24}N_2O$; (b) the tropanes—hyoscyamine, $C_{17}H_{22}NO_3$, and hyoscine, $C_{17}H_{20}NO_3$; and (c) the diterpenoid alkaloids—ajaconine, $C_{23}H_{32}NO_3$, and delcosine, $C_{24}H_{33}NO_7$. By means of chemical methods, it is possible to partially or completely oxidize these alkaloids. By using labeled compounds, it can be shown that the alkaloids in a lower state of oxidation are easily converted to the more highly oxidized ones by plants, but the reactions are not easily reversed.

2.5.1. The Tropane Alkaloids

In 1958 Romeike reported a successful cross between *Datura ferox* and *Datura stramonium*. *D. ferox* is a small, virus-susceptible plant that produces only minute amounts of alkaloids, but its alkaloid is the valuable drug hyoscine. The other parent, a vigorous and non-virus-susceptible plant containing at least ten times as much alkaloid, produces hyoscyamine, which could not readily be converted to hyoscine chemically. The difference is in the oxidation level, hyoscine being the epoxide of hyoscyamine. There are at least three reactions necessary to perform the oxidation. The F_1 plants inherited the high alkaloid level of *Datura stramonium* and the oxidation ability of *Datura ferox*. The yield of alkaloids per plant was 700 mg in the hybrids, as compared with *Datura ferox,* which produced less than 100 mg/plant. The percentage of hyoscine was over 90% that found in *Datura ferox.* The segregation ratio in F_2 was 3:1, which shows a dominance of hyoscine. In further experiments using labeled compounds and grafts between both species, Romeike (1961, 1962) established that hyoscyamine is the first alkaloid to be synthesized and is transformed into hyoscine only in plants with the correct enzyme system (three enzymes are suggested). The species of *Datura* involved in this hybridization study are actually very closely related so that a heterosis (hybrid vigor) was observed (Figure 2.10a and 2.10b).

2.5.2. The Quinolizidine Alkaloids

Kazimierski and Nowacki (1961a, b) reported a spontaneous hybrid in the genus *Lupinus* involving the California species *Lupinus arboreus* and a Mexican species *Lupinus hartwegi.* The species differ in alkaloid composition. Whereas *Lupinus arboreus,* the female plant which produced the spontaneous hybrid, accumulated only sparteine, *Lupinus hartwegi* accumulated only lupanine. The biological conversion of sparteine into lupanine was reported earlier, in 1958, by Nowacki, who found that in alkaloid-poor *Lupinus angustifolius* sparteine is easily converted into lupanine when injected into the plant. The reaction probably requires only a few biosynthetic steps. One of the steps may be the formation of dehydrosparteine, which Piechowski and Nowacki (1959) were able to obtain in *in vitro* experiments using acetone powders from a number of plants. The last

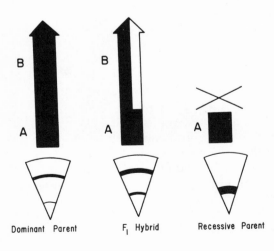

B

A

Dominant Parent

B

A

F₁ Hybrid

A

Recessive Parent

Figure 2.10a. Schematic diagram of a cross between two plants where it is feasible for one to convert *A* into *B* when the other lacks this ability. F_1 progeny have an alkaloid content like the dominant parent. Underneath are chromatograms.

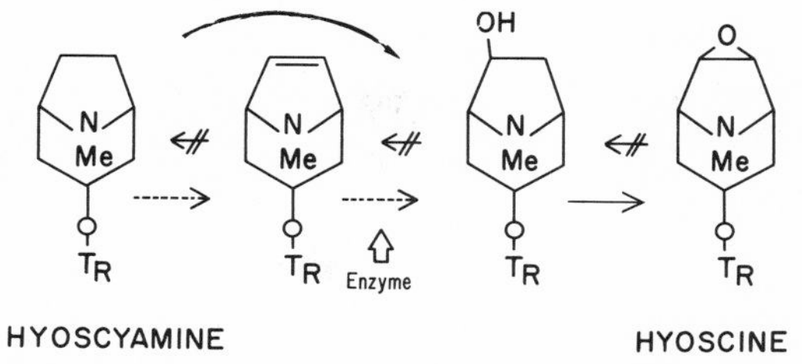

Figure 2.10b. Hyoscyamine (compound *A*) is converted in three steps to hyoscine; however, only one reaction is governed. The segregation ratio in the F_1 progeny is 3:1 (hyoscine to hyoscyamine plants). The hypothetical unsaturated intermediate has never been found, but in feeding experiments is transformed into hyoscine. T_R is tropic acid. (Our interpretation of Romeike's experiments, 1958, 1961, 1962.)

step is probably very specific and can be accomplished only in certain leguminous plants of the *Sophoreae* and *Genisteae* tribes. *Lupinus hartwegi* is one of these types. The F_1 generation was very vigorous and reasonably fertile (43%). The alkaloid composition of the hybrid was quite similar to that of *Lupinus hartwegi*, i.e., formation of the product, lupanine, was dominant, and the level of trace alkaloids was higher.

In the same year Nowacki *et al.* (1961) reported the formation of hybrids in American lupine species. The Washington lupine (*L. polyphyllus*), the Russell hybrid (*L. nootkatensis*), and all of the above-mentioned species were crossed with *L. arboreus*. The alkaloids of the F_1 plants were either lupanine alone or a mixture of lupanine and hydroxylupanine accompanied by traces of angustifoline and other unidentified compounds. In all cases accumulation of the major alkaloid of *L. arboreus*, viz., sparteine, was a recessive trait, and the F_2 segregations gave results (Table 2.3) that could be explained only if the pathway converts sparteine into lupanine and subsequently into hydroxylupanine (Figure 2.11). The results were confirmed by more crosses of Californian shrubby lupines performed by Nowacki and Dunn (1964). In all cases sparteine formation was recessive, and in F_1 progeny the alkaloid either disappeared completely in crossing with plants bearing lupanine and hydroxylupanine or the total sparteine level was much decreased. Only when crossing *Lupinus elegans* with *L. arboreus* was the inheritance intermediary; thus, it seems conceivable that the enzyme performing the transformation of sparteine into lupanine in *Lupinus elegans* is much less efficient or is present in a lesser amount. Therefore, the hybrid with only one-half a set of genes (heterozygous) (Figure 2.12) for the reaction accumulated sparteine to the extent of 40–70% of the total alkaloid level in certain periods of vegetative regrowth.

Most of the lupine genetic experiments were performed using seeds for analysis. Seeds genetically represent the next generation, so with F_1 plants the seeds are F_2, but as described in Hagberg's experiments in 1950 the

Table 2.3. Inheritance of Ability to Store Alkaloids in Interspecific Hybrids of Lupines[a,b]

Hybrid	Sparteine	Lupanine	Hydroxy-lupanine	Angusti-foline
A. *L. arboreus*	++++	S	−	−
L. nootkatensis	S	+++	++	−
F$_2$ 214 plants	S	+++	++	−
F$_2$ 73 plants	S	++++	−	−
F$_2$ 117 plants	++++	−	−	−
B. *L. arboreus*	++++	S	−	−
L. polyphyllus	S	+++	+++	+
F$_2$ 210 plants	S	+++	+++	+
F$_2$ 174 plants	S	+++	−	−
F$_2$ 75 plants	+++	−	−	−
F$_2$ 10 plants	+++	−	unknown	−
C. *L. arboreus*	++++	S	−	−
L. hartwegi	−	++++	S	−
F$_1$ *arboreus* × *hartwegi*	S	++++	−	−
F$_2$ 370 plants	−	++++	S	−
F$_2$ 70 plants	++++	−	−	−
D. *L. polyphyllus*	−	+++	+++	+++
F$_1$ *arboreus* × *polyphyllus*	−	+++	+++	+
E. *L. hartwegi*	−	++++	S	−
L. elegans	+++	++	−	−
F$_1$ *elegans* × *hartwegi*	−	++++	S	−
F$_2$ 51 plants	+++	++	−	−
F$_2$ 148 plants	−	++++	−	−
F. *L. arboreus*	+++	S	−	n
L. excubitus	+	++	+	n
L. chamissonis	++	+	−	n
L. albifrons	−	+++	+	n
L. longifolius	−	+++	+	n
F$_1$ hybrids of following parents:				
2 × 4	−	+++	+	n
2 × 5	−	+++	+	n
2 × 4	++	++	+	n
1 × 5	−	+++	+	n
4 × 5	−	+++	+	n
5 × 1	−	+++	+	n
5 × 4	−	+++	+	n

[a]From Nowacki and Dunn (1964), and Nowacki (1968).
[b]Legend: −: not present; +, ++, +++, ++++: increasing amount of alkaloid detected; S: trace; n: not investigated.

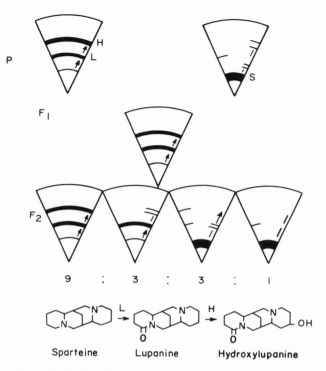

9 : 3 : 3 : 1

Sparteine → Lupanine → Hydroxylupanine

Figure 2.11a. Schematic diagram of a cross between plants with two different genes governing two reactions in a biosynthetic pathway. *P*—parent plants (left form) which produce lupanine and hydroxylupanine and (right form) which produce only sparteine. F_1 progeny are similar to the dominant parent form. In the F_2 progeny, the following segregation occurs: nine like the dominant parent, three with lupanine, and four like the recessive parent. The four plants which accumulated only sparteine are of two types. Three of them can convert artificially introduced lupanine to hydroxylupanine and one, like the recessive parent, cannot perform this reaction. The chromatogram arrows denote active enzymes; headless arrows denote inactive enzymes. Fragments on the left side of the chromatograms denote alkaloids in trace amounts.

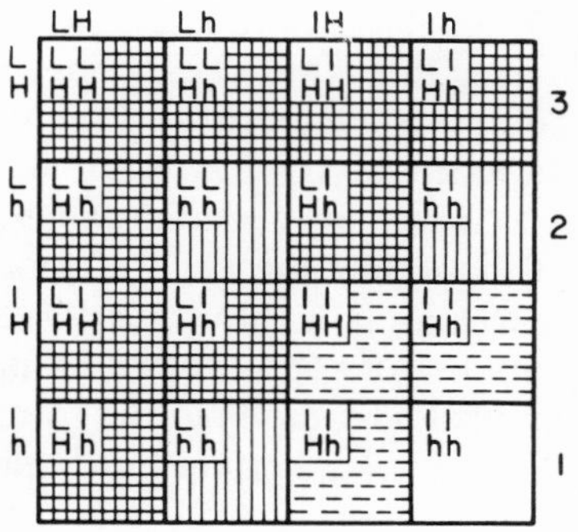

Figure 2.11b. Segregation pattern in F_2 progeny. Crosshatched fields represent plants like the dominant parent. Vertical lines represent plants that produce only lupanine. Horizontal lines and a clear field represent those plants that produce sparteine only. The plants represented here by the horizontal broken lines are able to transform introduced lupanine into hydroxysparteine.

	LH	Lh	lH	lh	
L H	LL HH	LL Hh	Ll HH	Ll Hh	3
L h	LL Hh	LL hh	Ll Hh	Ll hh	2
l H	Ll HH	Ll Hh	ll HH	ll Hh	
l h	Ll Hh	Ll hh	ll Hh	ll hh	1

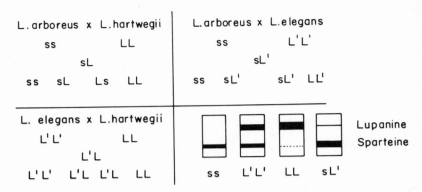

Figure 2.12. Genetic formulas of crosses of *Lupinus arboreus* with *L. hartwegi* or *L. elegans* with *L. hartwegi*. ss—sparteine-synthesizing form; LL—lupanine-synthesizing form; L′L′—lupanine-synthesizing form with lower efficiency. Schematic chromatograms correspond to the respective genotypes (Nowacki *et al.*, 1966).

alkaloid makeup of the seed in influenced not only by its genetic makeup, but also by the mother plant. So in a sparteine-producing plant all seeds accumulate sparteine even when they are of hybrid origin, but after germination the plants will produce lupanine. The results obtained by means of genetic experiments were simultaneously confirmed by experiments involving labeled compounds.

2.6. Inheritance of the Ability to Store *O*-Methylated Alkaloids

Plants belonging to the genus *Papaver* and the closely related genus *Argemone* produce alkaloids with different levels of methylation of oxygen atoms. The best known species of *Papaver* is the opium poppy, which accumulates morphine and small amounts of various other alkaloids: codeine, thebaine, papaverine, narcotoline, etc. These latter alkaloids can comprise as much as 60% of the total alkaloids produced. Crosses of varieties of poppy which differ in percentage and composition of the accompanying alkaloids were performed in various countries with the aim of increasing the yield of total alkaloids and improving the percentage of morphine (a highly sought-after drug). The search was successful and the varieties of poppy grown for pharmaceutical purposes at present are higher in alkaloid yield than the former ones. However, in only a few cases was the detailed inheritance information published (Moldenhauer, 1961; Dános,

1965). The lines used for hybridization always contained approximately the same amounts of morphine but contained different minor alkaloids in different amounts. Since the poppy seeds are virtually alkaloid-free, the analyses were performed on green plants and/or on mature capsules. It was hard to compare the alkaloid distribution and content of the hybrids since the development of the partners and especially the hybrids differed substantially, so the chemical results were somewhat inconclusive.

The inheritance pattern of minor alkaloids is a tedious subject to study. On chromatograms they are obscured by the presence of the major alkaloids and artifacts produced during the extraction procedure.[2] Some plants accumulated 20 or more chromatographically distinguishable compounds. Biosynthetically the minor alkaloids are considered metabolic intermediates or side products produced as results of rapidly occurring reactions which form the final accumulated alkaloids. The minor alkaloids may also be formed by unspecific enzymatic reactions. It was found (Piechowski and Nowacki, 1959) that crude extracts from alkaloid-free plants—i.e., from plants bearing different groups of alkaloids—from animal tissues and from bacteria can perform a considerable number of reactions with alkaloids. It is conceivable, therefore, that native enzymes in the plant, which are relatively nonspecific with respect to substrate (e.g., alcohol dehydrogenases, transmethylases, etc.), produce small amounts of alkaloids by using precursors of the main biosynthetic route. Since most of the alkaloids (and probably some of their intermediates) are dissolved either in the vacuole or in the milky latex fluid, they probably are exposed to a number of commonly occurring enzymes. Having this in mind, one will not be surprised that the level, as well as the distribution, of minor alkaloids is as changeable during the vegetative period and as easily influenced by environmental factors as is observed. Also, the state of the plant may be influenced by certain types of infections. Therefore a well-planned search to elucidate the genetics of minor alkaloids has been mostly inconclusive.

In conclusion, it appears that the impressive experiments on minor alkaloids in *Papaver somniferum* confirmed only the minor value of these compounds for genetic investigations (Table 2.4). On the other hand, crosses involving major alkaloids are important for such investigation. This kind of cross was performed between the morphine-bearing *Papaver somniferum* and the thebaine-accumulating *P. bracteatum* (Böhm, 1965). As was expected from the biosynthetic experiments of Stermitz and Rapoport

---

[2]With adequate prepurification of alkaloids it is now possible to use the combined gas chromatograph–mass spectrometer–computer to determine the precise mass of these minor alkaloids (Waller, 1972; Sweeley *et al.,* 1974). A method of rapidly assaying milligram quantities of alkaloid using combined gas chromatography–mass spectrometry has been developed by Millington *et al.* (1974), Games *et al.* (1974), and Hargreaves *et al.* (1974).

Table 2.4. Trace Alkaloids in Some Varieties of *Papaver somniferum* and the F_1 Generations of Intervarietal Hybrids[a]

	Names of parent and hybrids							
Trace alkaloids	Fe	Fe/C	C/Fe	C	Fe	Fe/P	P/Fe	P
I. In %								
Narcotine	0.348	0.478	0.383	0.236	0.456	0.332	0.316	0.204
Papaverine	0.303	0.280	0.253	—	0.383	0.347	0.202	—
Narcotoline	0.191	0.172	0.211	—	0.248	0.267	0.148	0.339
Protopine–cryptopine	0.280	0.135	0.138	—	0.139	—	—	—
Thebaine	0.157	0.159	—	—	0.145	0.130	0.208	—
Laudanine–laudanosine	0.528	0.360	0.314	0.122	0.369	0.320	0.222	0.125
Codeine	0.355	0.375	0.469	0.550	0.499	0.350	0.416	0.325
II. In mg/plant								
Narcotine	1.120	2.861	2.151	0.958	2.216	2.679	1.934	1.077
Papaverine	0.977	1.679	1.420	—	1.861	2.800	1.236	—
Narcotoline	0.614	1.027	1.187	—	1.205	2.154	0.906	1.790
Protopine–cryptopine	0.902	0.807	0.778	—	0.675	—	—	—
Thebaine	0.504	0.952	—	—	0.705	1.321	1.273	—
Laudanine–laudanosine	1.701	2.153	1.766	0.494	1.793	2.582	1.359	0.660
Codeine	1.070	2.243	2.634	2.233	2.425	2.824	2.546	1.715

[a]Courtesy of Danos (1965).

(1961) (who had shown that thebaine is the first base to be formed and that codeine and morphine are derived from successive irreversible *O*-demethylation steps), the hybrid obtained by Böhm accumulated only the demethylated products. Unfortunately the hybrid was sterile. Similarly, an even more interesting hybrid in the genus *Papaver* resulting from a cross between *P. orientale* and *P. somniferum* obtained in Hungary was sterile (Tétényi *et al.*, 1961). The Hungarian cross hybrid is worthy of mention since the parent species contained alkaloids not only at different methylation levels, but, in addition, formed by different biosynthetic sequences (Figure 2.13a). The *P. somniferum* alkaloids are derivatives of (−)-reticuline, while the *P. orientale* alkaloids are derivatives of (+)-orientaline (Barton *et al.*, 1965; Battersby *et al.*, 1965) (Figure 2.13b). Both orientaline and reticuline are converted in a series of similar reactions to yield, respectively, oripavine and morphine (Figure 2.13c). The main difference is in the first reaction leading from norlaudanosoline to the respective parent compound used in each of the reaction sequences. The enzymes are probably rather nonspecific, so both series of compounds can be transformed readily to their final respective alkaloids.

Figure 2.13a. The conversion of thebaine to codeine and morphine in *Papaver somniferum* plants. Chromatograms *A* and *D* are plants which produce only traces or no thebaine but accumulate codeine and morphine. *B* and *C* are *P. bracteatum* plants which accumulate only thebaine. *A* × *B* and *C* × *D* are F₁ hybrids that produce akaloids similar to the *P. somniferum* parent. 1—root; 2—stem; 3—capsule; Mo—morphine; Co—codeine; Th—thebaine; E—alpinigenine (Böhm, 1961). Courtesy of the author and the journal.

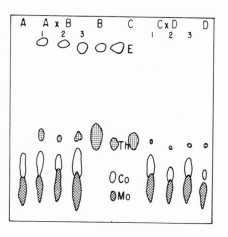

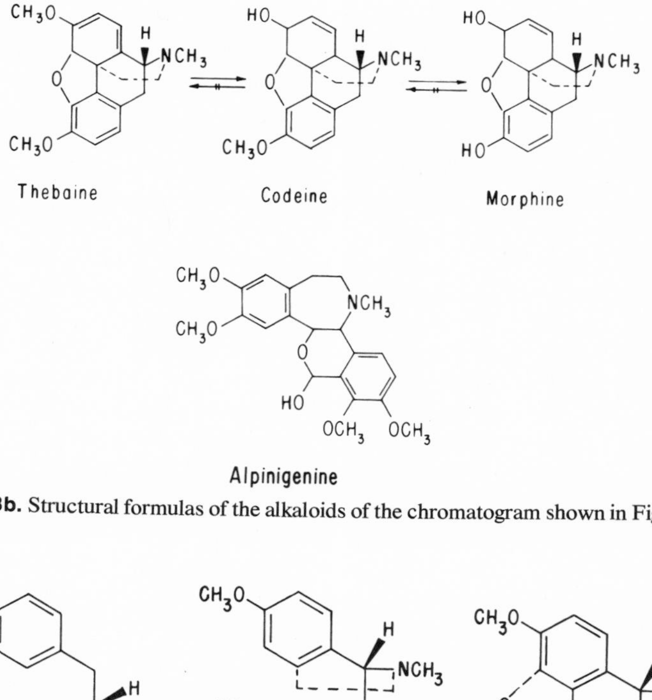

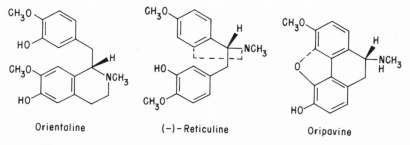

Thebaine   Codeine   Morphine

Alpinigenine

Figure 2.13b. Structural formulas of the alkaloids of the chromatogram shown in Figure 2.13a.

Orientaline   (−)-Reticuline   Oripavine

Figure 2.13c. Structures of orientaline, (−)-reticuline, and oripavine.

2.7. Inheritance Pattern for Nitrogen Methylation

As previously mentioned, the best example of the inheritance pattern of nitrogen methylation is nicotine and nornicotine (Figure 2.6a). For some time there was a certain discrepancy between the results of biosynthetic investigations and the genetic studies. This confusion was due to the first experiments concerning the origin of the pyrrolidine ring of nicotine. Results of tracer experiments published by Dawson *et al.* (1960), Dewey, Byerrum, and Ball (1955), and others showed that ornithine is incorporated into nicotine after decarboxylation and deaminative oxidation. On the other hand the genetic experiments indicated that an *N*-demethylation of nicotine occurred. The first experiments on the inheritance of nicotine and nornicotine were performed on interspecific crosses, and all the predictions concerning this type of cross were verified. In *Nicotiana* Neumann and Schroeter (1966) found α-*N*-methylornithine. Mizusaki *et al.* (1972) and Leete and Medekel (1972) found that α-methylaminobutyraldehyde was the precursor of the pyrrolidine ring of nicotine and that its incorporation occurred without loss of the *N*-methyl group.

The genetic experiments, however, were improved by Koelle (1966b) in a series of publications which firmly not only established that nicotine actually undergoes *N*-demethylation but, in addition, determined the number of genes involved in this transformation. As it may be expected in a tetraploid plant (actually an alloploid), the sets of genes were double.[3] Thus

[3]Only when the genetic matter is common to both species.

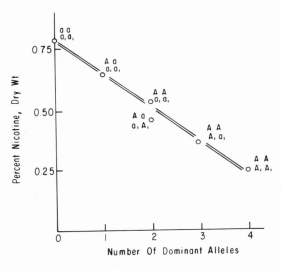

Figure 2.14a. The accumulation of nicotine. Data are from an average of 21 analyses of plants grown during the summer months. The gene *dossis* effects: *a'*—recessive defective gene and *A'*—dominant active gene for demethylation of nicotine to nornicotine.

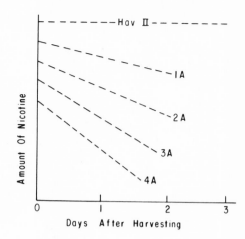

Figure 2.14b. The demethylation of nicotine during 3 days after harvest; the gene *dossis* effect is still visible.

Nicotiana tabacum has two sets of alleles, $A_1A_1A_2A_2$, which governed the transformation of nicotine into nornicotine. The actual number of dominant genes was easily recognized by determining the proportion of nicotine to nornicotine that occurred, as well as the ratio of conversion (Figures 2.14a,b,c and 2.15). Koelle (1975) showed that the nicotine content is positively correlated with the dose of the conversion genes. The relation was not linear, and the nicotine content decreased with an increased level of the gene. This means that with an increase of the gene dose, nicotine content in living leaves will not decrease to zero, and the steady-state situation remains unchanged. Thus the nicotine content offers a good example for the dependence of genes controlling the substrate supply and for the function of genes in the steady state.

Figure 2.14c. Koelle's explanation of the different chemical compositions observed in F_1 progeny resulting from crosses involving different varieties of tobacco. 1—Intermediate phenotype: the hybrid with a single dominant gene is capable of transforming only 50% of the nicotine. 2—Dominant phenotype, where a single gene can transform all of the substrate (nicotine). 3—The reverse dominance, where the homozygous plant is unable to transform all available nicotine, and the heterozygous plant, therefore, behaves as if nicotine were a dominant substrate (Koelle, 1965b, 1966). Courtesy of the author and the publisher of *Z. Pflanzenzücht.*

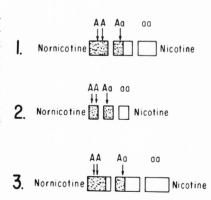

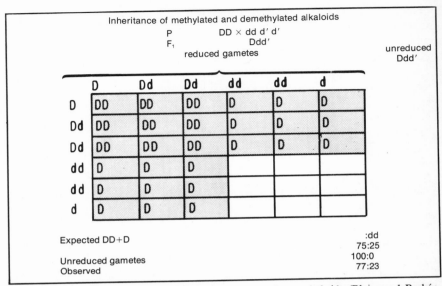

Figure 2.15. Segregation in *Nicotiana glauca* × *N. tabacum* hybrids (Blaim and Berbéc 1968). The ability to demethylate nicotine is shown in *D*—the dominant allele from *N. glauca*. Plants which bear the *D* gene produce nornicotine; expected—75%, observed—77%. Courtesy of the authors and the journal.

2.8. Inheritance Patterns Related to the Complexity of the Molecular Formula

Some plants accumulate alkaloids with very different chemical structures which are nevertheless related biosynthetically. In some cases the biosynthetic route may be branched, but in some it is rather clear that the more complicated structures arise from the simpler molecules. The latter condition is encountered in lupines. While the majority of alkaloids are of a coupled quinolizidine structure, two uncommon alkaloids, lupinine and epilupinine, are of a simple quinolizidine structure. The idea that lupinine is a precursor of sparteine, i.e., a bisquinolizidine, was first proven by experiments in which radioactive lupinine was fed to the plants and radioactive sparteine and its derivatives were isolated (Schuette, 1960; Nowacki *et al.*, 1961). In this study, varieties of *Lupinus luteus* from Palestine, Portugal, and central Europe were tested. There were plants that produced only one major alkaloid, either sparteine or lupinine; however, the hybrid plants which produced sparteine also produced a small amount of lupinine. Consequently, it was a typical example of intermediate inheritance. The F_2

generation of plants was separated into five groups based on chromatographic analysis of the alkaloids. In the first there were plants with only a
single major alkaloid, sparteine. In the next three groups there were plants
containing different amounts of sparteine and lupinine, while the fifth
accumulated only lupinine. The results are explained in the simplest way by
assuming that the conversion of lupinine into sparteine is governed by two
sets of alleles (Kazimierski and Nowacki, 1964).[4] There is no proof that *L.
luteus* is a tetraploid species, but this species, like all the Mediterranean
lupines, has a different chromosome number than is usual for the genus
Lupinus; apparently it is some kind of aneuploid, and therefore a double set
of alleles that control certain reactions probably exists (Gladstones, 1974)
(Figure 2.16).

In crosses of yellow lupines with lupinine as the major alkaloid with
plants bearing only sparteine, the F_1 plants had both alkaloids; and in F_2 a
segregation into five groups was observed, the extreme with only one
alkaloid comprising each about $\frac{1}{16}$ of the F_2 generation. The rest were

[4]Only if alkaloid genes are on the expected chromosomes. There are few (only 1?).

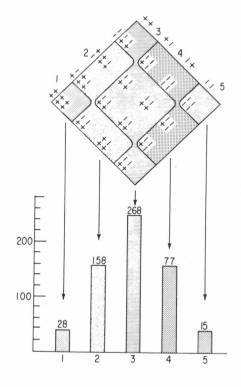

Figure 2.16. Tetrasomic inheritance,
expected segregation pattern when two
sets of genes are involved: + the dominant allele; − the recessive allele.
Observed segregation: 1 ++++ : plants
with sparteine only; 2 +++− : plants
with traces of lupinine and mostly sparteine; 3 ++−− : lupinine in high percentage, sparteine moderate; 4 +−−− :
traces of sparteine, low lupinine; 5
−−−− : no sparteine detectable and only
lupinine accumulated. Numbers over columns are the observed segregation; the
height of the columns is the expected segregation (Kazimierski and Nowacki,
1969). Courtesy of the journal.

plants with different quantities of lupinine or sparteine (Figure 2.9). The results indicate an action of two seeds of the plants which have alleles governing the conversion of lupinine into sparteine (Nowacki, 1968).

2.9. Conclusions

Genetic experiments can provide independent proof for the metabolic pathways used in alkaloid biosynthesis in plants. Although much remains to be done, the results confirm and extend those obtained by the use of isotopic tracers. Thus, the results of crosses between *Nicotiana glauca* and *N. tabacum* suggest that anabasine, as well as nornicotine, is synthesized via an *N*-methylated intermediate. The rules that apply to the inheritance of a specific compound clearly show the biosynthetic relationship of its precursors. In simple cases in intraspecific crosses with only one set of alleles, the segregation in F_2 is as conclusive as a biochemical experiment in which a labeled precursor is incorporated into the alkaloid with a low dilution factor. When no linkage between genes exists, the following segregation ratios occur:

1. *No biosynthetical relationship*
   Plant with compound A × plant with compound B:
   F_1 progeny—both compounds are present (sometimes at a decreased level).
   F_2 progeny—segregation ratios 9AB:3A:3B:1 without either compound.
2. *Branched pathway, A and B branches from a common pathway*
   F_1 progeny—both compounds are present, but the level of both is depressed.
   F_2 progeny—one set of genes differs in parents: segregation ratio is 1A:2AB:1B; two sets of genes differ in parents: segregation ratio is 1A:14(4 + 6 + 4 different proportions of A and B) AB:1B.
3. *Direct relation between precursor A and product B*
   F_1 progeny—product is dominant, overdominant, or transgressive; in exceptional cases it possesses intermediate dominance.
   F_2 progeny—segregation ratio is 3(1 + 2)B:1A.
4. *Direct relation of precursor A to intermediate B to end product C*
   Plant with compound A × plant with compound C:
   F_1 progeny—intermediate and end products are usually dominant.
   F_2 progeny—segregation ratio is 9C:3B:4A, and A is divided into three plants which can transform the introduced B into C, and one plant that cannot perform this reaction (Figure 2.11).

In interspecific crosses the segregation ratios can be so badly obscured by irregularities of gametogenesis that a proper evaluation of the results is impossible. Nevertheless, such results may yield data which can serve to confirm tracer experiments.

Unfortunately, quite interesting experiments in the breeding of plants that produce alkaloids are sometimes published without the necessary segregation data, so the proper discussion of these data is not possible; consequently, these types of genetic experiments are of only limited value for elucidating biosynthetic pathways.

3 | Environmental Influences on Alkaloid Production

3.1. Introduction

Although the level of alkaloid biosynthesis is gene-governed, there are remarkable fluctuations in the concentrations and the amounts of alkaloids produced per plant due to environmental influences (Mika, 1962). Environmental conditions affect the general growth of the plant as well as the formation of alkaloids. Because most alkaloids are formed in young, actively growing tissues, factors affecting the growth of this tissue, such as influence of light, supply of nitrogen, potassium, phosphorus, and other minerals, temperature, moisture of the soil, and height above sea level will affect the production of alkaloids. Given that alkaloid production involves several different metabolic pathways, many of which are not fully known, only a generalized approach seems appropriate. An environmental factor may alter the biosynthesis or degradation of alkaloids of various origins; this same factor will increase the alkaloid production in one species and decrease it in another species. Consequently, the present discussion will be an enumeration of the available data to give an outline of the problems that have been solved and to point out those that remain unclear because of contradictory findings.

3.2. The Influence of Light, Water, and General Climatic Conditions

Because the fixation of CO_2 and growth are directly light-dependent, one might expect remarkable differences in the production of secondary metabolites in plants grown under varying light conditions. It is evident that light cannot be an essential, direct factor for the control of biosynthesis of those alkaloids which are formed exclusively in the roots. However, the role of light as an indirect factor cannot be ignored, particularly during a short time, while the plants are growing under unfavorable light conditions. Schmid (1948) found that germinating *Nicotiana tabacum* seeds produced nicotine while germinating in light as well as in the dark, but in lower amounts; the amount of nicotine produced in darkness was higher than in light. The plantlets grown in the dark reached the maximum level of alkaloid production on the fifth day following germination, and then the level rapidly declined, whereas plants grown in light never reached the same high level of nicotine, and they showed no decrease in alkaloid production. The explanation for a higher level of nicotine in etiolated seedings as compared with those grown in light could be that the initial nicotine biosynthesis is proceeding from an intermediate already present in the seeds. The seeds germinating in the dark are respiring and losing carbohydrate reserves and they liberate certain amino acids from the storage proteins. Glutamic acid, an intermediate in nicotine biosynthesis, cannot be used for protein synthesis and is therefore converted to alkaloid production. On the other hand, a plant which is growing in light is supplied with new carbon from photosynthesis and nitrogen from the soil, so it is using all of the amino acids for protein synthesis; in addition some nicotine can be broken down to furnish nitrogen and carbon for fresh amino acid biosynthesis (Figure 3.1a).

An experiment similar to that of Schmid's was performed (1948) with results shown in Figure 3.1b (Weeks, 1970). Seeds were exposed to 8 hr of darkness per day for 144 hr and had rates of germination which were like those for seeds exposed to 10 hr of light per day at 27°C. At all sampling periods, except that of 48 hr, the seed germinated in the dark contained more total alkaloids than seeds with daily exposures to light. The

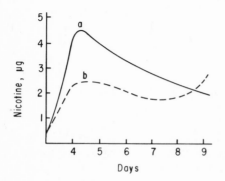

Figure 3.1a. Nicotine content of *Nicotiana tabacum* during germination; *a*—in the darkness, *b*—in the light. After H. Schmid (1948). Courtesy of *Schweizerische Botanische Gesellschaft.*

largest increase (300%) in total alkaloids occurred during the interval between the 96 and 120-hr sample points, whereas only a 4% increase in total alkaloids occurred in the next 24-hr period. The total alkaloid content of seeds grown at 144 hr was about three times that of the seeds germinated under daily exposure to light. Nicotine, nornicotine, and anabasine represented 90, 5, and 4% of the total alkaloid fraction at 144 hr. The remaining 1% was unidentified alkaloid-like substances. Nicotine was the predominant alkaloid in all samples except the 72-hr sample, where nornicotine accounted for 66% of the total alkaloids and nicotine accounted for 34%.

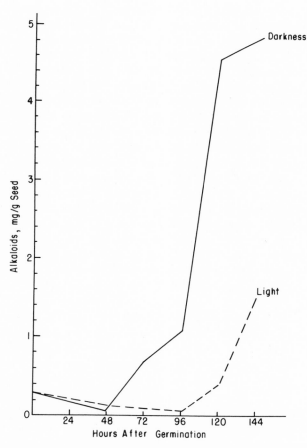

Figure 3.1b. Effect of time and light on alkaloid content of *Nicotiana tabacum* seedlings. Courtesy of W. W. Weeks (1970).

The unidentified alkaloid-like substance first appeared in the samples taken at 72 hr, and the quantity of this substance increased with time of germination and became maximal at 144 hr. In contrast, seed germinated in 10 hr of light per day contained the unidentified substance only in the 24- and 48-hr samples.

Seeds kept in the dark at 21°C following an initial 8-hr exposure to light during imbibition germinated at about the same rate as those exposed daily to 10 hr of light, but the accumulation of alkaloids during the first 48 hr after radicle protrusion was four times as great (6.5 *versus* 1.4 mg/g) in plants from seeds subjected to dark treatment as in those subjected to the light treatment.

Tobacco plants were grown on long or short photoperiods (Tso *et al.*, 1970) followed by 5 min of either red[1] or infrared[2] irradiation each day. Plants that received 16-hr photoperiods had a significantly higher concentration of total alkaloids than those that received 8-hr photoperiods. It was significant that higher total alkaloid content was found in plants that received red light rather than infrared radiation each day. The interaction among these variables (red light and far red radiation) produced the reverse effect on total alkaloid levels in that a higher concentration of total alkaloid was found in red light over the far red light treatment, but a higher total concentration of phenols was found in the far red light over red light treatment. This might be expected since the availability of phenylalanine and tyrosine, which are precursors of phenolic compounds in *Nicotiana* plants, is dependent on the longer day length.

The alkaloids of leguminous plants are synthesized predominantly in the aerial parts of the plant; therefore, one might expect a different reaction when the plant is exposed to darkness. Experiments with germinating *Lupinus* (lupine) seeds were performed by Nowotnówna (1928) and by Sabalitschka and Jungermann (1925); and the results were confirmed by Wallenbrock (1940). The synthesis of alkaloids by plants grown in the dark was notably retarded compared with those grown in light (Figure 3.2). An exception to this is presented by germinating seeds of *Ricinus communis* (castor bean), where the level of ricinine was remarkably higher in plants grown in darkness than in light. The results from 300 plantlets showed 273 mg of ricinine in normal green plants grown in light, while 3-week etiolated seedlings grown in the dark contained 422 mg (Weevers and van Oort, 1929). The steroidal alkaloids of *Solanum tuberosum* (potatoes) are formed in tubers exposed to light, whereas in darkness they do not form.

Well developed plants behave in a fashion similar to that of the

---

[1]2000 ft-c of illumination from cool white VHO fluorescent lamps.
[2]Adjusted to approximately $360\,\mu W^{-2}$ over the wavelength bands of 600–700 and 700–770 nm.

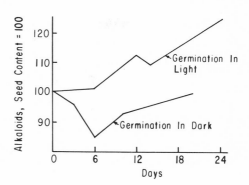

Figure 3.2. Alkaloids in *Lupinus luteus* seedlings germinating in the light and in the dark (Wallenbrock, 1940). Courtesy of *Acta Botanica Neerlandica.*

germinated seeds. A short-time light experiment in the dark usually caused a retardation of alkaloid synthesis in plants that produced alkaloids in the aerial part (Nowacki, 1958), while plants that produced alkaloids in the roots did not show decreased alkaloid formation while shaded (Ripert, 1921). Except for cultivation under sterile conditions, there is no way to grow plants or plant tissues for long periods of time in darkness. When supplied with carbohydrates while in darkness, young cotyledonless seedlings from different lupine varieties did not produce alkaloids, even though their dry weight increased tenfold (Nowacki, 1958). (Callus tissue of *Datura stramonium* produced alkaloids under those same conditions.) Varying the nature of sugars fed had no clear influence on the accumulation of alkaloids.

Long-term experiments with field-grown plants can be performed only within certain geographical limits that determine the day length as well as the intensity of light. For example, short daily exposure to dim light kills experimental plants, sometimes more rapidly than rearing them in complete darkness. The results of experiments using different times of isolation and/ or different intensities of light do not agree as well as those of the short-time experiments. *Nicotiana tabacum* when shaded produced less alkaloid than when exposed to full light (Stillings and Laurie, 1943), and *Atropa belladonna* (Unger, 1912) was shown to act in a similar manner; both of these species produce alkaloids in the roots (see following chapter). The lupines, which produce alkaloids in the aerial parts of the plant, accumulate less alkaloids when in full light than when in dispersed light and show an increase in the percentage of alkaloids when grown under dispersed light (Malarski and Sypniewski, 1923; Byszewski, 1959, 1960) (Figures 3.3a and 3.3b). These results were not those expected, for just the reverse had been predicted. *Nicotiana* species are strongly photoperiod-sensitive plants, and when cultivated under long-day conditions, they quickly proceed to the

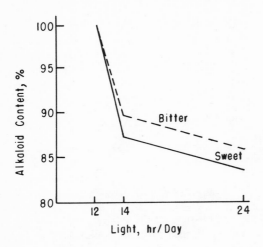

Figure 3.3a. The influence of time of illumination on the percentage of alkaloids in bitter and sweet lupine. The level found in a 12-hr illumination experiment is taken as 100% of the alkaloid content. The light intensity was constant (Malarski and Sypniewski, 1923). Courtesy of the authors and of the journal.

flowering stage of growth, a stage that usually ends alkaloid synthesis. On the other hand, plants produced under different light conditions are usually not comparable; not only are the developmental stages different, but also the sizes of plants are often tremendously different. In 1973 Nowacki *et al.* described a series of experiments on the effects of light on alkaloid synthesis in lupines. The percentage of alkaloids in the dry matter was found to decrease rapidly with prolongation of the time of exposure to light, but simultaneously the total amount of alkaloids increased because of the

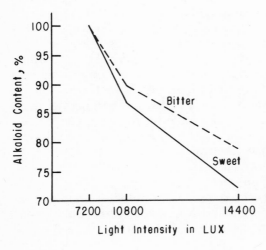

Figure 3.3b. Influence of light intensity on the percentage of alkaloids in bitter and sweet lupine. The amount of alkaloids produced by 7200 lux is taken as 100% alkaloid content (Byszewski, 1959, 1960). Courtesy of the author and of the journal.

overall better growth of the plants. Not only the *level* of alkaloids was changed in plants grown under different light conditions, but the *proportions* of alkaloids were markedly disturbed (Figures 3.4a, 3.4b, and 3.4c). The long-term experiments using different exposure times, while contributing less to our knowledge about the mechanism of alkaloid biosynthesis, are in some ways helpful in explaining the fact which has been known since ancient times, that the same species of plants is poisonous in one country and harmless in another.

Closely related to experiments with light intensity are those with plants grown at different geographical latitudes. During the 19th century, Charles Darwin stated that *Aconitum napellus* grown in Nova Scotia was not poisonous, while that which was grown in the Mediterranean was one of the most toxic plants. Sokolov (1952, 1959) found that the same variety of *Scopolia carniolica* accumulated over 1% alkaloids in the Caucasus, while in Sweden it accumulated only 0.3%; *Atropa belladonna* on the Crimean peninsula accumulated 1.3% alkaloids, while in Leningrad only 0.4–0.6% was accumulated. Similar data were obtained for a number of plants such as *Datura stramonium*, *Hyoscyamus niger*, and *Anisodus luridus*. The plants that were grown in the south were alkaloid-rich, while in the north they were alkaloid-poor. Besides light and temperature, there was at least one additional factor involved: the level of soil moisture. The soils in northern areas never dry to such extremes as can happen in the south. The effect was dramatically illustrated by the finding that the cinchona tree produces no quinine during the rainy season. Kuzmenko and Tikhvinska (1935) found that tobacco plants grown in a field with 90% saturation of soil-water capacity produced only traces of alkaloids, while the highest percentage of nicotine was found in those plants grown under the lowest level of water saturation (30%). Similar results were obtained by Sharapov (1956), who stated that the level of alkaloids in lupines was considerably less than normal during wet years, while during unusually dry years it was higher. The results of Sharapov (1961) were confirmed by Bachmanowa and Byszewski (1959).

Woodhead and Swain (1974), who investigated the biosynthesis of the alkaloid-like compound amaranthine encountered in Chenopodiaceae, have demonstrated the effect of light on the rate of biosynthesis of this pigment. The authors believe that the synthesis is controlled by phytochrome with a probable involvement at a higher level of light (Figure 3.5).

A number of experiments were performed with the opium poppy (*Papaver somniferum*). The highest alkaloid level in this species was noted when it was cultivated in moderately dry fields in the tropics. Il'inskaya and Yosifova (1956) found that the percentage of total alkaloids in the poppy capsule increased in the plants the farther south they were grown. Voseu-

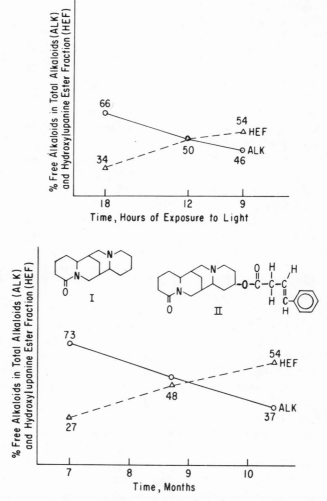

Figure 3.4a. The effect of exposure of plants to light on the composition of lupine alkaloids. Top: Artificial light: plants were grown in July, the plants exposed 18 hr were additionally illuminated with 4000 lux during the twilight hours, and the plants exposed 12 and 9 hr were covered to shorten the time of exposure. Bottom: Natural light: plants were planted at month intervals; hence, their growth rate was according to their location, Poznan, Poland. The curves represent the concentration of free and esterified alkaloids in the total plant in mg/100 g of plant material (dry weight). The numbers represent the milligram contribution of free alkaloids (ALK) and hydroxylupanine ester fraction (HEF). Formulas: lupanine (I) a free alkaloid, and cinnamoyloxylupanine (II) a HEF (Nowacki *et al.,* 1973). Courtesy of the journal.

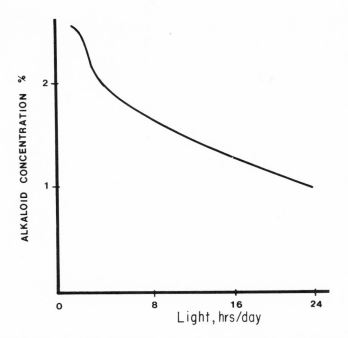

Figure 3.4b. Lupine alkaloids as influenced by time of illumination (Nowacki *et al.,* 1973). Courtesy of the journal.

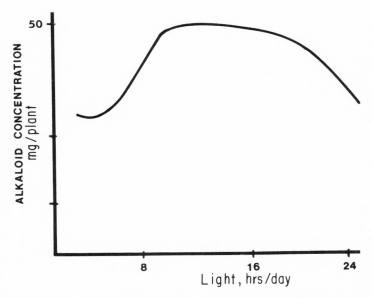

Figure 3.4c. Lupine alkaloids as influenced by time of illumination (Nowacki *et al.,* 1973). Courtesy of the journal.

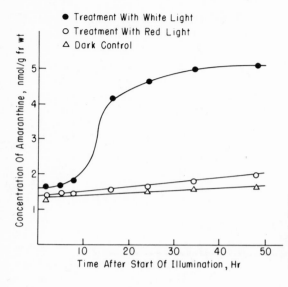

Figure 3.5. The increase in amaranthine in dark green *Amaranthus caudatus* seedlings after treatment with light (Woodhead and Swain, 1974; courtesy of Pergamon Press, copyright, 1974).

rusa (1960) has demonstrated the influence of not only the geographical location of the experiment station but, in addition, the variation of the climatic conditions within a cultivation year. Similar experiments performed in Hungary by Tétényi *et al.* (1961) and Vágújfalvi (1963a,b, 1964) established the effects of climatic conditions on the synthesis of alkaloids. In the cooler and more moist summers of Poland, the Hungarian variety of opium poppy, "Hatvani," accumulated morphine to a level of 0.36%, whereas the average level in hot and dry summers in Hungary was 0.41%. Matveyev (1959), while investigating over 800 varieties of poppies in different agriculture experiment stations throughout the Soviet Union, found the morphine level to fluctuate between 0.1 and 1.0%. Kleinschmidt and Mothes (1958) found that only in favorable years did the level of morphine in 20% of the investigated varieties get above 0.4%. In cool and rainy summers no varieties exceeded the 0.4% level. Duchnowska and Pawelczyk (1960) investigated a collection of opium poppies in Poland and found that in only 15% of the varieties the morphine level approached or exceeded 0.3%. There was clear evidence for the opium poppy that the amount of rainfall did exert a predominant influence, because the same varieties grown at the same latitude in Poland, as in the Russian Republic, produced more morphine in the dryer climate of Russia.

Experiments were performed with seedlings of *Nicotiana tabacum* germinated at different temperatures by Weeks and co-workers (1969, 1970, 1974). An increase in temperature from 21 to 27°C during the germination period gave the most rapid rate of alkaloid decomposition, whereas the most rapid rate of alkaloid biosynthesis corresponded with rapid germination (Figure 3.6). Radicle protrusion occurred after 72 hr, 24 hr earlier than at 21°C. The quantities of nicotine, nornicotine, and anabasine measured at the 48-hr sampling interval were lower at 27°C than at 21°C. However, at the 144-hr sampling, the total alkaloid content of plants from seed germinated at 27°C was twice that of plants started at 21°C. At the 144-hr sampling interval, the total alkaloid fraction was composed of 90% nicotine, 6% nornicotine, 4% anabasine, and trace amounts of anatabine. All the alkaloids decreased in quantity during the interval between the 96- and 120-hr sampling times. This decrease was observed in all replicates and cannot be fully explained with available data. The unidentified alkaloid-like substance was observed only at the 24- and 48-hr sampling periods. At 32°C radicle protrusion occurred at 48 hr, and by 72 hr an accumulation of nicotine and nornicotine was observed. However, the rate of accumulation of alkaloid in the seedlings for the first 24-hr period following radicle protrusion was lower at 32°C than at any of the other temperatures except 16°C. The total alkaloid fraction at 144 hr was composed of 93% nicotine,

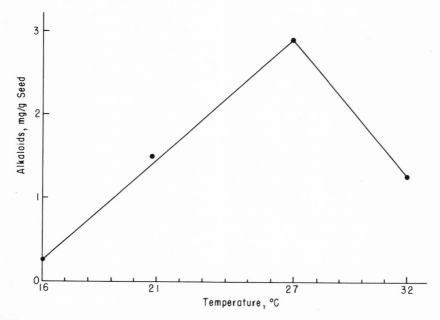

Figure 3.6. Effect of temperature on total alkaloid content of *Nicotiana tabacum,* variety Kentucky 16: 144 hr after germination (Courtesy of W. W. Weeks, 1970).

3.5% nornicotine, 3.5% anabasine, and traces of anatabine. The unidentified alkaloid-like substance was not detected in any samples grown at 32°C.

The concentrations of total alkaloids measured during the first 144 hr of germination at the four temperatures tested are shown in Figure 3.6. The nicotine content was highest at 27°C and was found in decreasing amounts at 21, 32, and 16°C. The percentage of nicotine in the total alkaloid fractions was 93, 90, 84, and 23% for 32, 27, 21, and 16°C, respectively. Although seedlings from the 32°C germination temperature had the highest percentage of nicotine content, total alkaloids from the seedlings grown at 32°C were lower than those at all other temperatures except 16°C. A comparison of the mean nicotine content by a least significance difference test indicated that the nicotine content of the seedlings at 144 hr from the 16 and 27°C germination temperatures differed from that found for all other germination temperatures. The nicotine content of seedlings from the 21°C treatment differed from all other treatments except at 32°C, where it was the same. The most favorable temperature for alkaloid synthesis during the first 144 hr of germination was 27°C.

Besides the investigation of the same varieties under different climatic conditions, the endemic form of a given species may accumulate different alkaloids at different levels. As an example, Proskurina (1955) isolated lycorine, tazettine, and haemanthidine from *Pancratium maritimum* collected along the Caucasian coast of the Black Sea, while Boit and Ehmke (1956) isolated lycorine, tazettine, and hippeastrine from Dutch specimens. Sandberg (1961) investigated the alkaloid composition of this same species grown in Egypt and on the islands of Rhodes and Corsica and found that the Egyptian form differed by the presence of two additional alkaloids, one uncommon to both forms and the other one different in each form. As long as ecotypes from various countries are not compared under identical conditions, it is not possible to assess the significance of the different causes for varying taxa (Figure 3.7). Varieties of fodder plants such as clovers, various grasses, and birdsfoot trefoil *(Lotus corniculatus)* are recognized as excellent forages in central Europe; but when they were introduced into semiarid areas of Australia, central Asia, and the United States, it was found, surprisingly, that the varieties turned out to be harmful in these new areas. Thus alkaloids of the pyrrolizidine type were found (Yunusov and Akramov, 1955, 1960) in species of *Festuca* and *Lolium* grown under climatic conditions of Uzbekistan, Australia, and the arid regions of the USA. *Phalaris tuberosa* grown in Australia contained a high level of N,N-dimethyltryptamine-type alkaloids (Yates and Tokey, 1965). When these varieties of grasses were introduced, they caused acute poisoning of sheep. Oram and Williams (1967) showed that the ecology affects the level of N,N-dimethyltryptamine alkaloids.

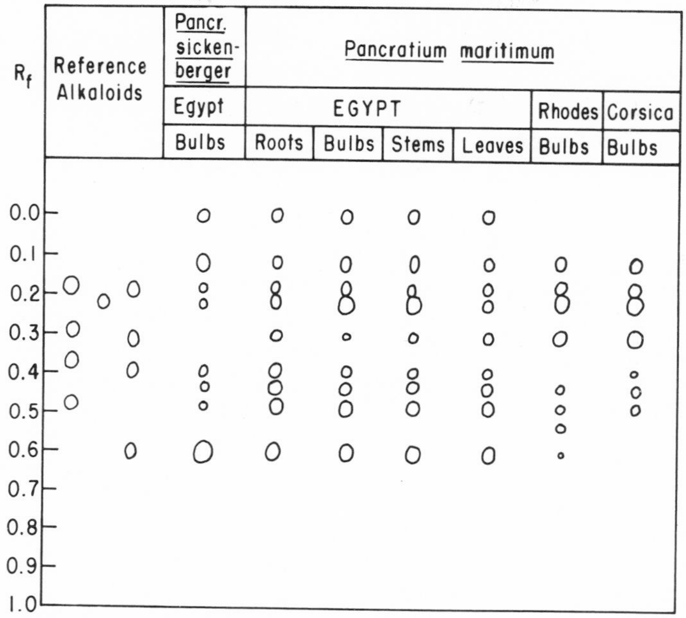

R_f	Reference Alkaloids	Pancr. sicken-berger Egypt Bulbs	Pancratium maritimum				Rhodes	Corsica
			EGYPT				Bulbs	Bulbs
			Roots	Bulbs	Stems	Leaves		
0.0		o	o	o	o	o		
0.1		o	o	o	o	o	o	o
0.2	O o O	o o	8	8	8	o o	8	8
0.3	O o		o	o	o	o	O	O
0.4	O o	o o	o o	o o	o o	o o	o	o o
0.5	O	o	o o	o o	o o	o o	o o / o	o o
0.6		o	O	O	O	O	O	o
0.7								
0.8								
0.9								
1.0								

Figure 3.7. Distribution of *Amaryllidaceae* alkaloids along the Mediterranean coast (Sandburg, 1961). Courtesy of the *Pakistan J. of Scientific and Industrial Research.*

3.3. Nitrogen, Potassium, and Other Mineral Nutrients

One of the most fundamental but least understood areas of plant physiology and plant biochemistry is the relationship between the quantities of the essential nutrient elements and the metabolic changes occurring in the plant. A living plant is a complex biological system composed of chemical constituents. Most minerals are essential building materials, and some form parts of indispensable catalysts. A deficiency in any element affects the normal metabolic system and thus disturbs the balance of the chemical constituents. Abnormal accumulation of certain compounds is associated with the development of a typical symptom usually recognized due to the abnormal supply of a specific element.

Since alkaloids are nitrogen-containing compounds, the availability of nitrogen is expected to play an important role in the biosynthesis and accumulation of alkaloids in plants; however, as in the case of light, some reservations are necessary. With respect to the biosynthetic pathways, the alkaloids can be divided into at least three groups. The first comprises those which liberate amino nitrogen from the precursor while biosynthesis is carried out. In this group belong all the alkaloids which are biosynthesized from the diamino acid family, the majority of the phenylalanine–tyrosine family, and some from the tryptophan group. Alkaloids of this group have a higher C:N ratio than their immediate precursors. During the process of converting an amino acid into an alkaloid, at least 1 mole of ammonia is liberated. The second group includes alkaloids with the same number of nitrogen atoms in the precursor as in the alkaloid. When condensation with isoprenoid and/or phenolic compounds occurs, the C:N ratio will of course increase. The third group comprises alkaloids which have a lower C:N ratio than the substrate. This group is sometimes recognized as the pseudoalkaloids, since it includes steroidal alkaloids, isoprene derivatives, and some others (Figures 3.8a, 3.8b, and 3.8c).

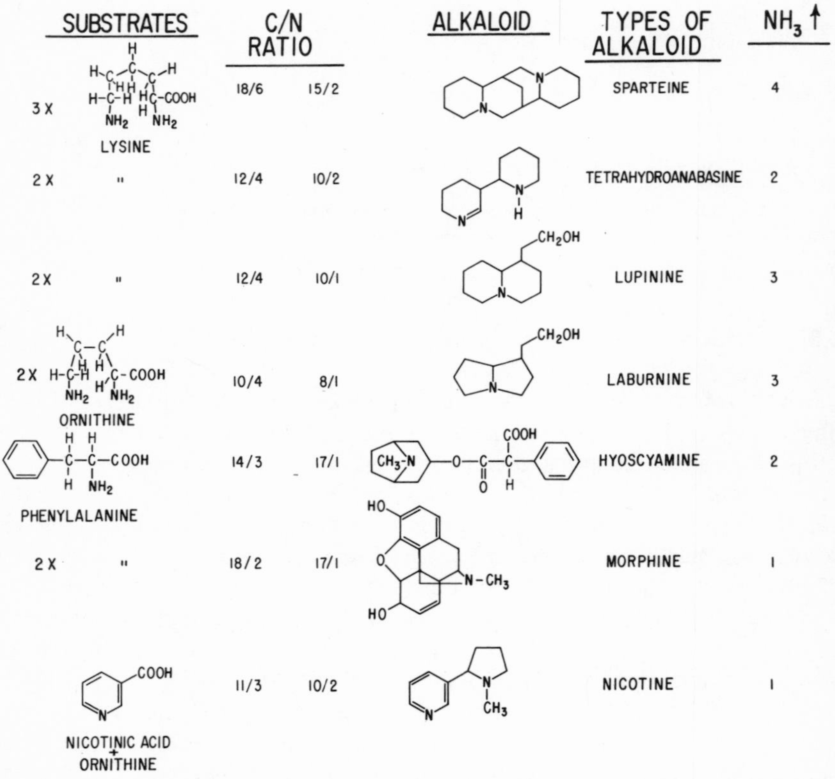

SUBSTRATES	C/N RATIO		ALKALOID	TYPES OF ALKALOID	NH₃ ↑
3 X LYSINE	18/6	15/2		SPARTEINE	4
2 X "	12/4	10/2		TETRAHYDROANABASINE	2
2 X "	12/4	10/1		LUPININE	3
2 X ORNITHINE	10/4	8/1		LABURNINE	3
PHENYLALANINE	14/3	17/1		HYOSCYAMINE	2
2 X "	18/2	17/1		MORPHINE	1
NICOTINIC ACID + ORNITHINE	11/3	10/2		NICOTINE	1

Figure 3.8a. Relationships in nitrogen content between alkaloids and precursors. The precursor is richer in nitrogen than the product alkaloid. Nitrogen is liberated (↑).

SUBSTRATE	C/N RATIO		ALKALOID	ALKALOID NAME	NH_3 BALANCE
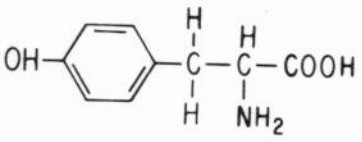 TYROSINE	9/1	14/1	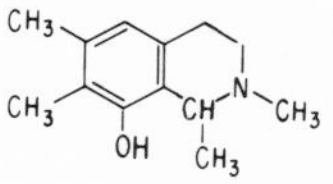	PELLOTINE	0
	9/1	29/2	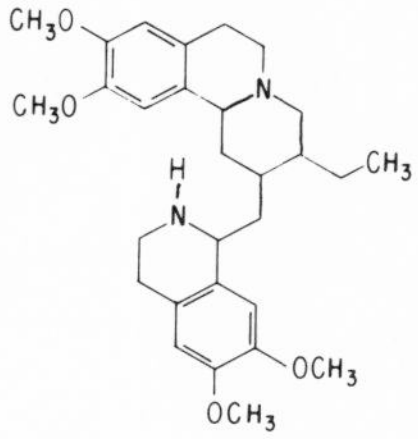	EMETINE	0
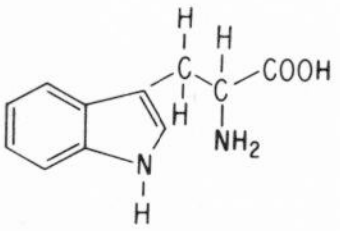 TRYPTOPHAN	11/2	11/2	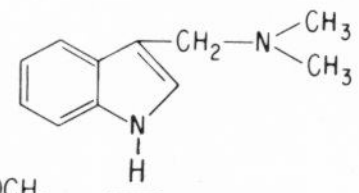	GRAMINE	0
"	11/2	20/2	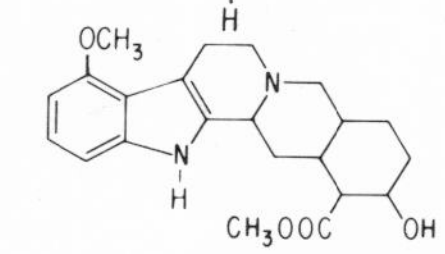	VENENATINE	0
"	11/2	17/2	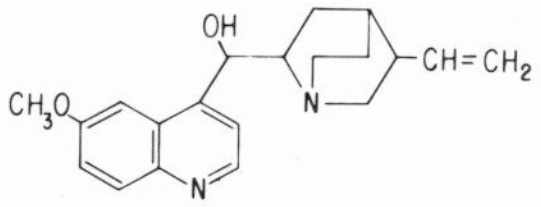	CINCHONINE	0
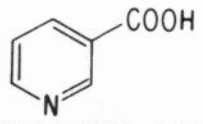 NICOTINIC ACID	6/1	10/1	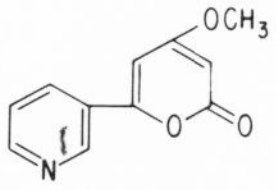	ANIBINE	0
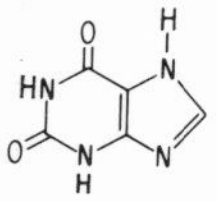 PURINE-2,6-DIONE	5/4	8/4	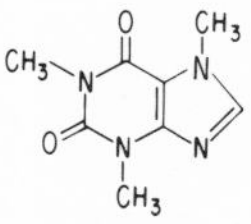	CAFFEINE	0
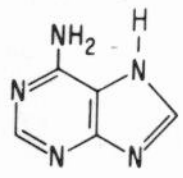 ADENINE	5/5	10/5	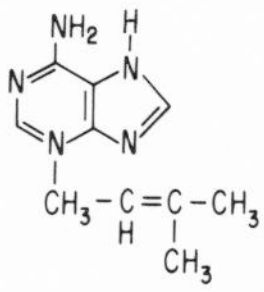	TRIACANTHINE	0

Figure 3.8b. Relationships in nitrogen content between alkaloids and precursors. The precursor has the same number of nitrogen atoms as the product. The C:N ratio is usually increased owing to condensation with a nitrogen-free compound.

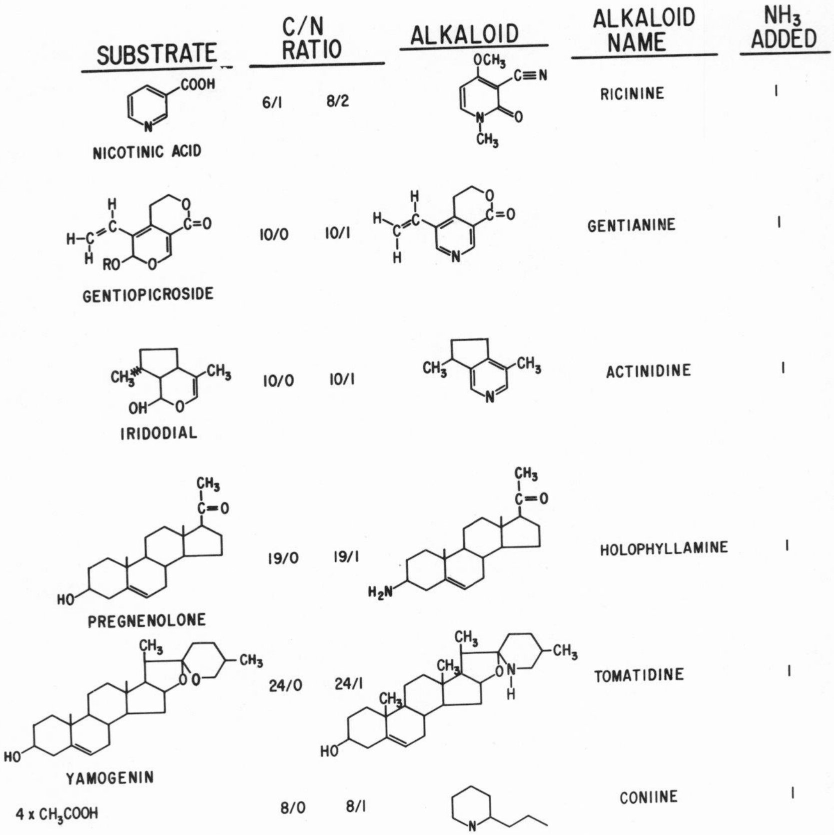

Figure 3.8c. Relationships in nitrogen content between alkaloids and precursors. The product has more nitrogen than the precursor. Nitrogen is a prerequisite for biosynthesis.

During the last 48 years evidence has been accumulated that all plants do not react in this way to increased nitrogen fertilization, i.e., with an increase in alkaloid biosynthesis (Nowacki *et al.*, 1975, 1976). It is possible to produce two- to tenfold increases of alkaloid content (in *Nicotiana* species, *Lupinus* species, and barley, *Hordeum* species) by treating the plants with high levels of nitrogen. Exceptions to this behavior are plants producing indole alkaloids and the *Solanum* glycoalkaloids. The effect of nitrogen fertilization on alkaloid production has not been the subject of extensive experiments.

It is conceivable that the levels of various groups of alkaloids will react differently to varying levels of nitrogen fertilization. One has to distinguish two situations: short-time experiments, sometimes with extremely differentiated levels of available nitrogen, and long (entire growing season) experiments. In the second type of experiment, extreme conditions will result in the death of the plant. There are few good examples of published short-time experiments concerning the role of nitrogen. One is that of Mothes, published nearly 50 years ago (1928). In this experiment, *Nicotiana* plants were grown under three different levels of added nitrogen. The alkaloid production per plant rapidly increased upon addition of the first level of nitrogen, but additional amounts of nitrogen decreased the total amount of alkaloids (Figure 3.9a). This was rather to be expected for an alkaloid like nicotine in which the C:N ratio markedly decreases in the final stage of alkaloid

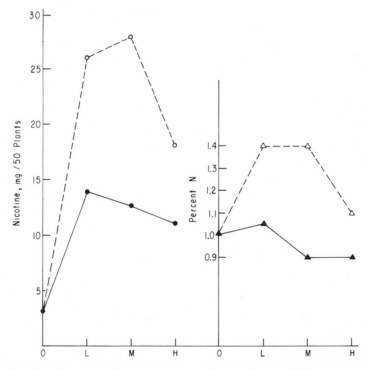

Figure 3.9a. Nicotine content in tobacco grown on low, medium, and high concentrations of ammonium sulfate in nutrient solution. Left—total amount of nicotine per 50 plants. Right—percentage of total nitrogen (Mothes, 1928). Courtesy of the author and *Planta,* published by Springer-Verlag, Heidelberg.

formation, from 11:3 to 10:2. Similar behavior would be expected for tropane and lupine alkaloids. Experiments performed by Nowacki *et al.* (1976) confirmed these expectations. In 1922 Vogel and Weber published data on alkaloid synthesis in lupines grown under sterile conditions without nitrogen and under symbiotic conditions with *Rhizobium*. The percentage of total nitrogen increased only fivefold. All the long-term experiments, which were quite numerous, yielded different results. Usually all plants reacted to nitrogen fertilization like the lupines inoculated with *Rhizobium*. With increasing levels of nitrogen fertilizer, the plants grew more luxuriously, and in most cases the extent of alkaloid synthesis paralleled the general growth of the plant. Stimulating growth by addition of fertilizer tended to increase the formation of alkaloids. The nitrogen doses administered to the plants were usually moderate, and the experiments were performed at agricultural experiment stations whose soils had a high level of fertility; consequently, there was always a certain initial level of nitrogen in the soil. Therefore, the control group, which was on a "zero" dose, rarely suffered nitrogen deficiency. Seldom was the other extreme, an overdose of nitrogen, reached, because most commonly the highest rates used were 200 kg N/ha with an average available soil nitrogen level of 60 kg N/ha. Most plants with an adequate supply of water can produce an increase of dry substance up to 1200 kg N/ha (Stuczynski and co-workers, 1969, 1970, 1971; Nowacki and Weznikas, 1975). The response to increased nitrogen depends upon the species and sometimes the variety used. Usually there is a simple increase in the overall growth, with up to 300 kg N/ha, and approximately the same chemical composition is maintained. After the optimal level is passed, the chemical makeup of the plant changes rapidly (Nowacki, 1969; Nowacki and Weznikas, 1975). Figure 3.9b shows the indole alkaloids of reed canarygrass as influenced by nutrient supply (Marten *et al.*, 1974). Ammonium sources of nitrogen caused a greater alkaloid concentration than did a nitrate source ($NaNO_3$), while grass supplied with both sources (NH_4NO_3) had an intermediate concentration. They concluded that nitrogen fertilization is likely to increase the alkaloid content in those strains where alkaloids are already normally high, but that practical levels of nitrogen (up to 200 kg/ha) will not increase alkaloid concentrations of those strains inherently low in alkaloids. Of interest to the geneticist and to the natural products chemist was the finding that none of the fertilizer elements (N, P, Mn, Co, Mg, S, Ca, B, Zn, Mo, K) changed the type of alkaloids (gramine, *N, N*-dimethyltryptamine and 5-methoxy-*N,N*-dimethyltryptamine) of the various clones and seed source, which indicated that this variable was under *genetic control*. Nowacki *et al.* (1976) recently found differences in the responses of alkaloid-producing plants grown with added ammonium nitrate. This demonstrated that plants which produce alkaloids originating from amino acids

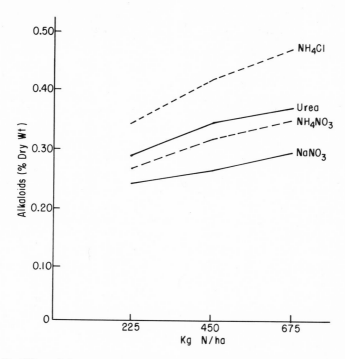

Figure 3.9b. Effect of nitrogen sources and rates in solution culture on alkaloid concentration in reed canarygrass (mean of four clones and two harvests). Courtesy of Marten *et al.*, 1974 and *Agronomy Journal.*

react to increased nitrogen fertilization with increased alkaloid production; the exceptions were *Vinca perenne* and *Vinca (Catharanthus) rosea*, both of which produced alkaloids derived from tryptophan (Figures 3.10a and 3.10b).

Waller *et al.* (1965) studied ricinine production by individual castor bean plants (Figure 3.11). The plants contained about 0.1 μmole in the ungerminated seed and about 1.2 μmol at 17 weeks of age. These plants had received the normal amount of fertilizer to ensure optimum production, and they were irrigated to avoid moisture stress. Immediately prior to flowering, the observed rate of ricinine synthesis was 36 μmol/day, which represented the highest rate in the nonflowering plants. The plants were not mature at 17 weeks of age, but they had reached a constant growth pattern, during which they continued developing seeds until the 21st week of age, at which time they began to senesce (see Chapter 6 for information on senescence). Although this whole experiment was not repeated in 1962 and

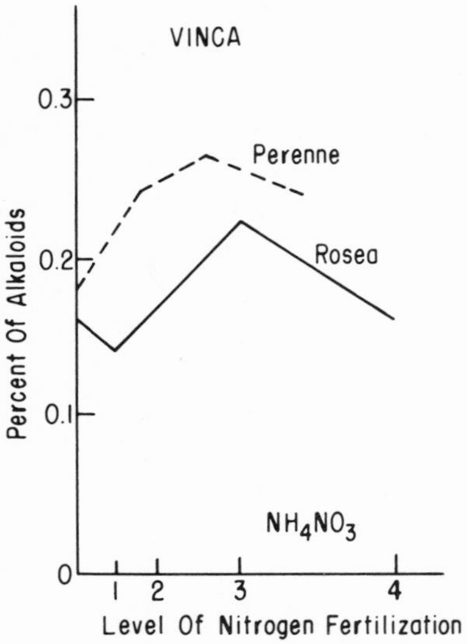

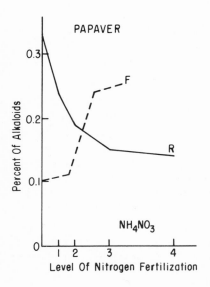

Figure 3.10a. Effect of nitrogen applied to soil on the alkaloid content of *Vinca perenne* and *Vinca (Catharanthus) rosea* in 1972 and 1973. Nitrogen applied one time at 0 week and at the levels: $1 = 1g; 2 = 2g; 3 = 4g; 4 = 8g\,NH_4NO_3$/pot.

1963, sufficient numbers of plants were analyzed to verify the 1961 findings. There was a noticeable variation in the chronological age of the plants at the same physiological state of development during this 3-year period. The plants grew most rapidly in 1961, as is indicated by the 40 days of age required to reach the flowering stage, whereas in 1963, 49 days were

Figure 3.10b. Effect of nitrogen applied to soil on the total alkaloid content in *Papaver somniferum* plants: *R*—rosette stage; *F*—flowering stage, 4 and 7 weeks, respectively. Nitrogen applied one time at 0 week, the same as in Figure 3.10a (Nowacki *et al.,* 1976). Courtesy of the journal, published by V. E. B. Gustav Fischer Verlag.

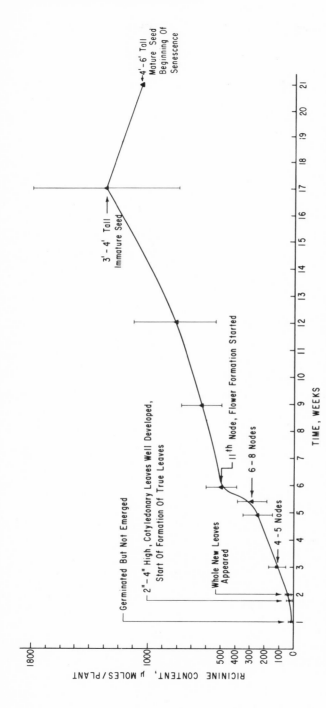

Figure 3.11. Ricinine content of the castor plant at different states of development. These plants were grown at the Oklahoma State University Agronomy Farm in Stillwater in 1961. The physiological stages of development are noted on the graph. The entire plants were used for the isolation of ricinine. The variations in analyses are indicated by vertical lines. The number of plants used in obtaining each point is as follows: ●—2 lots of 20 plants each; ■—2 lots of 5 plants each; ▲—2 plants (Waller *et al.*, 1965). Courtesy of *Plant Physiology*, published by the American Association of Plant Physiologists.

required. This might be expected to have had an effect on the synthesis and breakdown of ricinine, and a comparison of the results obtained in 1961 and 1963 confirmed this hypothesis. The castor bean plant, when fertilized with a shock level of ammonium nitrate in the middle of the growing season, first responded with a decrease in the percentage of ricinine, but after a few weeks, when a rapid growth and an overall rejuvenation of the plant had occurred, the level of ricinine was found to surpass that of the control. By adding ammonium nitrate in physiological doses simultaneously with the proper substrate for ricinine biosynthesis ([6-^{14}C]quinolinic acid or [7-^{14}C] nicotinic acid), results were obtained that indicated an inhibitory role of nitrogen on alkaloid biosynthesis. Transfer of the plant into darkness decreased the incorporation of nicotinic acid, while feeding with sucrose stimulated the conversion (Figure 3.12). Much more experimental evidence is needed to clarify the role of nitrogen.

The influence of nitrogen on the accumulation of alkaloids under field conditions was investigated by Smirnova-Ikonikova (1938), Zolotnicka (1949), Sukhorukov and Borodulina (1932), and Avramova (1955). In *Nicotiana, Atropa, Datura,* and *Papaver* the nitrogen promoted the overall development of the plants, and to a certain extent, increased the percentage of alkaloids (Figure 3.13).

In a rather obscure publication of the Hatano Tobacco Experiment Station, Yoshida (1973) determined the activities of enzymes catalyzing nicotine biosynthesis (see Chapter 6 for a further discussion of the biosynthesis) in the roots of tobacco plants. His experiments included plants that

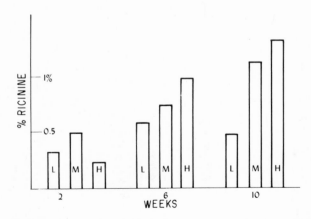

Figure 3.12. Effect of different levels of nitrogen applied to the soil on ricinine content: L— low level of nitrogen fertilization (none); M—medium level of nitrogen fertilization (250 kg N/ ha); H—high level of nitrogen fertilization (500 kg N/ha). (Nowacki and Waller 1972). Courtesy of the publisher.

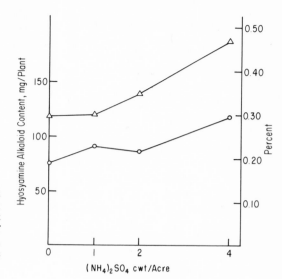

Figure 3.13. Influence of nitrogen fertilization on the alkaloid content in *Atropa belladonna* Δ—percent in shoots. ○—alkaloids, in mg of hyoscyamine per plant. (James, 1950). Courtesy of Academic Press.

had received complete nutrients and plants that were deficient in nitrogen, phosphorus, potassium, calcium, magnesium, sulfur, and boron. The activity of ornithine decarboxylase was increased by nitrogen deficiency, the putrescine N-methyltransferase decreased in the plants which were deficient in each nutrient, the N-methylputrescine oxidase decreased in the plants deficient in nitrogen, phosphorus, potassium, and sulfur, the NAD-pyrophosphatase decreased in the nitrogen-deficient plant, and the nicotinic acid decarboxylase decreased in the plants which were deficient in every nutrient tested except magnesium. In all cases the addition of nitrogen to the culture solution caused a rapid increase in the activity of the enzymes. It was assumed that the decrease in the nicotine in the nutrient-deficient plants is due to the decrease in activity of the enzymes.

Bush and Buckner (1973) found that tall fescue seedlings grown in the greenhouse increased in perloline (Figure 3.14) content as the level of nitrogen in the nutrient solution was increased; e.g., at 60 days, the level of

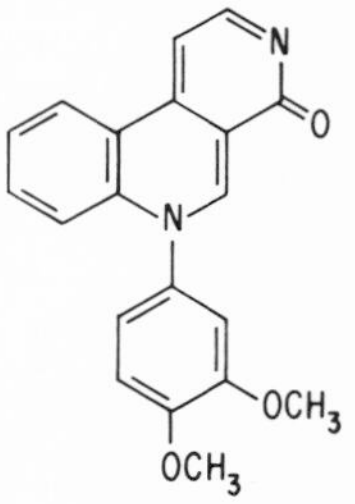

Figure 3.14. Structure of perloline.

nitrogen that the plants started from gave a perloline concentration of about 700 μg/g, but when the level of nitrogen was increased 10 times the concentration of perloline was 3000 μg/g.

Weybrew *et al.* (1974) conducted a study of the biochemical changes of the tobacco plant, *N. tabacum,* with respect to growth, maturation, and senescence over a 6-year period. They found that the external supply of nitrogen (1) times the onset of senescence, (2) influences the yield and, (3) influences the quality of the crop. Nitrogen metabolism is dominant during active growth, and since there is a complicated biochemical pathway that exists between nitrate reduction and glycolysis which is the direct precursor of starch biosynthesis, the starch accumulates only after the external reserve of nitrate is exhausted. Climatic factors, especially the distribution and total amount of rainfall, induce serious constituent imbalances with respect to nicotine, starch and sugar contents. Extended droughts or leaching rains have a pronounced effect on nitrogen reserves, and to an extent, the moisture reserve of the plant. A case in point is the 1971 season, which was notably dry: only 1.6 cm of rain fell between June 4 and July 10 and this was in the form of brief showers. On June 29, 3.2 cm of irrigation water was applied, and rainfall was fully adequate after July 10. The tobacco made phenomenal growth when the moisture deficiency was remedied; analysis of the leaves revealed 7% nicotine in some samples and 25% sugar concentration in other. Sequential sampling showed that nitrate reduction did not stop until July 27, which was two weeks later than normal. After July 27 there was insufficient time for the plants to accumulate starch; starch content in 1971 did not exceed 10%. Prolonged nitrogen metabolism reduced starch accumulation, and when the process is delayed as in 1971, the time remaining influences the yield.[3] Table 3.1 shows yield and compositional data from the summer of 1971.

The second important mineral nutrient, potassium, seems to be much easier to control, and the results are clear for alkaloids that are derived from amino acids. Smith (1965) found that the degrading enzyme N-carbamyl putrescine to putrescine is greater in leaves of potassium-deficient barley. Depletion of potassium usually results in an increase in the percentage of alkaloids in plants, while the total amount per plant is sometimes decreased. Results for tropane alkaloids were published by James (1950), Dafert and Siegmund (1932), and El-Hamidi *et al.* (1965). In lupines the results were similar to those obtained for tropane alkaloids, i.e., the potassium-depleted plants had a higher alkaloid level (Mironenko, 1965, 1975; Andreeva, 1952). Alkaloid-free plants, e.g., some species of *Trifolium,*

---

[3]Yield of flue cured (or Bright Leaf or Virginia) tobacco is calculated as kg/ha or lb/acre. This is the major component of most cigarettes; it is readily recognizable by the bright yellow or orange color of the cured leaves and it is distinguished by its relatively low nicotine content (around 2.5%) and high content of sugar (18%).

Table 3.1. The Effect of Season and Cultural Managements on the Yield and Composition of Flue-Cured Tobacco,[a] *Nicotiana tabacum* (variety C254, Oxford)[b]

Treatment and year	Yield[c] kg/ha	Total alkaloid[d] %	Reducing sugar[d] %
18 leaves/plant × 15,650 plants/ha (6 harvest × 3 leaves)			
1970	2374	4.12	12.0
1971	2367	4.98	7.4
12 leaves/plant × 23,485 plants/ha (3 harvest × 4 leaves)			
1970	2370	4.33	12.3
1971	2150	5.46	7.0

[a]Note that the harvesting of the low-topped tobacco had been completed by August 13, whereas two-thirds of the leaves on the check plants were still in the field on that date.
[b]From Weybrew *et al.*, 1974.
[c]All cured leaves.
[d]Weighted by thirds of plants.

Hordeum, and *Melianthus,* upon depletion of potassium, respond by accumulating amines such as putrescine and agmatine (Suzuki and McLeod, 1970; Sinclair, 1967; Koch and Mangel, 1972).

The effect of phosphorus is less spectacular. However, it is the reverse of that of potassium. With an increase of fertilization the level of alkaloids in the lupines rises both in percentage and in total amount (Ermakov *et al.*, 1935; Mironenko, 1965, 1975). Gentry *et al.* (1969) found that phosphorus and potassium added together greatly reduced perloline biosynthesis in tall fescue *(Festuca arundinacea),* whereas nitrogen increased perloline synthesis. Perloline and total alkaloid content were reduced as the plant approached maturity. The addition of phosphate either alone or together with nitrate had no significant effect on the perloline level (Bennett, 1963).

The effect of soil pH must be considered along with that of mineral nutrients. McNair (1942) found a clear relation of soil acidity and the percentage of alkaloids in plants in certain floras. The percent of alkaloid-rich plants was found to increase with the soil basicity. Less than 4% of the investigated species were alkaloid-rich in the strongly acidic soils, but with the basic soils, over 15% accumulated alkaloids above the normal level. McNair's finding probably cannot be explained as a simple adaptation to soil pH, but it is perhaps caused by other environmental conditions. Strongly basic soils are often encountered in arid areas, and the peculiar accumulation of alkaloid-rich plants in these areas is probably a result of environmental pressure (caused by predation and the slow-growth ratio as a preference of alkaloid-producing plants for basic soils.)[4] Timofieyuk (1929)

[4]This subject will be considered further in Chapter 4, which deals with the role of alkaloids.

found that lupine plants grown in soils with added sodium or potassium carbonate produced less alkaloid, and also the percentage was lower. On the other hand, *Cinchona* produced more alkaloids in alkaline soil. Stillings and Laurie (1943) found that *Atropa belladonna* produced more alkaloids in plants grown in the plots with pH 5.5 to 6.5 than in plots with higher pH.

Keeler (1975), in his essay on toxins and teratogens in higher plants, pointed out, in reference to the very high accumulation of toxic plants in arid and semiarid areas, the following:

> It is difficult for one familiar with the lush grass pastures of the midwest to appreciate the sparse range situation in the more arid western parts of the country and the toxic plant abundance on these ranges. Cattle often travel over 5 miles each day in search of feed between water holes. Under such circumstances, stumbling onto something green to eat must be refreshing even if it is toxic. Millions of cattle and sheep graze such winter ranges. And during summer months on the more lush mountain ranges, the hazards are not reduced although the plant genera are completely different. On these ranges there are even more toxic genera that present grazing hazards than on the arid ranges.

The effects of micronutrients are like those of macronutrients: they are not always alike. Mironenko (1965) found that the addition of such micronutrients as B, Mo, Mn, and Cu caused a depression in the alkaloid level in the seeds of three lupine species. Tso *et al.* (1973) tested 54 elements for their effects on the production of nicotine in plants *(Nicotiana tabacum)* grown in solution culture. Some elements such as Be, Cu, Pd, Pt, and Sm definitely increased the nicotine yield (over 25%), while other elements had less pronounced effects (Ni, Rb, Ag, Tl, Sn, U, V), and Zn decreased the yield of nicotine. Tso *et al.* (1972) give a complete and thorough explanation of the role of all the elements in tobacco plants.

All the environmental factors, light, water, macro- and micronutrients, pH of the soil, temperature, and probably others have still been insufficiently investigated. There was considerable interest in this type of research in the 19th century; mostly the alkaloids were isolated as salts of silicotungstic or phosphomolybdic acids. Because these form different complex salts with alkaloids encountered in the plant, it must be assumed *ipso facto* that the obtained data were only crude approximations.

3.4. External Influences

Recently the alkaloid content of lupines grown in southern California was compared with that of lupines grown in Poland (Figure 3.15). Great differences in the total level of alkaloids and in their composition were found. Results were obtained (Figure 3.16) for *Skytanthus acutus* M., which produces α, β, and δ isomers of skytanthine and dehydroskytan-

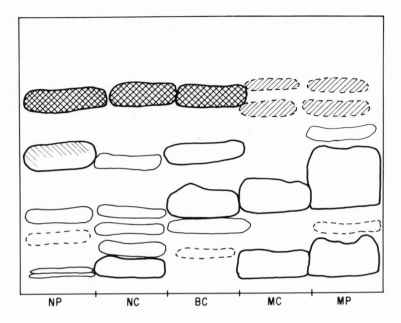

Figure 3.15. Chromatograms of alkaloids of *Lupinus nanus* (N) and *L. micranthus* (M) grown in California (C) during the winter–spring growing season and in Poland (P) during the spring–summer growing season. BC is *L. bicolor* grown in California, in 1972.

thine, which can serve as links to higher alkaloids.[5] Seeds from the studied plants were collected in the Atacama Desert in Chile and grown in Stillwater, Oklahoma (Auda *et al.*, 1967; Appel and Streeter, 1970a,b) (Figure 3.16b). This plant, owing to the shape of the seed pods, is commonly referred to as "Goat's horn" by natives of the area. Chile's Atacama Desert is devoid of plant life except the *Skytanthus* plant (Figures 3.16c and 3.16d). The Atacama stretches 600 miles into Peru, Bolivia, and Argentina and was chosen for study by the National Aeronautics and Space Administration because it resembles the terrain of Mars, particularly because it has no rainfall. In fact, the Atacama Desert gets its moisture from the early morning fog which drifts in from the Pacific Ocean, and *Skytanthus* is the only surface plant growing in the Atacama Desert. The alkaloid compositions of the plants grown in the Atacama Desert and in Stillwater are not known. Actinidine (Figure 3.17a) was found in Japanese *Actinidia polygama* plants in Japan and in Stillwater, Oklahoma in 1965 when they were first brought over (Figures 3.17b and 3.17c). However, the relative amounts of actinidine decreased from 1965 to 1968, finally reaching zero

[5]The α and β forms of this alkaloid have been reported to be artifacts generated during the process of isolation (Appel and Streeter, 1970a, b).

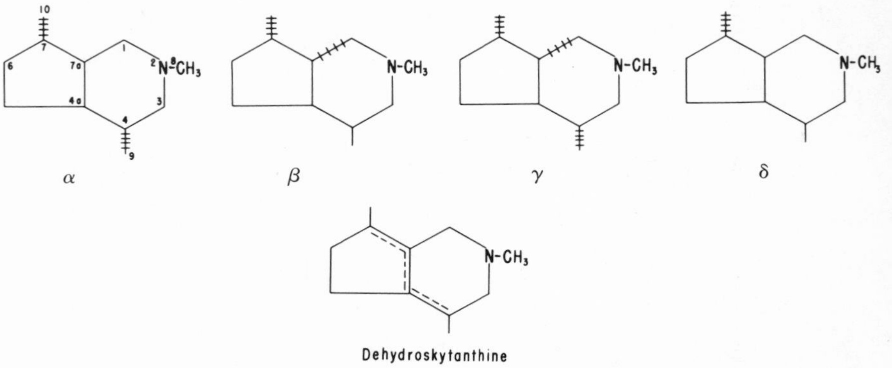

Dehydroskytanthine

Figure 3.16a. Structures of *Skytanthus* alkaloids. The Greek letters refer to isomeric skytanthines.

Figure 3.16b. *Skythanthus acutus M.* plants grown in the greenhouse of the Department of Horticulture, Oklahoma State University.

Figure 3.16c. *Skythantus acutus M.* plants grown in the Atacama Desert of Chile.

when analyzed by gas chromatography–mass spectrometry (Johnson, 1972). The *Actinidia* grown on rocky soil is known to thrive at an altitude of 3000 to 5000 m above sea level. It had lost its ability to produce actinidine in the Stillwater environment, due to the lack of rocky soil, an elevation of only 375 m above sea level, and different light and temperature conditions.

Figure 3.16d. Another view of *Skythanthus acutus M.* plants in the Atacama Desert.

Figure 3.17a. The structure of actinidine from *Actinidia polygama*.

Figure 3.17b. *Actinidia polygama* plants grown in the greenhouse of the Department of Horticulture, Oklahoma State University.

Figure 3.17c. *Actinidia polygama* plants grown near Kofu, Japan.

The alkaloids of the genus *Delphinium* may be classified into two distinct structural types: the atisine type, represented by ajaconine, and the lycoctonine type, represented by delcosine, acetyldelcosine, and delsoline (Figure 3.18). The common larkspurs are widely distributed in the western United States and in Canada where they are a common cause of cattle poisoning. Both types of alkaloids from *Delphinium ajacis* have been studied (Frost *et al.*, 1967). Plants were grown in the field on sandy loam soil at the Horticultural Farm in Stillwater with normal fertilization and

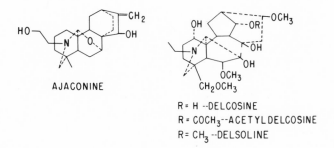

R = H --DELCOSINE
R = COCH₃--ACETYLDELCOSINE
R = CH₃ --DELSOLINE

Figure 3.18. Structures of some diterpenoid alkaloids from *Delphinium ajacis*.

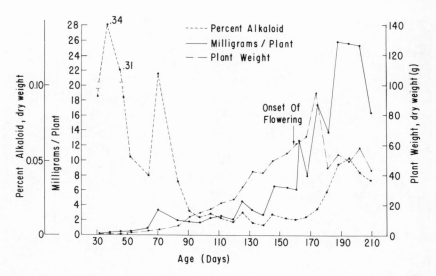

Figure 3.19. Variation of alkaloid content with age of *Delphinium ajacis* plants (Frost *et al.* 1967). Courtesy of The Chemical Society Publications.

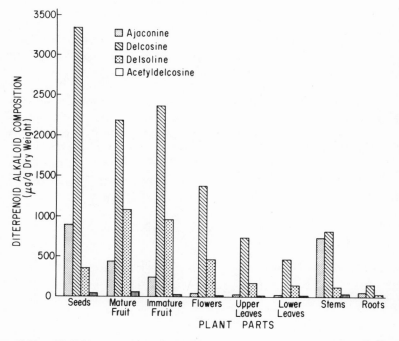

Figure 3.20a. Alkaloid composition of *Delphinium ajacis* plants (190 days old) (Waller and Lawrence, 1975).

irrigation (Figure 3.19). The analysis of the four diterpenoid alkaloids from plants grown from germination to 190 days old (Waller and Lawrence, 1975) is shown in Figure 3.20a; probe runs were then analyzed by repetitive scan in the high mass range of the mass spectrometer (where most of the important fragment ions occur). The total diterpenoid alkaloid content was lowest in the roots, lower leaves, and upper leaves; it was highest in the seeds, mature fruit, immature fruit, and the flowers. Thus all plant parts analyzed contained from 0.02 to 0.46% diterpenoid alkaloids. Delcosine was the predominant diterpenoid alkaloid in all tissues. Of the known alkaloids studied, acetyldelcosine was the only one found to be totally absent in tissue; none was found in the roots. No ions derived from the root alkaloids larger than delcosine ($M^+ = 467$) were observed. Ajaconine was highest in the stems and seeds, and it was lowest in the leaves and flowers (Figure 3.20b). Results from TLC, GLC, GLC-MS (mass spectrometry) and direct probe mass spectral data indicated the identity of four unknowns with apparent molecular ions (MW) of 481, 509, 523, and 537 in the tissues investigated (Figure 3.21). These are compounds that represent a modifica-

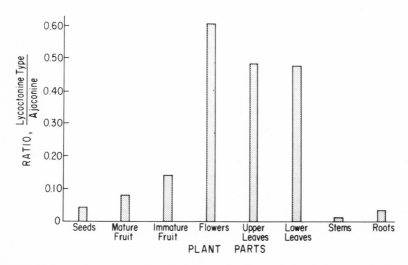

Figure 3.20b. The ratio of lycoctonine-type to ajaconine alkaloids in *Delphinium ajacis*. These results were obtained using the direct probe–mass spectrometer LKB-9000P and a crude alkaloid fraction. Major fragment ions that are characteristic of the compound and which have low cross contribution from the other compounds were selected. The cross contribution was determined as were the response factors from known amounts of pure compounds. Samples were then analyzed by repetitive scans in the high mass range (where most of the important fragment ions occur). The ion intensities of the interpretive fragment ions are plotted, and the area under each curve is determined and corrected for cross contributions and response factors to arrive at a total microgram estimation for each known compound. Unknowns are monitored by their characteristic fragment ions and estimated relative to the knowns (from Waller and Lawrence, 1975).

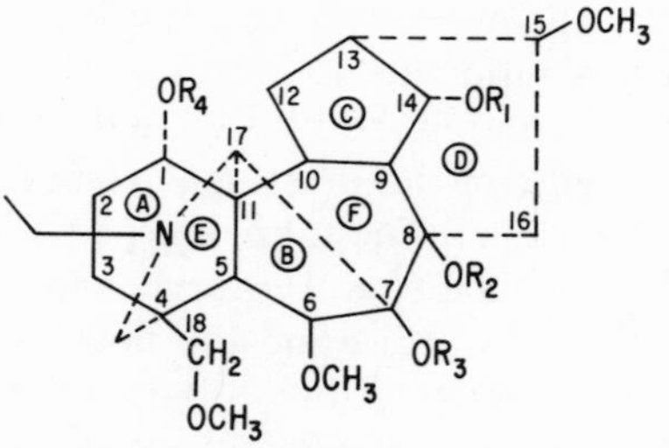

Unknown Diterpenoid Alkaloids from D. ajacis – Proposed Structures* and Names

Proposed Structures*	Proposed Name
lycoctonine skeletal type	
$R_2 = R_3 = CH_3$; $R_1 = R_4 = H$	M.W. = 481 Delosine
$R_1 = R_2 = R_3 = R_4 = CH_3$	M.W. = 509 Delcoline
$R_1 = COCH_3$; $R_2 = R_3 = CH_3$; $R_4 = H$	M.W. = 523 Dimethylacetyldelcosine
$R_1 = COCH_3$; $R_2 = R_3 = R_4 = CH_3$	M.W. = 537 Trimethylacetyldelcosine

* Other combinations of R are possible

Known Alkaloids	Accepted Name
$R_1 = R_2 = R_3 = R_4 = H$	M.W. = 453 Delcosine
$R_1 = CH_3$; $R_2 = R_3 = R_4 = H$	M.W. = 467 Delsoline
$R_1 = COCH_3$; $R_2 = R_3 = R_4 = H$	M.W. = 495 Acetyldelcosine

Figure 3.21. Unknown diterpenoid alkaloids from *D. ajacis*—proposed structures (lycoctonine skeletal type) (Waller and Lawrence, 1975).

tion of the lycoctonine-type skeleton in which (1) two of the four free hydroxyl groups are methylated (MW = 481), (2) all four free hydroxyl groups are methylated (MW = 509), (3) C-14 is *O*-acetylated and two of the three remaining hydroxyl groups are methylated (MW = 523), and (4) C-14 is *O*-acetylated and all three hydroxyl groups are methylated (MW = 537). It was suggested that these compounds be named *delosine, delcoline, dimethylacetyldelcosine* and *trimethylacetyldelcosine*. The molecular weights of all of these known and unknown lycoctonine-type alkaloids occur in a homologous series as 453, 467, 481, 495, 509, 523, and 537. These compounds are in highest concentration in the lower leaves of the plant and are present in all parts of the mature plant except the roots. Apparently the roots do not possess the enzyme capable of linking the acetate to delcosine, acetyldelcosine, dimethylacetyldelcosine, and trimethylacetyldelcosine. These TLC, GLC, and mass spectral data provide a good example of what can be done using modern instrumentation: more alkaloids can be found, and in significant quantities and these can vary with the environment.

3.5. Conclusion

It seems appropriate to point out that for both better understanding of the role the alkaloids are playing in plant metabolism, as well as for avoidance of unpleasant surprises in biosynthesis and catabolic experiments, research on factors affecting alkaloid synthesis should be promoted. Since the need for alkaloids as a source of drugs is not being met sufficiently, the search for the best conditions for alkaloid synthesis seems to be of greater importance than before. Also, a number of alkaloid-producing plants have been converted into fodder plants, such as lupines, crotalaria, and certain grasses; the residual alkaloid level can sometimes exceed safe limits and cause great losses in livestock by a peculiar convergence of factors. With the increased utilization of soils with unfavorable properties, this problem can be serious. In modern agricultural practices, there is a tendency to increase the level of mineral fertilization, which may, under certain conditions, force plants that normally do not produce alkaloids to start accumulating them or related substances. Alkaloid production provides the plant with protection when predators have a choice between alkaloid-free plants versus plants that produce alkaloids in abundance. These alkaloid-rich plants may cause metabolic disorders or otherwise be toxic to the predators.

4 | Sites of Alkaloid Formation

4.1. Introduction

Plant cells are less specialized than animal cells in their metabolic abilities. An animal cell can develop or lose some metabolic properties depending on the tissue; however, this process is mostly irreversible, e.g., there is no way known at present to reverse the development of a nerve cell or a liver cell into an embryonic one. This is true to some extent for plant cells also; usually a leaf cell performs leaf metabolism, while a root cell performs only root metabolism. Under certain conditions, it is possible to change the metabolic properties. It was shown by Steward (1964) that carrot root cells, when released from the limitations of their normal tissue environment, can undergo differentiation to form all possible types of cells, leading ultimately to an entire plantlet. Some leaves or pieces of stem can readily form roots. Therefore, owing to this metabolic versatility, one might expect that alkaloids can be formed in all cells of a plant. This may be the case in some plants, such as *Ricinus communis* (castor bean), but it is not universal.

To determine the site of alkaloid synthesis, a number of techniques have been developed. The most widely used of these are (1) grafting alkaloid-producing plants on alkaloid-free roots, (2) grafting alkaloid-free plants on roots from an alkaloid-producing plant, (3) administering isotopically labeled compounds to sterile tissue cultures grown *in vitro,* (4) feeding excised organs with isotopically labeled compounds, and (5) employing chemical microtechniques to identify the active site. Each of these methods is subject to criticism. The grafting methods, while still recognized as the best, are not free of experimental errors. It can happen (and it happens frequently) that the shoot forms adventitious roots which may not be visible when they grow into the stock. Even a very short root can show considerable alkaloid biosynthetic activity. Furthermore, it has not yet been proved whether a shoot growing on foreign rootstock behaves as it does when it is grown on its own. Experiments have shown that a foreign shoot may be either poisoned by the products of the root or have a toxic influence on the root. Some graftings were unsuccessful due to the toxic properties of the graft partners; for instance, it was never possible to grow *Laburnum* on *Lupinus,* or *Pisum* (peas) on *Lathyrus* (tangier pea) (Nowacki *et al.*, 1967). On the other hand, some quite strange combinations were successful: *Zinnia elegans* grew well when grafted on tobacco roots (Schroeter, 1955). In spite of the reservations mentioned, grafting, a technique used in the 19th century, has furnished much useful information that is valuable when in agreement with results obtained by other techniques.

The third method, involving *in vitro* tissue culture, often fails because it is difficult to compose a nutrient solution capable of supporting both proper growth and production of the same alkaloids that occur normally.

Feeding isolated organs with an isotopically labeled substrate also presents difficulties. All shortcomings mentioned for the other methods are valid for this one. In addition, the substrate may not be the proper one; in other words, it may be toxic. The simplest thing would be to feed a universal substrate such as carbon dioxide or acetate, but the isolated tissue might not possess all enzymes necessary to perform all the reactions involved in alkaloid synthesis.

There have been attempts to locate the site of alkaloid production by means of chemical microtechniques. These methods were based on the assumption that the site of accumulation is also the site of biosynthesis; however, conclusive results have not been obtained.

Since the site of alkaloid production varies from species to species and varies from compound to compound within species, producing more than one alkaloid, we will deal with this subject separately for certain genera.

4.2. Solanaceae Alkaloids

It is generally recognized that the most intensively studied group of alkaloids are ones encountered in genera of the family Solanaceae. The number of papers concerning the site of synthesis of these alkaloids is enormous, and no attempt to cite all of them will be made; however, the most important ones will be discussed. The *Nicotiana* alkaloids are synthesized predominantly in the youngest parts of the roots; consequently, they can be associated with the metabolism of the rapidly growing and dividing root cells. Feeding experiments using various substrates were successful only when the plants used for experiments had fast-growing roots. Dewey *et al.* (1955) and Leete (1955, 1957) have shown that a whole *Nicotiana*

Table 4.1. Growth and Nicotine Production by a Clone of Excised Roots of *Nicotiana tabacum* var. Turkish[a]

Age of culture (days)	Number of cultures	Mean dry wt (mg)	Mean length (mm)	Mean number of branches[b]	Mean nicotine content (μg)
7	28	0.32	57	3.4	9
11	22	0.89	163	7.5	26
14	23	1.71	284	11.8	55
17	12	3.21	564	33.2	110
20	7	6.34	1098	73.5	182

[a] From Solt, 1957a.
[b] Does not include main branch.

plant is able to incorporate ornithine and nicotinic acid into nicotine and nornicotine, and the latter acid also into anabasine. Solt (1957a, b), working with excised roots, confirmed that the roots were capable of synthesizing alkaloids while growing on sterile media (Table 4.1). A similar result was published by Mothes *et al.* (1957) in the same year.

In grafting experiments involving *Nicotiana rustica* or *N. tabacum* and a nicotine-free stock like potato (Meyer and Schmidt, 1910) or tomato (Schmuk *et al.*, 1941a,b; Dawson, 1944, 1951; Lashuk, 1948; Ill'in, 1948, 1955; Solt and Dawson, 1958; Mothes, 1955, 1969; Kuzdowicz, 1955), it was found that the scions were alkaloid-free or had a considerably reduced alkaloid content (Dawson, 1942) (Table 4.2). (The reader is referred to articles by Dawson (1948) and Dawson and Osdene (1972) for more infor-

Table 4.2. Accumulation of Nicotine in Reciprocal Grafts of *Nicotiana tabacum* (Tobacco) and *Lycopersicum esculentum* (Tomato)[a]

Location of sample of tomato scions	Nicotine, mg per scion (mean of four)			
	Number of days after preparing grafts			
	0	31	58	84
Leafy tobacco stocks	0	55	258	571
Leafless tobacco stocks	0	38	135	501
Leafy tomato stocks	1.3	2.6	4.1	2.5

[a] From Dawson, 1942.

mation.) The total accumulation by the scion is a smoothly continuing process, and the distribution of the alkaloid between the various parts of the aerial shoot is similar to that in the intact tobacco plant (Table 4.1). In the same experiments, tobacco scions grafted to tomato rootstocks contained nicotine only in the lowermost leaves (which had been present in juvenile form in the original tobacco scion) and then only in amounts comparable with those of scions used at the time of grafting. *N. glauca* produces both nicotine and anabasine, and the above-mentioned authors have simultaneously stated that anabasine is formed in grafts, regardless of which part of the graft is from *N. glauca,* scion or stock. It can be concluded that the *Nicotiana* alkaloids with the pyrrolidine ring are synthesized predominantly in the root, while anabasine, a piperidine alkaloid, is produced in both the roots and the aerial parts of the plant. Some results contrary to those mentioned above were obtained by Mashkovtsev and Sirotenko (1956), who confirmed the results of Pal and Hath (1944) that a tobacco scion can produce some nicotine when grown on alien (tomato) roots. Still, most of the evidence available at present is quite convincing that at least nicotine is formed predominantly in the root. It can be demethylated in the shoot to nornicotine: therefore, the site of nicotine synthesis must be different from that of nornicotine. It would appear that the enzyme for N-demethylation is not present in the roots.

Several authors have studied nicotine production (i.e., biosynthesis) in callus tissue cultures (Speake *et al.,* 1964; Benveniste *et al.,* 1966; Furuya *et al.,* 1966, 1971; Tabata *et al.,* 1968, 1971; Shiio and Ohta, 1973; and Heinze, 1975). The biosynthesis of nicotine is dependent upon the formation of organized tissue within the callus. Nodule-like structures similar to roots were observed in our laboratories using tobacco variety Maryland-872, which produces 96% of its alkaloids as nicotine. Shoot formation stimulated nicotine production in the callus, and nicotine may have been transported from the callus to the shoot. Nicotine production and tissue differentiation were dependent upon concentrations and types of growth regulators in the culture medium (Tables 4.3 and 4.4). The vegetative buds and leaves (shoots) contained about five times as much nicotine as callus without buds or leaves, which is in agreement with the results of Tabata *et al.* (1968).

Similar results have been published for other Solanaceae alkaloids, namely, those of the tropane group. Heine (1942) grafted *Datura* scions upon *Nicotiana rustica* and found that nicotine accumulated instead of tropane alkaloids. This finding was confirmed by Hills *et al.* (1946) with *Nicotiana–Duboisia* grafts. The tobacco scion on *Duboisia* root was found to accumulate tropane alkaloids. Surprisingly, the tobacco scion accumulated both hyoscine and the unesterified tropine. In all grafts between plants

Table 4.3. Nicotine Analysis[a] of *Nicotiana tabacum*, Maryland-872 (Heinze and Waller, 1975)

Explant[b] material from tobacco tissues	Nicotine content (μg/g fresh weight)		
Pith[c]	406 ± 43		
Tissue cultures of callus subculture no. 3[c]			
	1 week	4 weeks	6 weeks
Pith	1770	80	1465

[a] Average of three examples in which nicotine was determined spectrophotometrically after thin-layer chromatography.
[b] Refers to section of plant used.
[c] Pith tissue was removed with a cork borer, sectioned, sterilized, and placed on the media. Callus induced on one medium was subsequently subcultured on the same medium at eight-week intervals. Subcultures were so designated.

producing the *Nicotiana* type of alkaloids and those producing tropane alkaloids, the alkaloid which was accumulated was that of the root. The synthesis of these alkaloids in the aerial parts of the plant seems to be negligible, yet not entirely excluded. A great number of experiments were performed (Figure 4.1) using a tropane-alkaloid-producing plant as one partner and a solanine-producing species as the other partner.

Histologic peculiarities of the Solanaceae make it relatively easy to combine different species of these plants in a graft of chimera. The Solanaceae were a good choice for this type of experiments because some species of this plant family contain plants which produce different types of alkaloids and still others are alkaloid-free. Thus the *Nicotiana* and *Datura* genera are capable of producing alkaloids derived from ornithine, while

Table 4.4. Nicotine Analysis of Subculture No. 1 of Callus (Pith)[a] (Heinze and Waller, 1975)

Callus tissues	Nicotine content,[b] (μg/g fresh weight)
Callus without development of leaves	14 ± 2
Callus with the developed leaves removed	71 ± 10
Leaves that were removed from the callus (above)	335 ± 20

[a] See Table 4.3 for explanation of subculture.
[b] Average of three samples in which nicotine was determined spectrophotometrically after thin-layer chromatography.

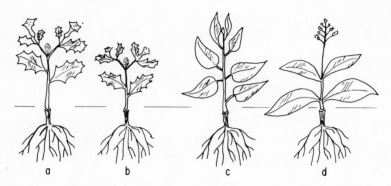

Figure 4.1a. Grafting experiments with Solanaceae. Grafting on *Datura stramonium* roots: *a—Datura stramonium* control on *D. stramonium; b—Datura tatula* on *D. stramonium; c— Cyphomandra betacea* on *D. stramonium; d—Nicotiana tabacum* on *D. stramonium.* In all cases the aerial part (scion) accumulated tropane alkaloids like *Datura stramonium,* except in *b,* where the alkaloid hyoscyamine was converted into hyoscine.

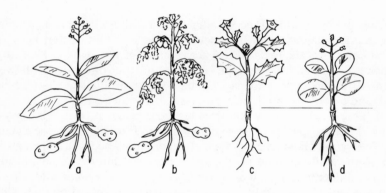

Figure 4.1b. Grafting experiments with Solanaceae. Alkaloid-rich plants grafted on alkaloid-poor roots: *a—Nicotiana tabacum* on *Solanum tuberosum; b—Lycopersicon esculentum* on *S. tuberosum* (control); *c—Datura stramonium* on *Cyphomandra betacea; d—Nicotiana glauca* on *L. esculentum.* Grafts *a, b, c,* were alkaloid-free; graft *d* produced only anabasine.

Solanum can accumulate glycoalkaloids and *Cyphomandra* is practically alkaloid-free. Therefore, various reciprocal graftings can elucidate which main part of the plant (root or shoot) is capable of producing alkaloids. Fortunately, the alkaloids have sufficiently different properties to enable easy identification of the partner from which they came. The alkaloids are translocated across the graft very easily, with the majority remaining in the other part of the chimera without undergoing any further change. There is

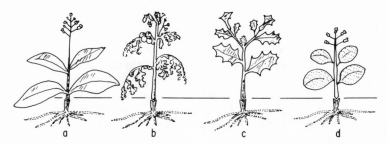

Figure 4.1c. Grafting experiments with Solanaceae. Graftings on *Nicotiana tabacum* roots: *a—Nicotiana tabacum* on *N. tabacum* (control); *b—L. esculentum* on *N. tabacum; c— Datura stramonium* on *N. tabacum; d—N. glauca* on *N. tabacum*. In *a, b, c,* the aerial part regardless of species accumulated only *Nicotiana tabacum* alkaloids, nicotine and nornicotine. In *d,* anabasine, an alkaloid characteristic of *Nicotiana glauca,* was produced also.

of course some conversion of alkaloids translocated across the graft, but this transformation is negligible, since the heterogeneous graft partner lacks the enzymes necessary for the degradation of a foreign alkaloid.

Different *Solanum* species (*Solanum tuberosum,* potato, *Solanum nigrum,* nightshade), and *Lysopersicum* (*L. esculentum,* tomato) were also grafted. The first grafting experiments were performed in the 19th century by Strasburger (1885, 1906).

The site of biosynthesis of tropane alkaloids is well documented in a series of publications by Romeike (1953, 1956, 1959, 1964). Romeike grafted alkaloid-producing species *Datura ferox, D. stramonium, D. innoxia,* and *Atropa belladonna* on alkaloid-free roots of *Cyphomandra betacea.* The reverse grafts were also performed. Grafts between plants that produce different sets of alkaloids—for example, grafts of *Datura ferox,* which produces hyoscine, with *Datura stramonium,* which produces predominantly hyoscyamine—were also performed. The first experiments of Romeike confirmed the data presented by Evans and Partridge (1953), but the explanation offered was different. The difference was actually due to an oversight by the earlier authors of the possibility that an aerial part can transform alkaloids produced in the roots. In all cases where the roots belonged to alkaloid-producing species, the scion accumulated alkaloids; however, when the stock was alkaloid-free, the scion was also alkaloid-free.

The scion of *Datura ferox* transformed hyoscyamine into hyoscine (see Chapter 2, on genetics of tropane alkaloids, p. 69). Since the *Datura stramonium* scion was not able to transform alkaloids, it accumulated the same type as was produced in the root; similarly, the *Cyphomandra* scion could not effect transformation: it accumulated only hyoscyamine. Therefore, it seems justified to state that the alkaloid formed in the root of all *Datura* and *Atropa* species is hyoscyamine. It can be converted to hyoscine in *Datura ferox* scions. Meteloidine is probably produced in roots, and it can also be synthesized in the aerial parts, although more poorly. These grafting experiments are in good agreement with biosynthetic investigations. In 1943, Cromwell stated that injection of putrescine and arginine permitted him to obtain a higher content of hyoscyamine in *Atropa* and *Datura;* moreover, he observed a putrescine-oxidizing enzyme in the roots. Similar results were obtained by Haga (1956), who supplied *Atropa* roots with ornithine and arginine. In 1951, Diaper *et al.* supplied *Datura stramonium* with labeled putrescine but was unable to isolate radioactive alkaloids. Three years later, in the same laboratory, Leete *et al.* (1954) demonstrated [2-^{14}C]ornithine was incorporated into hyoscyamine but not into hyoscine. Results obtained later in different laboratories on tropane-alkaloid synthesis will be discussed in the following chapter; however, they all confirmed the data obtained from grafting experiments.

A method that should be mentioned was applied in a number of laboratories. *Nicotiana* and *Datura* alkaloids were used with the objective of trying to establish the site of alkaloid synthesis. This method is the analysis of "bleeding sap." Hemberg and Fluck (1953) stated that *D. stramonium* transports alkaloids into the aerial part, but Warren-Wilson (1952) was able to show that a downward migration of alkaloids occurred in *Atropa*. Mothes and Engelbrecht (1956b) have confirmed the finding of Dawson (1941), that alkaloids in *Nicotiana* are rapidly transported from the roots to the shoots and leaves. The upward transport is observed in xylem tissue, while the downward is in phloem tissue. Using chemical microtechnical methods, Chojecki (1949) established that nicotine was distributed all over the tobacco plant. He found that the leaf epidermis was the richest tissue, while the axial cylinder of the root was the poorest in the alkaloid. Elze and Teuscher (1967) have shown that submerged single cells of *Datura* are able to take up added alkaloids from the medium and accumulate them. The level reached in the cells was much higher than the concentration of alkaloids in the solution. Plant species best accumulate those alkaloids that they do not produce, e.g., papaverine, which was accumulated in an unexpectedly higher level in *Datura* than the species accumulates its own alkaloids. The accumulation was inhibited by sodium azide and 2,4-dinitrophenol. It may be concluded that the *Solanum* alkaloids are synthesized predominantly in roots and moved, with the transpiration current, through the xylem tissue into the aerial part of the plant. The alkaloids are taken up by active participation of the protoplasm into the leaf epidermis cells as well as into other parenchymatous-like tissues of the stem and the spongy mesophyll of the leaves. They can be converted to other alkaloids as well as to nonalkaloidal substances in the aerial part of the plant. It is interesting to note that the Solanaceae alkaloids can be degraded both in the plant which produces them and in other species; that is, nicotine can be degraded by *Lycopersicum esculentum, Atropa bella-donna,* and *Cyphomandra betacea* and also by the tobacco root parasite *Orobache crenata.* The alkaloid-poor *Nicotiana alata* degraded one-half of introduced nicotine within 20 hr, while *Nicotiana rustica* did not degrade nicotine in 40 days (Neumann and Tschoepe, 1966). Atropine is degraded by some Solanaceae even more rapidly. Japanese workers (Kisaki and Tamaki, 1966; Wada *et al.*, 1959) found that *Nicotiana* alkaloids are degraded in the leaf as well as the root, and that optical isomers undergo different metabolic fates. Prokoshev *et al.* (1952) investigated, by means of grafting, the site of synthesis of the glucosidic alkaloids of Solanaceae. Grafts were performed between *Solanum tuberosum* (potato), which produces solanine, and *S. demissium* and *S. punae,* which produce solanine and demissine. Two species of tomato, *Lycopersicum esculentum* and *L. hirsutum,* which produce tomatine, were also used. It was concluded that the glucoalkaloids were synthesized in leaves and not in roots. This finding was corroborated using tracer methods by Guseva and Paseshnichenko (1958), who fed rootless potato sprouts with labeled acetate and were able to isolate the radioactive glucoalkaloid.

4.3. Lupine Alkaloids

While the Solanaceae alkaloids are synthesized predominantly in roots and converted to other compounds or degraded in the aerial parts of the plant, the opposite seems to be true for lupine alkaloids, which are synthesized and accumulated in the aerial parts of the plant (White and Spencer, 1964; Nalborczyk, 1968). In 1940, Smirnova and Moshkov reported the alkaloid distribution in a graft between *Pisum sativum* (pea) and a bitter lupine *(Lupinus).* The stock as well as the scion were allowed to produce shoots, and it was found that both partners of the graft contained alkaloids. Alkaloids were also found in the pea seeds. This result is also different from those with Solanaceae. In the case of tomato scion growing on alkaloid-producing roots, the alkaloid was distributed throughout the tomato plant except in the fruit. Its absence there is apparently due to its inability to enter the fruit and not to its degradation there; when mature or immature tomato fruits were injected with alkaloids, there was no sign of a breakdown (Hills *et al.*, 1946). The legume fruit accumulates a high level of alkaloids when it has the opportunity to obtain them. In 1956, two laboratories reported similar results on the site of alkaloid synthesis

(Peters *et al.*, Mothes and Engelbrecht, a, b). In 1958, Mironenko and Rikhovska reported similar results with different species of lupines. Mironenko and Spiridonova (1959) and Mironenko (1965) added further examples. In the meantime, grafting experiments by Kazimierski and Nowacki (1960), and Nowacki and Nowacka (1966) used plants differing not only in the percentage of alkaloids accumulated in dry substance but also in qualitative composition. The material used was the following: *Lupinus albus*, bitter and sweet varieties; *L. angustifolius*, bitter and sweet varieties; *Lupinus pilosus, L. mutabilis, L. arboreus, L. hartwegi,* and *L. polyphyllus*, all bitter varieties. In all cases, the growing apices of 2-week-old plants were cleft-grafted on 3-week-old stock plants. In some cases, adventitious sideshoots from the stock were removed as they appeared, and in other experiments the sideshoots were allowed to grow to the same approximate size of the grafted scion. During the vegetation period, assays for alkaloid content were made. The content in the scions of plants without sideshoots was approximately the same as that in normal plants. The evidence is that all lupine alkaloids are synthesized in green parts of the plants. Anatomical analyses performed at the site of fusion failed to reveal any adventitious roots on the scion (which otherwise could have been regarded as the organ synthesizing alkaloids). Very interesting results were obtained when two different shoots grew (the grafted and the original) on a single root. In this case, the bitter shoots produced alkaloids which were transported all over the plant. When the sweet partner was of a species containing the wild-type genes capable of oxidizing alkaloids (i.e., *L. angustifolius*) and the bitter partner produced only sparteine *(L. arboreus)*, the alkaloids that accumulated were of the oxidized type. It is conceivable and in agreement with biochemical experiments involving transformation of labeled sparteine into lupanine and hydroxylupanine, that, while the bitter shoots produced sparteine, the sweet shoots transformed it into the oxidized derivatives (Figures 4.2a and 4.2b). The results may be compared with those of *Nicotiana* and *Datura* grafting experiments, where alkaloids produced by one partner in the root were transformed by the grafted scion into more highly oxygenated derivatives. The difference in lupines lies in the inability of the root to synthesize alkaloids, so that a union of two shoots on a single root was necessary to corroborate the idea of oxidative conversion. To date, there are no experimental data on the biosynthesis of dehydrogenated quinolizidine alkaloids such as anagyrine, thermopsine, and cytisine. The grafts between lupines and *Cytisus, Laburnum, Thermopsis,* and *Baptisia* species were all unsuccessful. This is probably due to the highly toxic effect of cytisine on plants that do not produce it (Pöhm, 1955, 1957, 1966). This adverse reaction to cytisine seems to be a peculiarity of metabolism. It is possible to graft distantly related plants such as peas

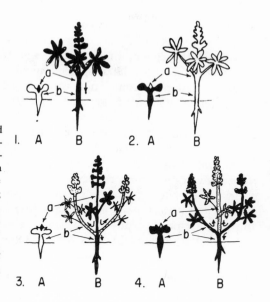

Figure 4.2a. Scheme of graftings and bittering of plants. Black area indicates bitter plants; dotted area indicates bittering of shoots; white area indicates low-alkaloid plants. (A) = at the beginning of the experiment; (B) = at the end of the experiment. 1. Bitter scion on a low-alkaloid stock; 2. low-alkaloid scion on bitter stock; 3. bitter scion on a low-alkaloid stock with a sideshoot of the stock left; 4. low-alkaloid scion on bitter stock with a sideshoot left.

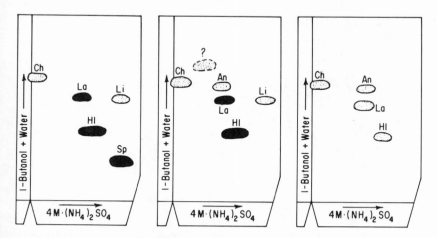

Figure 4.2b. Diagrams of chromatograms of alkaloids of grafted plants and controls. *Left*—chromatogram of the alkaloids of seeds of *Lupinus mutabilis* growing on its own roots. *Middle*—chromatogram of alkaloids of the seeds of blue lupine growing on *Lupinus mutabilis* stock with a sideshoot. *Right*—chromatogram of alkaloids of blue lupine (*L. angustifolius*) growing on its own roots. La—Lupanine; Sp—sparteine; Hl—hydroxylupanine; An—angustifoline; Ch—quinine, added as a reference alkaloid: dotted area—traces of alkaloid.

or beans on lupines, but no success is obtained when such closely related plants as lupines and *Cytisus* are grafted. A similar effect was observed by Tschiersch (1962) in grafting canavanine-producing plants with canavanine-free plants. Closely related species failed to grow when the graft components differed in the canavanine-producing character. Distant species grow well together when both are canavanine-free.

The cytisine phenomenon needs further investigation. In a few European botanical gardens a peculiar shrub can be encountered. It is *Cytisus adamii* (chimera), a plant that has an outer layer of tissues derived from *Laburnum anagyroides*, while the more central tissues belong to *Cytisus purpureus*. *Laburnum* produces cytisine. Depending on the season of the year, more or less *N*-methylcytisine and traces of laburnine are encountered. *Cytisus purpureus* is an alkaloid-poor plant. It accumulates traces of the L-sparteine derivatives—lupanine, anagyrine, cytisine, and probably some dipiperidine alkaloids as well as some biological amines (methylated

Figure 4.3. Schematic distribution of alkaloids in a lupine plant according to Mironenko (1965). The concentration of black corresponds to the concentration of alkaloids. Only the youngest plants are alkaloid-rich; in older plants a decrease of alkaloids occurs.

tyramines). No experimental data are available for this peculiar chimera, and it would be very interesting to know which kind of alkaloid it accumulates and how high the level is. Such work could also prove which layer of tissue actually produces an alkaloid. Microchemical analysis of lupine tissues confirmed the assumption that the alkaloids are transported into phloem tissue and that they can be easily detected in root vascular tissues even when the root is not synthesizing alkaloids (Tomaszewski, 1957). Roots of some perennial lupines accumulate alkaloids in high concentration. Electron-microscopic examination of *Baptisia* cells confirmed the supposition that lupine alkaloids are accumulated in the vacuoles (Cranmer, 1965). It should be mentioned that biosynthesis experiments with lupine alkaloids formed from various precursors were performed with rootless shoots (Schuette and Nowacki, 1959). Schuette's (1961) experiments revealed that lupine alkaloids can be synthesized in all aerial parts of the plant from [1,5-^{14}C]cadaverine. Mironenko *et al.* (1959a,b) established that the aerial part can synthesize lupine alkaloids when supplied by $^{14}CO_2$ or NaH$^{14}CO_3$; the incorporation was not influenced by the root. The highest incorporation was in leaves and stems, whereas the lowest was in old leaves (Figure 4.3).

Birecka *et al.* (1960) found that aerial parts without roots are able to produce alkaloids from $^{14}CO_2$, while Mothes and Kretschmer (1946) found that root cultures of lupines were unable to synthesize alkaloids on sterile media.

4.4. Papaveraceae Alkaloids

Due to a physiological pecularity of the Papaveraceae (latex production), it is not possible to graft *Papaver* species; therefore, the only evidence available on the site of alkaloid synthesis is from biosynthetic experiments performed on intact plants, isolated organs, or tissue cultures. In 1959, 1960 Kleinschmidt and Mothes reported on the biosynthesis of alkaloids from tyrosine in the isolated leaf, capsules, and the latex of *P. somniferum*. All these parts were able to synthesize alkaloids, but the latex was the best. This finding was confirmed by the same authors with the nonspecific substrates glucose and sucrose. Both sugars were converted to alkaloids, but now alkaloid biosynthesis was lower in the latex than in the leaf. This difference was attributed to lack, in the sap, of the enzymes responsible for converting sugars into tyrosine, the direct intermediate used in Papaveraceae alkaloid biosynthesis.

Callus cultures of *Macleaya cordata* are able to synthesize alkaloids. Neumann and Muller (1967) found that the callus cells differentiate into two types. Some are distinguishable under the light microscope as yellow to orange colored. These cells accumulate the Papaveraceae alkaloids protopine, sanguinarine, and allocryptopine. Since the alkaloid sanguinarine is orange, it was readily distinguishable in the cells. Histochemical methods revealed that the alkaloids were encountered only in specialized cells referred to as "alkaloid cells." The feeding of labeled phenylalanine, followed by precipitation of alkaloids *in situ,* extraction of other compounds, and radioautography revealed that, while all the proteins in all cells were radioactive, only in some vacuoles was a radioactive precipitation of alkaloids found. The exact localization of alkaloids in the cells was achieved by means of electron-microscope radioautography.

4.5. Ricinine in *Ricinus communis*

The site of alkaloid synthesis varies in different groups of plants. While some plants synthesize alkaloids predominantly in the roots, others perform this reaction in the aerial parts, and still others produce alkaloids in all organs. One plant that can synthesize the simple alkaloid ricinine from nicotinic acid (which is itself synthesized by a number of other substrates) is the castor bean plant. As was revealed by Waller and Nakazawa (1963), leaf disks from seedlings grown in sterile conditions can synthesize alkaloids even when depleted of the endosperm. Root cultures grown under sterile conditions (Hadwiger and Waller, 1964) also synthesized ricinine when supplied with the proper (labeled) substrates. It was also observed that ricinine could have a "sparing effect" on the nicotinic acid required for growth, a finding that permits speculation about the possible role of the alkaloid in growth. It was demonstrated by Yang and Waller (1965) that the incorporation of label from $[4-^{14}C](\pm)$-aspartic acid into ricinine by seed clusters of castor plants was somewhat less than in young seedlings; however, the incorporation of $[7-^{14}C]$nicotinic acid was considerably greater. The amount of label from both precursors in ricinine produced by seed clusters differed in at least one other respect from that of seedlings: it began to decrease after a short period of time. This indicated that degradation was taking place in the seed clusters and that the degradation was more rapid than that previously demonstrated in *R. communis* seedlings. It is evident that ricinine synthesis in the seeds is slower than in the stem, leaves, or pods, and it is possible that under normal conditions quite negligible synthesis takes place in the seeds and that most of the alkaloid that accumulates there is the result of transport into the seeds from other organs.

The *R. communis* plant seems to have the ability to synthesize ricinine in all parts, but it is probable that some organs are more efficient in their ability to perform this reaction. To test this idea, an experiment was performed with rootless plants fed with [7-^{14}C]nicotinic acid. The harvested plant was divided into three parts: the mature but still dark green leaves, the young leaves with the still growing part of stem, and the stem (Nowacki and Waller, 1973). The alkaloid content as well as the total and specific radioactivity were determined. While the highest percentage of ricinine was found in mature leaves, the highest specific activity was detected in the growing part of the plant (Figures 4.4a and 4.4b). When plants were used that had some old leaves that were turning yellow, such leaves had only traces of inactive ricinine. The alkaloid is translocated from aging leaves into younger ones (Skursky, Burleson, and Waller, 1969; Lee and Waller, 1972; Waller and Skursky, 1972). Similar behavior may be observed in a number of other plants. This means that all organs are capable of synthesizing alkaloids, but the most active growing portions are the most efficient.

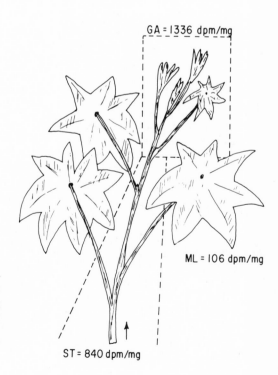

GA = 1336 dpm/mg

ML = 106 dpm/mg

ST = 840 dpm/mg

Figure 4.4a. Specific activity of ricinine in rootless plants fed with [7-^{14}C]nicotinic acid: ST—stems; ML—mature leaves; GA—growing area, ricinine isolated (Nowacki and Waller, 1973). Courtesy of Akademie Verlag.

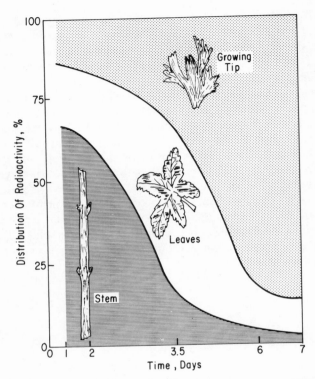

Figure 4.4b. Radioactive ricinine infiltrated into *Ricinus communis* shoots; distribution of the label versus time (Nowacki and Waller, 1975). Courtesy of the authors and the Society of Translocation and Accumulation of Nutrients in Plant Organisms, Warsaw.

A series of competitive feeding experiments were performed with nicotinonitrile, ricininic acid, inactive ricinine, and 2-, 3-, and 4-hydroxypyridines to determine if these compounds could inhibit the transformation of nicotinic acid into ricinine (Nowacki and Waller, 1972). The 2- and 4-hydroxypyridines had no significant effect, the 3-hydroxypyridine was poisonous to the plant, nicotinonitrile suppressed the conversion of nicotinic acid by over 50%, and ricininic acid completely inhibited the conversion. Since, according to Schiedt *et al.* (1962), ricininic acid is not an efficient precursor, it may be serving as a specific inhibitor (most of it was recovered from the plant unchanged). During this last experiment, it became apparent that not all pyridine compounds are similarly distributed throughout the plant. When ricinine or nicotinic acid were fed to rootless plants, they were at first passively transported with the transpiration current, but, after some additional time, they were translocated from the lower part of the stem and older leaves into the growing tip. Nicotinonitrile and the hydroxypyridines remained at the site at which they were placed during the uptake of ricinine (as did ricinine when fed to plants poisoned with *p*-aminobenzaldehyde) (Figure 4.5). In the case of poisonous substances, necrosis occurred at first in the lower leaves, then on the edges of

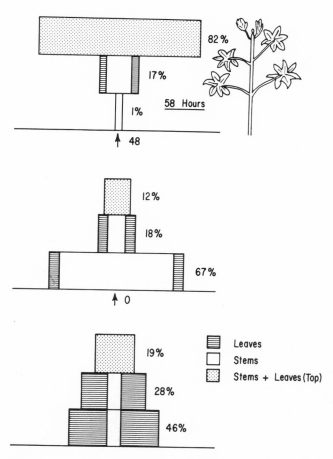

Figure 4.5. Translocation of ricinine in young rootless plants after it was added to rootless plants. *Middle*—distribution of radioactive ricinine after 10 hr (time required for [8-¹⁴C]-ricinine to be taken up). *Top*—distribution of radioactive ricinine after an additional 48 hr. *Bottom*—distribution of ricinine in plants poisoned with *p*-aminobenzaldehyde after 70 hr. Numbers indicate percentages of radioactivity recovered (Nowacki and Waller, 1972). Courtesy of the journal.

Figure 4.6. *Ricinus communis* (castor bean) leaf fed a foreign alkaloid (sparteine). Alkaloid accumulated in intercostal field and close to the edges of the leaf will not be transported away from the leaf. Some of the poorer precursors, such as nicotinonitrile, behave similarly. Shown is a drawing; for original see Nowacki and Waller (1972).

these leaves, and finally in the intercostal fields (Figure 4.6), as indicated by the effect of added sparteine (a lupine alkaloid) on the castor bean plant. The young leaves remained unaffected, and only traces of the toxic substances were found there. In view of these data, the poor incorporation of nicotinonitrile, ricininic acid, and other structurally related compounds is probably due to the fact that they are not efficiently transported to the growing part of the plant, where the biosynthesis of ricinine occurs. The possibility exists that quinolinic acid and nicotinic acid, which are known to be synthesized in the plant by the condensation of a C_3 unit and a C_4 unit, may not represent the only pathway used for ricinine biosynthesis. Perhaps some small compounds containing two nitrogen atoms are involved; such a parallel pathway could arise from the condensation of a C_3 compound and asparagine or β-cyanoalanine. Conclusive proof must await chemical degradation of the biosynthesized ricinine. *To be emphasized is the concept that alkaloids are actually side products of intermediary metabolism reactions and that the rate of their biosynthesis is directly proportional to the synthetic activities of the tissue.*

4.6. Other Plants

In spite of extensive research on alkaloid biosynthesis, the active site remains unknown for many plants. Some data are available that confirm, in part, the ideas mentioned above. Shibata and Imaseki (1953, 1956) fed shoots of *Ephedra distachya* with [^{15}N]ammonium sulfate and found after one week that the isotope was in the alkaloid. The same authors three years later fed [^{15}N]phenylalanine in the same way and verified that the shoot was able to synthesize alkaloids from the applied precursor. The alkaloids of

Cinchona are produced predominantly in the aerial part of the plant, since the scions of *C. ledgeriana* are rich in alkaloids regardless of the root on which they are grafted. In this species, the level of alkaloids and the qualitative composition invariably remain the same, whether the plants are grown on their own roots or are grafted on roots of *C. succirubra* (Moens, 1882). Moerloose (1954) found that the leaves and bark produce radioactive alkaloids when exposed to $^{14}CO_2$; the rate of growth was related to alkaloid production.

4.7. Alkaloid Accumulation in Seeds

Seeds of a number of alkaloid-producing plants are virtually alkaloid-free. This is true for *Nicotiana, Hyoscyamus, Papaver,* and many other species. In contrast, seeds of the leguminous plants contain a high content of alkaloids; some species of lupines, castor beans, and *Crotalaria* can accumulate alkaloids up to 6% of the dry matter. So a fundamental question exists: Do the seeds produce alkaloids or are the seeds merely the site of alkaloid accumulation? Since the seeds are actually new plants growing on the mother plant, an experiment was designed by Hagberg (1950) crossing sweet and bitter lupines, with bitter as the pollen parent. The plants grown from the seeds resulting from this cross were tested. It was found that within a month after the seeds were sown, all the F_1 plants were sweet, while the F_1 plants from a reciprocal cross grown under similar conditions were bitter. The leaves of the older F_1 plants and the seed produced by these plants were bitter. All of the germinating F_2 plants were bitter, but after some time (more than a month) a segregation ratio of 3:1 was observed. The pollinated sweet lupines (in Pulawy, Poland) were crossed with pollen grains from bitter plants. It was found that the resulting seeds were heterozygous and genetically bitter owing to the dominance of alkaloid synthesis over the sweet trait. The seeds were always sweet. Yet *Pisum sativum* (peas) grafted on bitter lupines were able to accumulate alkaloids. Later in an experiment in Pulawy, Poland, with a sweet variety of *L. angustifolius,* which was also recessive to anthocyanin production, the flowers were fertilized with pollen from a bitter anthocyanin-bearing form. The seeds were externally like the homozygous ones, but all of the hybrids contained anthocyanin pigments on the surface of the cotyledons under the seed cover. All hybrid seeds were sweet, just as in Hagberg's experiment. The F_1 heterozygous lupine plants produce two kinds of seeds: (1) homozygous-sweet, and (2) both homozygous-bitter and heterozygous-bitter. One-half the seeds produced by the F_1 plants were analyzed, and all were found to be bitter. The other half were germinated, and the plants were analyzed at 6 weeks of age. A segregation ratio of 3:1 for bitter:sweet was observed. There seemed to be a tendency in lupines seed to accumulate alkaloids, regardless of their genetic makeup, but no considerable synthesis occurred. The distribution of alkaloids in a seed is quite different in various lupine species. While most American lupines have a similar level of alkaloids in cotyledons, root, and primordial leaves, the Mediterranean species tend to accumulate less alkaloid in the embryo proper (Nowacki, 1958).

The alkaloid-free seeds of *Nicotiana, Papaver,* and barley begin alkaloid synthesis with the onset of germination, and the production is more efficient when germination is carried out in the dark (Schmidt, 1948). Young barley sprouts that are still utilizing the endosperm nutrients for their main energy supply produce tyramine, methyltyramine, hordenine, and gramine. After the endosperms were detached, they could still convert an added substrate (Leete *et al.,* 1952) such as tyramine (but not tyrosine) to alkaloids. These results show the endosperm to be necessary for the decarboxylation of tyrosine. James (1953) and James and Butt (1957) obtained similar results. Rabitzsch (1958) found that whole barley seedlings can convert tyrosine to tyramine. Massicot and Marion (1957) added further evidence showing that even phenylalanine could be converted into hordenine as long as the seedlings were attached to the endosperm. Unlike the alkaloid-free seeds that start alkaloid production immediately after the beginning of germination, the alkaloid-rich seeds of the Leguminosae lose their alkaloids during the first week (Nowotnówna, 1928) following germination. This phenomenon was also observed in *Strychnos nux-vomica* (Sabalitschka and Jungermann, 1925). In *Laburnum* seedlings, the accumulated cytisine is converted into *N*-methylcytisine (Pöhm, 1955), and the content of both alkaloids decreases. New alkaloid synthesis did not start until 1 week to 10 days after germination; embryos without cotyledons did not synthesize alkaloids. Even when detached cotyledons increased in size, they could not synthesize alkaloids (Reifer and Kleczkowska, 1957); thus, it must be concluded that the entire plant is required for alkaloid synthesis. Glowacki (1975) investigated the protein composition of lupine species by means of immunoelectrophoresis and established that the beginning of alkaloid production coincides with the disappearance of the majority of the storage proteins in the seed.

Waller and Skursky (1972) showed that [3,5-^{14}C]ricinine administered to senescent leaves of the castor bean plant, *Ricinus communis,* was translocated to all other tissues of the plant; the developing fruit and especially the seeds were found to be labeled the most rapidly. For further discussion of the subject, see page 206.

4.8. Conclusions

Unlike some callus tissue cultures, which tend to synthesize alkaloids only when aging (Neumann and Muller, 1967), the intact plant or grafted chimeras produce alkaloids mostly in young, rapidly growing tissues. These tissues vary among species, with some plants synthesizing alkaloids only in roots, others in aerial parts, and still others in all parts. The young seeds and young fruits are usually rich in alkaloids in all alkaloid-producing species of plants; but the entire fruit of tomato, for example, loses alkaloids during maturation. Tobacco and poppies lose alkaloids only in seeds, and the fruit remains alkaloid-rich. Leguminous plants tend to accumulate alkaloids until the end of maturation. Alkaloid-free seeds start alkaloid synthesis simultaneously with germination, whereas alkaloid-rich seeds delay alkaloid synthesis. Processes which accelerate the growth usually simultaneously increase alkaloid production (Maciejewska-Potapczykowa and Nowacki, 1959).

5 | The Role of Alkaloids in Plants

5.1. Introduction

The proposed roles of alkaloids in plant metabolism, plant catabolism, or plant physiology are (1) end products of metabolism or waste products, (2) storage reservoirs of nitrogen, (3) protective agents for the plant against attack by predators, (4) growth regulators (since structures of some of them resemble structures of known growth regulators), or (5) substitutes for minerals in plants, such as potassium and calcium (according to Justus von Liebig (1840, 1841). Of these items, (2) and (5) appear to be the least promising, and are not considered further here, but the other items, as well as new concepts, will be discussed in this chapter. There are several reviews of the *raison d'être* of secondary compounds (Errera, 1887; James, 1950; Mothes, 1953, 1966, 1969; Fraenkel, 1959, 1969; Bu'Lock, 1961, 1965; Mothes and Schuette, 1969; Geissman and Crout, 1969; Weinburg, 1971; Luckner, 1972; Robinson, 1974).

5.2. Alkaloids as Growth Stimulators and Inhibitors

The structural similarity of some alkaloid structures to plant growth hormones stimulated the idea that at least some alkaloids can play a role as plant growth regulators. Although the idea is quite old, the methodological obstacles were enormous, and the role of alkaloids in growth regulation was proven in only a few cases. The first difficulty a researcher meets while attempting to investigate the role of alkaloids as growth regulators is the choice of the experimental plant material. If he or she decides to use an alkaloid-rich plant, the plant has alkaloids already present and is growing;

thus, an addition of alkaloids can either exceed the optimal alkaloid level and prove to be inhibiting, or it may still be below the critical point and be stimulating. It can, as well, have no effect. On the other hand, if he or she decides to use an alkaloid-free plant, the experiment will be artificial, since the plant normally develops without alkaloids; so it will react to the added alkaloid in the same general way as to other physiologically active, but not natural, substances. Therefore, if an alkaloid added to an alkaloid-free plant influences the growth, nothing is proved except that the compound is not without effect in the particular plant. If a compound does give a similar reaction in a number of different species of plants, its role can be extrapolated to the plant from which it originates. Still, some uncertainty will remain. As a rule, alkaloid-rich plants grow more slowly than related alkaloid-poor species, but this rule has a number of exceptions. In addition, it is quite common that alkaloid-rich species grow under unfavorable conditions and that the abundance of alkaloids is caused by other factors which have no effect on the growth, but this abundance does preserve the plants from herbivorous animals. The slow growth rate is an adaptation to the unfavorable environment, which does not permit more luxurious growth. A plant in an optimal environment, if grazed, can quickly regenerate and produce seeds, while a plant growing in an arid area, especially if a high percentage of the foliage is destroyed, cannot recover. Growth and predation are so interwoven that it is difficult to judge which is the cause and which is the result. When comparing plants belonging to one family, we quite often encounter slow-growing species and genera that are alkaloid-rich growing in unfavorable conditions; in contrast, relatives growing luxuriously in better environments are alkaloid-poor. For example, the alkaloid-rich Genisteae are usually encountered in arid areas, while the alkaloid-free Vicieae are encountered in moist areas (see Chapter 1, pp. 22, 94). One explanation is that the alkaloid-rich plant, finding less competition for sun, does not need to grow so rapidly to avoid being overshadowed, and therefore a slow-growing xerophytic type is selectively favored. Other plausible explanations such as the availability of soil moisture and the nutrient levels could also be important.

The best known example of an alkaloid which inhibits cell division in plants is colchicine. Added in minute amounts, this alkaloid interferes with the formation of the cell carokinetic spindle; instead of a division of the cell into two daughter cells, a restitutive cell is formed with a doubled set of chromosomes. The alkaloid, while very active on cells of most species of plants, produces no effects in *Colchicum autumnale,* the most common source of this compound. Alkaloids of *Senecio* and *Crotalaria* can cause chromosome breakage in a number of organisms, mostly animals, but are harmless to the species producing them. Some alkaloids, when added to water in which seeds of alkaloid-free plants are soaked, inhibit germination. These types of allopathic effects are quite common and are caused not only by alkaloids but also by a number of other secondary metabolites. The alkaloid prevents the germination of seeds of foreign species, thus preserving living space for its own progeny. However, it was found recently that some plants can absorb alkaloids from the soil without serious effects. Not only can foreign alkaloids from decaying plants be absorbed, but a number of synthetic alkaloidlike chemicals, as well as some microbial products, can also be taken in.

The foreign compound is sometimes converted into derivatives, while at other times it is only accumulated. In cases where it has been proved that an alkaloid is poisonous to a foreign plant, a postulation can be made that the alkaloid-producing species has evolved a hormonal system that can effectively operate in spite of the alkaloid present. Waller and Burstrom (1969) have shown that certain diterpenoid alkaloids from *Delphinium ajacis* exhibit growth-inhibiting effects on pea cambium growth. The diterpenoid alkaloids are related in chemical structure to the gibberellins in that the A/B ring junction is similar in the two groups of compounds and is antipodal to that of most naturally occurring steroids. These results clearly indicated that delcosine and delsoline were growth-inhibiting for both phloem and xylem tissue (Sastry and Waller, 1971). Particularly significant was the inhibitory or delaying effect on the initiation of cambium by

Table 5.1. Effect of *D. ajacis* Diterpenoid Alkaloids[a] on Pea Cambium Growth[b]
[Courtesy of Waller and Burstrom (1969) and *Nature*]

Compound[c]	Surface in transverse section, mm^2		Percent inhibition	
	Phloem	Xylem	Phloem	Xylem
Initially	0.147	0.033		
Control	0.301	0.065		
Control + delcosine	0.229	0.051	24	22
Control + delsoline[d]	0.202	0.041	33	37
Control + ajaconine	0.301	0.065	nil	nil
Standard error	±4%	±6%		

[a]Concentration: 5×10^{-4}M
[b]This test is quite sensitive, as indicated by the alkaloid concentration used for the inhibition studies.
[c]These are different treatments and not different results for the same treatment.
[d]This series of tests also delayed or inhibited initiation of the cambium.

delsoline. In contrast, ajaconine had no significant effect on growth (Table 5.1). This growth inhibition can be explained in several ways. Delcosine and delsoline may compete with the gibberellic acids for enzyme active sites, which changes their catalytic function. Such interaction may be at or near the active site or at secondary allosteric sites. Another mode of action might be feedback control of gibberellic acid(s) production by delcosine and delsoline (Figure 5.1). Each group of compounds probably originates from a common diterpenoid pyrophosphate intermediate, the structure of which is unknown. In the usual method of feedback control, the end product controls its production at one of the initial steps in biosynthesis (such as the branch point). It is conceivable that the diterpenoid alkaloid control could be exerted at some other step on the gibberellic acid biosynthesis branch of the pathway, as indicated in Figure 5.1 (see Chapter 6, p. 238).

In further studies of the relationship of diterpenoid alkaloids and gibberellic acids (GA), Lawrence and Waller (1973a, b, 1975) showed that the *Delphinium ajacis* diterpenoid alkaloids did inhibit the action of GA_3 in cucumber hypocotyl bioassays (Table 5.2). Comparison of the effects of diterpenoid alkaloids with those of abscissic acid (ABA) (an inhibitor of GA action) and 2-isopropyl-4-trimethylammonio-5-methylphenyl piperidine-1-carboxylate chloride (AMO-1618) (an inhibitor of gibberellic acid, GA_3) synthesis were completed. The results indicated that certain responses occurred with tests requiring the presence of GA_3 when inhibited α-amylase synthesis in germination seeds (Table 5.3) induced the elongation of the hypocotyl and stem internode of lettuce seedlings in the light and dark, and induced cucumber hypocotyl elongation; however, they did not consistently mimic the effects of either ABA or AMO-1618. These results suggest that diterpenoid alkaloids affect plant development through their influence on GA biosynthesis and/or transport within the plant. Perhaps diterpenoid alkaloids inhibit microtubule formation (Shibaoka, 1974), cellulose synthesis (Hogetsu *et al.,* 1974), or auxin translocation (Basler, 1975). It must be realized that these nitrogenous bases can be isolated from plants of the *Aconitum* and *Delphinium* genera of Ranunculaceae, the *Garrya* genus of Garryaceae, and *Inula royleana* of the Compositae; thus, although they might exert an effect on the plants endogenous to gibberellic acid on ABA, it cannot be considered general in the plant kingdom.

The remarkable translocation of ricinine from senescent leaves into maturing seeds provokes questions about the possible function of this alkaloid in seed germination (Skursky and Waller, 1972). Rapid *de novo* synthesis of ricinine in the second week of germination has been reported (Schiedt *et al.,* 1962). In this study, *N*-demethylricinine was detected as a normal constituent of *R. communis* (castor beans) during dormancy and in the very first stages of germination. Shortly after germination began, the

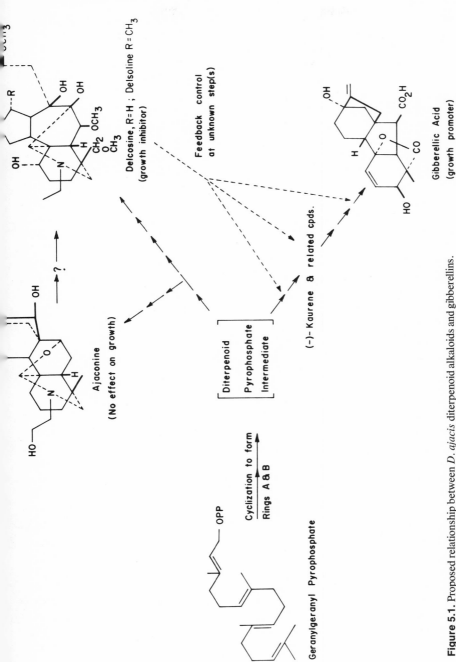

Figure 5.1. Proposed relationship between *D. ajacis* diterpenoid alkaloids and gibberellins.

Table 5.2. The effect of *D. ajacis* Diterpenoid Alkaloids on Normal and Gibberellic
Acid$_3$ (GA$_3$)—Induced Cucumber Hypocotyl Elongation
[Courtesy of Lawrence and Waller (1973a,b, 1975)]

Treatment[a]		Length of marked segment of hypocotyl after 72 hr		
	μg/plant	No GA$_3$	+GA$_3$ 1 μg/plant	Percent inhibition of GA$_3$-induced growth[b]
Control		26.8 ± 0.8[a]	36.4 ± 1.3	
Ajaconine	50.0	27.4 ± 2.2	30.5 ± 2[c]**	61
Ajaconine	0.5	27.4 ± 2.1	32.7 ± 1.0[c]*	27
Delcosine	50.0	28.8 ± 3.0	31.6 ± 1.9	50
Delcosine	0.5	25.4 ± 1.5	30.8 ± 1.3[c]**	58
Acetyldelcosine	50.0	27.1 ± 0.6	31.6 ± 1.1[c]*	50
Acetyldelcosine	0.5	26.0 ± 1.2	30.5 ± 0.7[c]**	61
Seed extract[e]	50.0	21.7 ± 9.4[c]	28.8 ± 1.1[c]**	79
Abscissic acid	10.0	23.5 ± 0.8[c]	28.5 ± 1.6[c]**	82

[a]One-week-old plants were treated with 20 μl of 50% ethanol and 0.01% Tween 20 containing compounds as indicated. Solutions were applied to the bracts subtending the third internode prior to its elongation.
[b]Calculated as (A-X) / (A-B) $\times$ 100, where X = treatment + GA$_3$ internode length, A = control + GA$_3$ internode length, and B = control − GA$_3$ internode length.
[c]Treatment mean is significantly different from control mean in that column at the 5%* and 1%** levels, respectively.
[d]Mean ± standard error of the mean of eight plants.
[e]Distilled water seed extract of five *D. ajacis* seeds per plant (note: the diterpenoid alkaloids are leached out of the seed with water, but it is not known which diterpenoid alkaloid or how much is extracted; also, other compounds such as endogenous GA$_3$ and ABA are water soluble).

concentration of *N*-demethylricinine temporarily increased, apparently being related to the biosynthesis of ricinine; but this alkaloid almost completely disappeared quite rapidly (Figure 5.2). There is not enough evidence to permit a proposal of any hypothesis yet. Efforts are in progress to determine if *N*-demethylricinine is a biologically active compound, with ricinine as the inactive form. In maturing seeds, *N*-demethylricinine is first detectable when pigmentation of the seed coat (testa) begins, i.e., when the development of the seeds has almost reached the final stage of maturity.

The increase of ricinine and *N*-demethylricinine during the first day of germination (Figure 5.2) might be caused by the release of ricinine from some bound form. The idea of "protein-bound ricinine" was first suggested by Robinson (1969), but convincing experimental data are yet to be obtained.

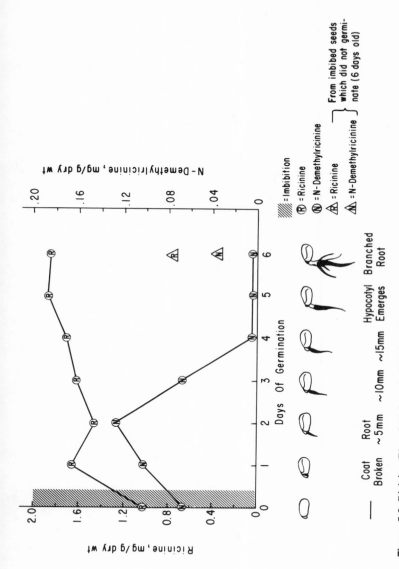

Figure 5.2. Ricinine (R) and *N*-demethylricinine (N) content of castor bean seedlings during germination. Each analysis was made on a composite sample of 20 seeds or seedlings following testa removal (Skursky and Waller, 1972). Courtesy of the journal.

Table 5.3. Effect of Diterpenoid *D. ajacis* Alkaloids on α-Amylase Synthesis in Barley (*Hordeum vulgare* var. Himalaya) Seeds
[Courtesy of Lawrence and Waller (1973a,b, 1975)]

Additions to incubation medium	μg/ml	Units of α-amylase[a]					
		Embryo with half-seed[b]				Embryo less half-seed[c]	
		Incubation medium			Seed homogenate	Incubation medium	Seed homogenate
		Expt. 1	Expt. 2	Expt. 3	Expt. 2		
None (control)		53 ± 11[a]	67 ± 3	69 ± 2	142 ± 15	135 ± 6	88 ± 1
Ajaconine	10	31 ± 2	55 ± 4	54 ± 6	137 ± 5	135 ± 16	86 ± 2
Ajaconine	1		61 ± 2	64 ± 5	151 ± 1	132 ± 12	89 ± 2
Ajaconine	0.1	32 ± 2	53 ± 5	65 ± 6	163 ± 8	94 ± 1	80 ± 3
Delcosine	10	43 ± 12	62 ± 2	65 ± 5	159 ± 0	113 ± 7	79 ± 5
Delcosine	1		68 ± 3	62 ± 4	161 ± 15	127 ± 9	91 ± 8
Delcosine	0.1	31 ± 2	57 ± 8	50 ± 4	147 ± 5	109 ± 2	91 ± 0
Acetyldelcosine	10	34 ± 2				109 ± 18	75 ± 3
Acetyldelcosine	1					119 ± 13	84 ± 2
Acetyldelcosine	0.1	32 ± 7				129 ± 14	91 ± 1

Treatment	Conc.					
Seed leachate[e]		19 ± 1				
ABA	0.1	20 ± 1				
AMO-1618	1.0mM	75 ± 3	75 ± 5	141 ± 3		
GA$_3$	0.001				890 ± 38	168 ± 2
Ajaconine + GA$_3$[f]	10				1033 ± 55	195 ± 2
Ajaconine + GA$_3$	1				961 ± 79	171 ± 2
Ajaconine + GA$_3$	0.1				953 ± 92	162 ± 3
Delcosine + GA$_3$[f]	10				925 ± 83	131 ± 15
Delcosine + GA$_3$	1				969 ± 58	162 ± 1
Delcosine + GA$_3$	0.1				843 ± 78	100 ± 2
Acetyldelcosine + GA$_3$[f]	10				1031 ± 92	168 ± 5
Acetyldelcosine + GA$_3$	1				1117 ± 93	191 ± 1
Acetyldelcosine + GA$_3$	0.1				752 ± 54	159 ± 8

[a] One enzyme unit equals 100 μg starch hydrolyzed/ml/min for 10 barley half-seeds.
[b] Seeds incubated in treatment solution for 48 hr.
[c] Seeds incubated in treatment solution for 24 hr.
[d] Means ± standard error of the mean of four replicates.
[e] Distilled water leachate from approximately 25 *D. ajacis* seeds/ml incubation medium.
[f] GA$_3$ concentration of 0.001 μg/ml.

In a preliminary experiment, it was shown that [7-^{14}C] nicotinic acid was significantly incorporated into both ricinine and N-demethylricinine during the first day of germination, which indicated that alkaloid biosynthesis is one of the earliest metabolic processes initiated with growth and differentiation in *R. communis*.

Experiments performed on the nicotinic acid–ricinine relationship in sterile cultures of *R. communis* established clearly that (1) the relationship exists, and (2) the metabolism of ricinine can be spared by the presence of higher concentrations of nicotinic acid than normally found in the tissue (Waller and Nakazawa, 1963). This sparing action of nicotinic acid on ricinine utilization suggests a vitamin–alkaloid metabolic relationship not previously found in a plant system.

Certain alkaloid precursors such as indole derivatives, purines, and nicotinic acid are powerful growth stimulators, and some, such as certain phenylalanine–tyrosine derivatives, are inhibitors. The biosynthetic con-

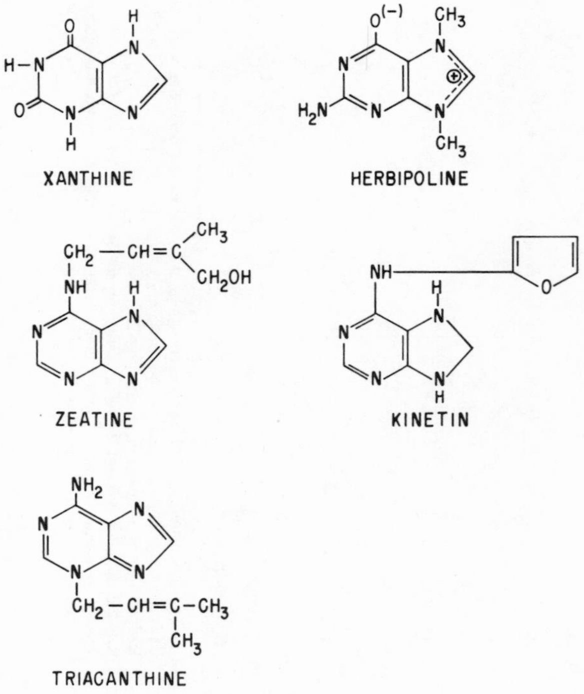

Figure 5.3. Some pseudoalkaloids derived from purines. They are alkaloid-like compounds with growth-stimulating properties.

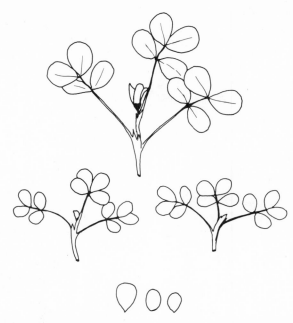

Figure 5.4a. Effect of lupine alkaloids on peanut plants. *Top*—control plant; *middle*—plants grown from seeds with added alkaloids. left—hydroxylupanine, right—lupinine; *bottom*—average leaflets from these plants in the same order.

version of these compounds into alkaloids would indicate a deactivation process (Figure 5.3). Most of the plants in which alkaloids were introduced in physiological amounts (up to 1% of the dry substance of the plant) exhibited no effects at all; but even when an inhibition of growth is observed or a poisoning results, this is no indication of a regulating role, because the level is well over the level of proper growth-regulating hormones (Figures 5.4a and 5.4b). A useful tool for experimentation was developed with the breeding of plants with genetically altered alkaloid levels. Varieties of *Datura, Papaver,* and *Atropa* were produced which contained as much as three times the normal alkaloid contents (Mothes and Romeike, 1954). These plants sometimes grew more slowly and were more susceptible to unfavorable environmental conditions than normal plants. In 1930 it was found that alkaloid-poor mutants of lupines were at first glance as vigorous as the wild type (Sengbusch, 1934). Careful examination of the yield of seeds and straw indicated, however, a lower reproduction coefficient for the mutant (Aniol *et al.,* 1968, 1972; Nowacki and Kazimerski, 1971; Kazimerski and Nowacki, 1967). In this case, it appears that the

Figure 5.4b Effect of lupine alkaloids on several plant species; *right*—control plant; *left*—plants grown from seeds with added alkaloids (a mixture of sparteine and lupanine). Species are (1) *Lupinus luteus*, (2) *Brassica olearacea* (cabbage), and (3) leaves of *Brassica olearacea*.

mutation causes a weakening of the plant by unbalancing the fine equilibrium between different compounds. Despite intensive breeding programs, the alkaloid-poor (sweet) forms are still inferior, even compared to unselected alkaloid-rich (bitter) ones. This is, in some ways, a confirmation that the alkaloid-poor mutation has an additional pleitropic side effect, which is that the original sweet mutants are weaker individuals. On the other hand, the alkaloid might act as a stimulant or growth regulator. Repeated backcrossing of sweet with bitter forms does improve the performance of the former. Great differences are encountered in the number of seeds produced (Norwacki and Kazimierski, 1971; Zachow, 1967), since in the sweet forms, decreased seed production is the most striking feature. Mackiewicz (1958) analyzed the viability of pollen grains of sweet and bitter plants; he was able to confirm that the sweet plants actually have a lower percentage of stainable (and presumably viable) pollen grains. In the favorable environmental conditions, germinating sweet plants sometimes developed better than the bitter plants, but later in the ontogeny these grew poorer than the wild type. In 1957 Maysurian reported an isolation of sweet *Lupinus polyphyllus*. This mutant was remarkable because the seeds did not germinate when placed on petri dishes in plain water. When alkaloids extracted from the seeds of bitter *Lupinus angustifolius* were added, the seeds from *Lupinus polyphyllus* germinated well. Since *Lupinus angustifolius* has a fairly similar alkaloid spectrum to *Lupinus polyphyllus,* no clues were given as to which particular alkaloid was necessary for the initiation of germination. However, seed progeny of the alkaloid-free plant germinated quite well. From seeds that were received from Maysurian in 1957, plants were reared in Poland and compared to the bitter genotypes, but it was impossible to ascertain the reason for the inferiority of the mutant, since the bitter plants available for comparison were collected in west Poland and the mutant originated from a collection in the vicinity of Moscow (USSR). Even if the original stock came from the same part of North America, the approximately 100 years of propagation in Europe could have caused the differences. Therefore, the two strains were crossed, and the F_2 segregants were investigated. The difference of vitality of sweet and bitter plants was so striking that even without testing for alkaloids it was possible to tell the plants apart. Since *L. polyphyllus* is a predominantly cross-pollinated species, the sweet plants were spatially isolated from the bitter *L. polyphyllus* to prevent cross-pollination with the bitter forms. The space between the bitter and sweet plants (approximately 400 m) was sown with different North American lupines. Usually *L. polyphyllus* does not hybridize easily. In this particular case, however, the number of interspecific spontaneous hybrids was amazingly high, which prompted the experimenters to investigate the pollen viability. It was found that some plants did not produce any viable pollen grains. Lamprecht (1964a, b), while investigating some aberrant mutations causing formation of morphological character peculiar to different species (he proposes to call these types of mutations exomutations, or interspecific mutations), found that an exomutation is always accompanied by reduced fertility of the mutant and that the segregation ratio is usually distorted by a deficit of the recessive homozygotes. An alkaloid-poor mutant in lupines can be regarded as a typical exomutation, because the mutated form is lacking character of not only the species but also of the tribe's metabolical peculiarity to synthesize and accumulate quinolizidine alkaloids. The mutations in other lupine species are not so drastic, and the level of the residual alkaloids is usually higher than in the sweet *L. polyphyllus*. Applying lupanine solution to leaves of sweet *L. albus*, Nowotny-Mieczynska and Zientkiewicz (1955) were able to show a growth-stimulating property of the applied alkaloid. On the other hand, Nowacki (1958) was not able to find any significant differences between sweet and bitter *L. angustifolius* growing on soil supplemented with sparteine and that grown on soil to which the alkaloid spectrum typical to the *L. angustifolius* alkaloid mixture was added. The plants had absorbed at least a part of the added alkaloids, but the increase of dry matter production was too low to be attributed to the effect of alkaloids. In unpublished experiments in laboratories in Poland, some seeds of sweet lupines were soaked with salts of pure alkaloids in levels from 0.01 to 1% of dry substance of the seeds. The results were confusing, but some conclusions can be drawn: hydroxylated alkaloids like lupinine and hydroxylupanine were slightly stimulating to growth, while sparteine and lupanine were inhibiting, as were most species of foreign alkaloids such as nicotine, quinine, cytisine, and a number of other compounds. While the sweet mutants were developing more poorly than the bitter forms, mutants with increased alkaloid levels also suffered (Mothes and Romeike, 1954). Alkaloid-free seeds such as peas and kidney beans reacted likewise. Pöhm (1966) found that cytisine was quite poisonous to *Phaseolus* seedlings, while *N*-methylcytisine was harmless. Tomatoes and belladonna grafted upon *Nicotiana* stock suffered, and the leaves developed brown necrotic spots (Mothes and Romeike, 1954) (see Chapter 4, Figure 4.1).

Table 5.4 shows some examples of known biochemical effects of alkaloids. It is important to recognize that these processes do not occur in all plants. It is quite possible that these alkaloids influence their own metabolism in the manner shown in Table 5.4.

An alkaloid common for a plant family or a similar taxonomical unit most probably originated together with the group of the species. Since the age of most plant families is traceable back to the upper Cretaceous era, the alkaloids must be of the same age. Thus in approximately 80 million years

Table 5.4. Some Other Biochemical Effects of Alkaloids in Plants[a]

Alkaloid tested	Primary action	Later effect	System used	Reference
Theophylline	Inhibits 3',5'-phosphodiesterase	α-Amylase released	Barley endosperm	Duffus and Duffus (1969)
Quinine	Intercalates in DNA helix	Phenylalanine-ammonia lyase induced	Pea pods	Hadwiger and Swochaw (1971)
Caffeine	Binds to part of operon	Adenylsuccinate lyase activity increased	*Bacillus subtilis*	Nishikawa and Shiio (1969)
Diaminosteroid alkaloids	Complexes with DNA	Replication inhibited	Bacteriophage	Mahler and Baylor (1967)
Colchicine	Cell carokinesis	Cell is formed with a double set of chromosomes	Most plants	Eigsti and Dustin (1955)
Colchicine	Binds the purified microtubular protein	In chromosome movement during mitosis, etc.	Cells of a general type	Sherline *et al.* (1975); Olmsted and Borisy (1973)
Veratrum alkaloids	Inhibits growth	Effect on DNA stability	Rye, oats	Olney (1968)
Nicotine	Antiauxin	—	—	Ramshorn (1955)
Nicotine	Inhibits chlorophyll synthesis	—	—	Hassall (1969)
Gramine	Plant competition	—	Barley	Overland (1966)

[a]Patterned after Robinson (1974).

of selection, the character has proved of some advantage to the species comprising the family, and a balance of metabolites has been created. Therefore, it is hardly possible to change the alkaloid level without seriously impairing the vitality of the plants. Still, when mutants with altered alkaloid levels are produced, they can grow under special conditions. The age of alkaloid-rich families is probably not the same; therefore, the time necessary to develop the character was different. In cases in which the alkaloid character is very old and all members of the plant family are alkaloid-rich, it is hardly to be expected that alkaloid-poor, viable mutants will be produced. Therefore, despite repeated attempts, no alkaloidless mutants have been found in *Papaver* and in some other alkaloid-rich genera.

5.3. Alkaloids as Plant-Protecting Compounds

It is quite certain that alkaloids are not growth-regulating compounds—at least in the sense of normal growth regulators—a fact that requires the scientists who are interested in the role of alkaloids to search for some other role. Some alkaloids that are randomly distributed all over the Angiospermeae and that often occur simultaneously in some individuals of a given species and not in others are without selective value. However, since alkaloids restricted to a certain natural systematic unit and present in all plants of the unit, without exception, establish themselves only due to selective pressure, they do play a role in the plants by increasing their fitness. A number of alkaloid-free plants can grow and produce seeds when supplied with alkaloids; thus these plants, though not producing alkaloids, can tolerate them. This is important to the further discussion of the role of alkaloids in plants. Since most alkaloids are not toxic for most plants, a mutation leading to alkaloid synthesis would be without effect on the plant and could be established as a neutral character merely by genetic drift. The chance that genetic drift can eradicate completely one neutral character and impose another is small. Therefore, it is justifiable to assume that alkaloids, without exception, in a natural higher taxonomical unit are or were of significant selective value for the plants containing them. As early as 1887, Errera held the notion that alkaloids protect plants against herbivorous animals and possibly pathogenic microorganisms. Since then, with accumulating knowledge on alkaloid physiology in animals, fungi, and bacteria, the idea has been enlarged. Still more recently, James (1950) and Mothes (1960, 1975) came to other conclusions. James, after noting that alkaloid-rich plants suffered from fungus infection when eaten by insects, writes, "In view of such facts, it is not possible to ascribe any great part to alkaloids in controlling the attacks either of animals or plant parasites." Mothes' statement is similar: "To conclude, it might be remarked that much has been learned about the metabolism of alkaloids, but their functions in the plant, if any, are still largely unknown." While the role of alkaloids in plant metabolism is still a mystery and has never been satisfactorily explained, it is almost inconceivable, in view of the differences in their chemical constitution and their restricted distribution, that they have the same function in all plants.

5.4. Protection against Fungi and Bacteria

Some alkaloids, such as quinine and other *Cinchona* alkaloids, have been used for centuries as drugs for killing protozoa while having little effect on the mammals ingesting such alkaloids. There have been attempts to extrapolate the poisonous effects of alkaloids to other microorganisms (protists), but it has been difficult to do. It was found that alkaloid-rich plants are affected by parasitic fungi and by pathogenic bacteria as much as alkaloid-free species. *Phytophthora* will attack, with the same deadly effects, alkaloid-rich *Nicotiana* species as readily as the less alkaloid-rich tomatoes and potatoes. *Cladosporium* develops in solutions containing a number of alkaloids as well as it does in media without alkaloids. A number of pathogenic and saprophytic fungi thrive on media with lupine alkaloids as well as they do in controls (Nowacki, 1958). *Fusarium* isolated from infected lupine plants grew on media containing sparteine, quinine, and a mixture of lupine alkaloids (up to concentrations that are physiological in the plants' natural environment (Table 5.5). The alkaloids in low concentration even stimulated growth. One of the saprophytic fungi, *Aspergillus niger*, not only was grown in alkaloid-rich media but also utilized the alkaloids. Similar results were obtained by Blaszczak and Statkun (1958), Zalewski *et al.* (1959), Nowacki (1961), and Massingill and Hodgkins (1967). Still, in some species, resistant forms can be clearly attributed to chemical differences. As an example, Virtanen (1958) found that *Fusarium*-resistant varieties of rye (*Secale cereale*) owe their resistance to a presence of an alkaloid-like (glycoside) fungistatic compound. Two substances, benzoxazolinone extracted from rye and 6-methoxy-2(3)-benzoxazolinone (Virtanen, 1958; Virtanen *et al.*, 1957), cause the resistance of corn (*Zea mays*) and wheat (*Triticum* spp.) toward certain fungi (Figure 5.5). They occur as their glycosides in the plants from which they are released by glycosidase which becomes active during extraction. Present evidence suggests that they are not important in resistance because neither the glycosides nor the aglycones are very toxic. Since most alkaloids do not affect parasitic microorganisms, there are probably a number of parasites that attack only alkaloid-free species; consequently, it seems that many alkaloids have not in the past played a role of protecting compounds and that parasites acted as selection factors only in some rare instances. Savile (1954), comparing the distribution of *Puccinia* and other related pathogens, pointed out that

Table 5.5. The Development of Parasitic and Saprophytic Fungi on Media with Added Alkaloids (Average from Five Cultures)[a]

	Dry weight of mycelium, mg				
Medium	*Fusarium culmorum*	*Rhizoctonia solani*	*Aspergillus niger*	*Aspergillus terreus*	*Penicillium glaucum*
Without alkaloids	305.4	389.0	308.7	335.0	301.3
0.01% Sparteine	297.5	391.6	336.3	316.7	292.7
0.1% Sparteine	284.0	396.5	296.0	315.2	286.5
1% Sparteine	383.5	386.7	319.9	311.3	286.7
0.02% Alkaloid extract	296.2	376.0	313.8	308.4	288.5
0.02% Alkaloid extract	291.2	401.8	322.1	304.3	290.5
0.1% Quinine	307.3	396.6	331.4	338.5	297.8

[a] Nowacki (1958).

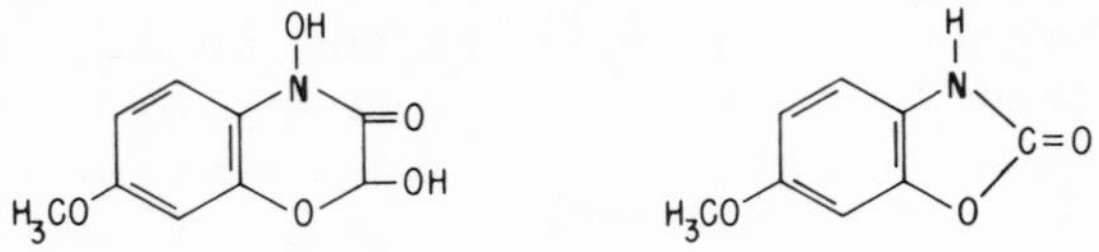

2,4-DIHYDROXY-7-METHOXY-1,4   COIXOL
-BENZOXAZINONE

Figure 5.5. Fungistatic compounds with alkaloid-like structure from *Zea mays* and *Coix lacryma-jobi,* both plants belonging to the family Maydeae.

the Pucciniaceae are older than the Angiospermeae, and, since the number of possible mutations in an organism with so short a life cycle and so enormous a number of spores can readily exceed the number of mutations of a higher plant, a development of alkaloid-based immunity seems to be hopeless. Even if a mutation of a higher plant has caused accumulation of a fungistatic alkaloid so that the mutant has a certain degree of immunity and replaces the susceptible original form, it will probably never achieve the final goal of total immunity. This is because, in the meantime, a mutation will occur in the infecting fungus, which will in turn be able to attack the temporarily immune form. The result will be a steady state involving the immune form, the original form, and both forms of the parasite. The difficulties in breeding rust-resistant varieties of cultivated plants are the best illustrations of this problem. The race between the development of immune forms of crop plants and the mutations of the parasite is a case in point, for every time a resistant form is produced, a mutation of the parasite diminishes its usefulness.

Phymatotrichum omnivorum parasitizes and destroys roots of a wide variety of higher plants from many families; Taubenhaus and Ezekiel (1936) first suggested that alkaloids are responsible for the resistance to this parasite. Greathouse (1938, 1939) investigated *Mahonia trifoliolata* and *M. swaseyi,* two species resistant to *P. omnivorum,* and found them to be high in berberine in a continuous zone of cells beneath the periderm of the roots in greater concentrations than that which prevents growth of the parasite in culture. Similar evidence is that the alkaloids of a number of other resistant plants prevent the growth of *P. omnivorum* in culture at concentrations lower than those found in resistant tissues, e.g., sanguinarine prevents growth at 2 to 5 ppm and chelerythrine does so at 50 ppm. Both substances occur in plant *(Sanguinaria canadensis)* tissue at much higher concentration and are considered important in the plant's resistance. Of 62 alkaloids tested from a variety of plants, sanguinarine is the most toxic. In general, it is found that resistance of a plant species to a parasite is related to the toxicity of the alkaloids. Wood (1967) observed that *P. omnivorum* was more susceptible to alkaloids whereas the vascular wilt fungi were the least susceptible of those tested. In the monocotyledons, which generally have resistance to the parasite, the high resistance of members of the Amaryllidaceae is believed to be due to their high content of alkaloids (Greathouse, 1938, 1939; Greathouse and Rigler, 1940a, b).

Invasion by *Fusarium* spp. in potato tubers may be prevented by alkaloids. The role of solanine in diseases caused by *F. avenaceum* and *F. caeruleum* is attributed to the fact that it occurs mainly in vacuoles of living cells and because *F. avenaceum,* unlike *F. caeruleum,* quickly kills and penetrates cells which it approaches. This means that *F. avenaceum* is rapidly exposed to the alkaloid; whereas *F. caeruleum* grows at first between cells of the tuber in intercellular spaces, so does not kill the cell at once (McKee, 1955). The concentration of solanine (glycoside of solanidine, see Figure 6.7) and chaconine (glycoside of solanidine, see Figure 6.7) increases near mounds produced by *F. caeruleum* and this increase is accompanied by increase in resistance (McKee, 1961). Solanine in concentrations of 500 ppm is a factor in the resistance of green tomato fruit to *Colleotrichum phomoides*. Tomatine and tomatidine (see Figures 6.7, 6.24) are thought to play a part in resistance to wilt caused by *Fusarium oxysporum f. lycopersici* (Kern, 1952).

Since in most cases the resistance is caused by factors other than alkaloids, such as differences in cell wall structure, accumulation of phenolic compounds, or an ability to form suberin layers around the infection point, it seems that alkaloids have never played an important role in protecting plants against fungal and bacterial infections. Most alkaloids are without effect on the development of adapted, as well as foreign, fungi in experiments *in vitro*. Therefore, it seems justified to assume that in cases in which an alkaloid-rich plant is not infected by a certain parasite, the reason for its immunity is not the presence of alkaloids.

5.5. Alkaloids and Virus Resistance

Resistance to a virus in a certain plant may be caused by two major factors: (1) the cell medium of the plant is capable of providing no suitable material for replication of the viral nucleic acids and protein, and (2) the plant is avoided by the vector, usually an insect, because of certain chemical or mechanical properties. We shall deal with the first possibility. The host specificity of most plant viruses is known; still there is a broad spectrum of species and genera that may be attacked by the same virus. Very often the species and genera belong to a single family, but exceptions are common. The virus develops quite well on alkaloid-rich and alkaloid-poor plants, but in some cases only plants with a defined alkaloid group suffer. While infecting *Nicotiana tabacum* and *N. glauca* with most strains of TMV (tobacco mosaic virus) is successful, the symptoms of the disease are very different in these species. *N. tabacum* rapidly becomes sick, the infection spreads all over the plant, and the typical mosaic pattern is formed. In contrast, *N. glauca* develops only restricted necrotic spots and the plant remains apparently healthy; the development and production of seeds is not retarded. A similar case was found in lupines of the Mediterransan species. *L. digitatus, L. albus,* and *L. luteus* were victims of the viral infection that causes the narrow leafiness in yellow lupines. The infected plants were found to have a severalfold increase in free arginine content. The plants usually died before being able to produce seeds, while the same virus produced no obvious symptoms in *L. mutabilis,* a South American species that has a different set of alkaloids (Nowacki and Waller, 1973) (Figure 5.6). In both cases, the *Nicotiana* and *Lupinus,* the virus was able to infect not only the species belonging to the given genera but a number of other not closely related host plants as well. In the case of tobacco mosaic virus (TMV), some Solanaceae plants were damaged when infected. Strangely, only *N. glauca* seems to be resistant. In the case of the lupines, the virus used developed not only on lupines, but on a number of Papilionaceae: peas, beans, clovers, and alfalfa. There is no evidence that alkaloids were actually involved in the resistance, but this is quite possible. The evidence that certain alkaloids are active mutagenic, carcinogenic, and antitumor agents is sufficiently great to suggest that the alkaloid present in the cell may interfere with the replication of the virus nucleic acid molecules. One might wonder why the alkaloids are not interfering with the exact replication of plant DNA (Figure 5.7). This case seems to be sufficiently explained in Fowden's (1974) experiments and reports, in which he found no incorporation of unusual amino acid homologues (or alkaloids, in the broader alkaloid definition) into proteins in the plant species that produces them, whereas unrelated plants do incorporate the unusual compounds instead of the proper ones. Dawson and Osdene (1972) have hypothesized on the possibility of nicotine and anabasine being bonded to a polynucleotide as an integral part of the macromolecules DNA and RNA. Although they present powerful arguments for the occurrence of a DNA or *t*-RNA containing nicotine and anabasine, *no alkaloid* has been isolated from either type of macromolecule. A system for discriminating between alkaloids and nucleic acid bases or, in other cases, between alkaloids and protein amino acids, must exist in plants which tolerate a great concentration of alkaloids. The mutagenic action of a number of alkaloids in cells of plants, bacteria, insects, and mammals is well known, but in no instance

Figure 5.6. Appearance of yellow lupine plants *(Lupinus luteus)* used in experiments 21 days after infection. Left—healthy; right—infected (Nowacki and Waller, 1973). Courtesy of the journal.

does an alkaloid produce a mutation in the plant where it is native. This can be partially attributed to the fact that the native alkaloid can be deposited outside the active centers of the cells, as it usually is. The alkaloids are mostly deposited in vacuoles. In contrast, the alkaloids introduced by artificial means can be distributed differently in the cells. For example, in cells of alkaloid-rich plants there may be alkaloids which are always on the way from the centers of active metabolism to the dumping areas, the vacuoles, and when the cell is infected by a virus, enough alkaloid is available to confuse the virus replication system. This concept needs

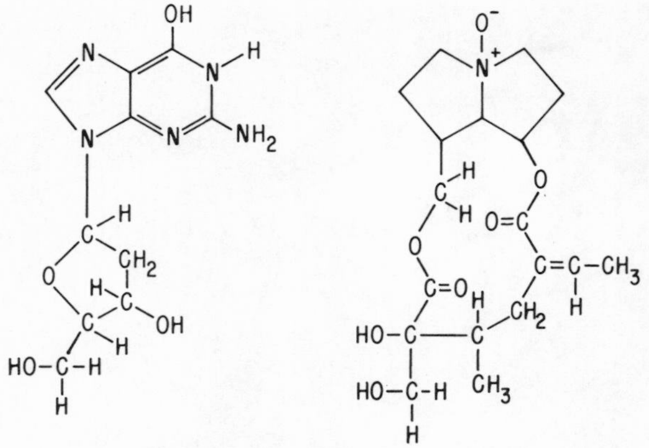

Figure 5.7. Structural formulas of a guanyldeoxyriboside (encountered in DNA and RNA) and a teratogenous alkaloid of the necine ester type.

experimental verification, of course, but the evidence is promising that in genera in which a number of species are virus-infected and others are apparently immune, there will exist simultaneously in the infected and immune species different alkaloid spectra.

Evidence of activities of alkaloids in the DNA reproduction mechanism is impressive. Some alkaloids are used as drugs for inhibiting the development, e.g., of animal cancerous tumors, including humans (Kupchan and By, 1968; Kupchan, 1975), some are known for hallucinogenic properties (Schultes, 1975; Wall, 1975; Hoffman *et al.*, 1975), antimicrobial effects (Mitscher, 1975), neurotoxins (Ressler, 1975), and teratogenic (Kuc, 1975) and allergenic behavior (Mitchell, 1975), and some are used in folk medicine (Duke, 1975).

The other aspects of resistance, the unpalatability of the plant for the vector, will be dealt with in the next section of this chapter, together with further evidence for plant–insect coevolution.

5.6. Plant–Insect Coevolution: The Role of Alkaloids

The causes for replacement of the paleophytic flora of the Pennsylvanian (Carboniferous) by the mesophytic flora of the Permian period are reasonably self-evident, because the cryptogam plants are inferior in reproductive capacity to the Gymnospermeae. The second replacement, which occurred in the Jurassic and Cretaceous periods, is less easily understood. Indeed, the Gymnospermeae have persisted up to the present time and are, in some areas, still the dominant flora (e.g., in the northern temperate forest, the Taiga). The origin of present "higher plants," the Angiospermeae, was for Charles Darwin an "abominable mystery." According to Cronquist (1968), "On the basis of present information, it seems that the origin of the angiosperms as we know them was fortuituous." Since the replacement of the mesophytic flora by the cenophytic cannot be attributed solely to the dubious improvement in the reproductive organs, some other explanations are necessary. The angiosperms now constitute up to 95% of all living vascular plant species. As the dominant flora, they provide food for the majority of phytophagous animals, but we are not certain that they thus provided food while they were replacing the gymnosperms. Two basic changes in the animal kingdom took place after the angiosperms had established themselves. The first and most remarkable was the general change of insect families; some underwent a thorough modification as new classes arose. The Cretaceous period was the time when most of the reptile orders perished. Surprisingly, the now surviving reptiles, with few exceptions, are carnivorous or insectivorous. The conspicuous colored and odoriferous flowers of the angiosperms, which were provided with nutritious exudates such as nectar, have been recognized for a long time as attractants for insects. The response of the insects was an evolution of the adapted species, genera, and classes of flower-pollination insects. It is common knowledge that the pigments and flavoring substances owe their origin and existence most likely to their function as attractants for insects. While the angiosperms have the same basic chemical composition as the algae and cryptosperms involving proteins, carbohydrates, nucleic acids, and fats, they, unlike the older classes, contain a number of "secondary metabolites." There are practically no plants that are entirely free of glucosides, terpenes, cyanogenetic compounds, mustard oils, alkaloids, and other representatives of what has been called "metabolic trash." Only the 0.1% of angiosperm plants that have become cultivated for human food are, to a certain degree, free from poisons, thanks mostly to plant breeding efforts; their wild ancestors were often not so innocent. The wild and primitive forms of the lima bean, *Phaseolus lunatus,* are rich in cyanogenetic compounds. The wild peas, *Pisum fulvum* and *P. elatius,* have a high concentration of bitter-tasting substances, and the ancestors of cereal grains accumulated a number of unpalatable or even poisonous compounds. The abundance of indifferent metabolites in angiosperms may be the factor which made possible their Jurassic displacement of the gymnosperms. Ehrlich and Raven (1964, 1967) and Raven (1973) write: "Angiosperms have, through occasional mutations and recombination, produced a series of chemical compounds not directly related to their basic metabolic pathways but not inimical to normal growth and development." Some of these compounds, by chance, serve to reduce or destroy the palatability of the plant in which they are produced (Fraenkel, 1959). Such a plant, protected from attacks of phytophagous animals, would in a sense have entered a new adaptive zone. During evolution some radiation might follow, and eventually what began as a chance mutation or recombination might characterize an entire family or several families. Phytophagous insects, however, can evolve relatively rapidly in response to physiological obstacles, as is shown by recent human experiences with commercial insecticides. It is suggested that some of the secondary metabolites present in early angiosperms may have afforded them an unusual degree of protection from the herbivorous animals of that time. Behind such a chemical shield the angiosperms may have developed and become structurally diverse. Owing to a simultaneous, even if delayed, adaptive evolution of insects, the shield was pierced long ago. The secondary metabolites no longer protected the plants

Figure 5.8a. Larva of European euphorbic moth feeding on a poisonous *Euphorbia* in Poznan,. Poland.

Figure 5.8b. Colorado bettle in a potato plant in Poznan, Poland.

against specialized vermin; therefore, in future evolution, new mutations were sometimes favored which did not waste essential metabolites to produce secondary compounds without protective properties. The youngest families therefore seem to be rather free of poisons (e.g., gramines). They are not digestible for most monogastric herbivorous animals because of a high percentage of cellulose and fibers in the leaf. Present alkaloid-rich plants are hosts for a number of phytophagous insects which apparently suffer no injuries from the alkaloids they ingest. Examples of insects feeding on alkaloid-rich plants are shown in Figures 5.8a through 5.8d (Nowacki and Waller, 1973b). This finding prompted some scientists to conclude that alkaloids have no protective properties. Yet the same alkaloids, when fed to unspecialized insects, prove to be violent poisons, e.g., nicotine fed to *Periplaneta americana* (Yamamoto, 1965). Thus, by accumulating alkaloids, plants are actually decreasing the number of potentially damaging animals that will eat or graze on them.

Leptinotarsa decemlineata, the common Colorado potato beetle, is encountered on a number of Solanaceae. But the development of the larvae can be observed only on some plants. All plants containing nicotine are

Figure 5.8c. Example of a displaced insect: caterpillar of a European butterfly feeding on a bitter American lupine plant in the lupine collection in Poznan, Poland.

Figure 5.8d. Moth of *Protoparce sexta* grown on *Datura* in Poznan, Poland.

toxic, and growth on plants containing *Datura* alkaloids is retarded as demonstrated by Hsiao and Fraenkel (1968), who reared the larvae on a basic diet with addition of alkaloids (Figure 5.9). The alkaloid solanine, which is encountered in all species supporting growth of the larvae, is growth-inhibiting only in concentrations exceeding those usually found in plants, and even in this case the weight gain is approximately 90% that of larvae on the alkaloid-free basic diet. Other *Solanum* alkaloids (capsaicine and tomatine) at supernormal levels inhibit the growth very drastically.

The occurrence of specialized predators indicates the role of alkaloids as plant-protecting agents; when the alkaloid becomes in the last resort useless agaist the plant predators, it even in some cases provides immunity for the alkaloid-rich insects from predation by birds and mammals.

The process of coevolution of the plant and the predatory insect may be either simple or complicated. A rather simple case would be as follows: A species of plant is broadly distributed, it has a spectrum of predators, and a balanced system exists. The number of herbivorous insects developed on

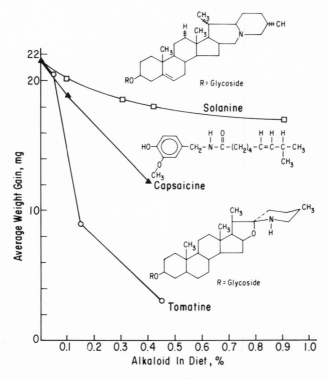

Figure 5.9. Effect of plant chemicals on feeding and growth of Colorado potato beetle (Hsiao and Fraenkel, 1968). Courtesy of the authors and of the Entomological Society of America.

the plant is so high that an average decrease of 40% of the seed production is observed. In some years the insects increase enough in number that they practically eradicate the plant population. Next, because of the lack of alternative hosts, nearly all the insect population dies off; as a result, the plant increases in number, and the insect population follows in this increase. This pendulum follows precise cyclic periods. The plant has a mutation ratio of 1 mutant for every 10 million produced gametes. One mutation can cause a conversion of an amino acid into an alkaloid. As in most mutants the character may be recessive, and only homozygous plants will produce sufficient alkaloids to be recognized as alkaloid-rich plants. The alkaloid produced by these rare mutated homozygous plants may be a complete surprise for the predatory insect. Each larva, after eating a single piece of leaf, becomes violently sick and dies. The plant free of predation produces seeds according to the full capability of the genotype, instead of the average 60% for the seriously maimed normal ones. It can, however, be assumed that since it is a mutant, it will produce fewer seeds than the normal form when not injured by insects. Due to the enormous size of the population and the very low mutation rate, the accumulation of the mutated form will proceed slowly, but steadily. As long as the mutant comprises only a small fraction of the population, it does not specifically influence the insects developing on the normal plants. Since it is assumed that the insect cannot distinguish the mutant from the normal plants, the female will oviposit on both types. We are dealing with millions of individual plants and thousands of years. A mutation can originate in the insect which causes it to be immune to a physiological innovation such as the alkaloids. The advantage the mutant has had for some time is gone, and the mutant will remain in the population or be lost. This mutation proves to be a failure. But if the alkaloid-resistant form of the insect has not developed during the first stages of the numerical growth of the mutants, the mutant may now comprise 10% of the population. While the reproduction coefficient of the entire population is still 1, the mutant is reproducing itself with a coefficient of 1.6, and the increase of the mutant is very large. Thus it will reduce the percentage of the original form in the population. Now when the number of the mutated plants reaches about 50%, the mutant starts to protect the normal form, serving as a trap for the insect. A great number of eggs are lost every year because the female has deposited them on the poisonous plant. Selection in the insect will favor two new forms, one avoiding the poisonous plant and the other able to survive on the alkaloid-rich diet. All mutations of the insects will slow down the elimination of the old form to a certain degree, but in the end the mutant will be superior and the original form will disappear because the number of predators feeding on the mutants will be lower; this may be due to insects which had not mutated or to those that had learned to avoid the poisonous plant. But the mutant had inherited the insects of the original form which were able to mutate to the new source of food before the old one had disappeared. A replacement of an old form by a poisonous mutant can take place without the simultaneous adaptation of the insects. The plant will be immune for some time and will spread in the ecological niche, replacing other species not so fortunate in mutations. It can happen that a certain insect is deprived of its usual host plants and the females will oviposit on plants which somewhat resemble the normal food source (in smell, taste, hairiness, etc.). Most of the larvae will then perish, but some can prove mutable and survive on the new food source. The advantage the alkaloid-rich plant had in the past will then disappear (Figure 5.10). Therefore, it is logical that alkaloids are mostly restricted to smaller taxonomical units because of the insects which have an evolution rate that is rather greater than that of the higher plants; they can always match the mutations which the plants are producing. This also explains the diversity of plants. Over 190,000 plant species are known to exist, but they are matched with at least a half-million herbivorous arthropods.

The chemical arms race is continually leading to deadlocks, increasing only the chemical diversity of plants and simultaneously the devices the insects develop to enable them to live on food unusable by other species. The larvae of the blue butterfly, *Plebeius icarioides,* are restricted in feeding habits to various species of the plant genus *Lupinus* (Downey and Dunn, 1964), which are rich in alkaloids. In all laboratory tests so far conducted the insect readily feeds on any species of lupine including paleoarctic and neotropical species with which extinct insect populations have never been associated. The larvae will not accept other types of plants and will succumb without lupines.

The alkaloid level in some American species is approaching 3% of dry matter, and the alkaloids, although all of the same origin and basic formula, have very different physiological properties. On the other hand, some North American lupines and the cultivated European species contain only traces of alkaloids. Thus this butterfly must be recognized as a species which is tolerant to the alkaloids present in the host plant but does not need them as constituents of the food.

The alkaloids are mostly unnecessary for the normal development of the insect, but in some cases the insect accumulates them, and then becomes unpalatable for animals feeding on the insect. Such accumulators are the insect *Helopeltis,* which feeds on *Cinchona* bark and comes to contain amorphous cinchonine-like alkaloids, and *Attacus atlas,* which also accumulates cinchonine. A good example of insects prospering from poisons acquired from plants was described by Brower *et al.* (1968).

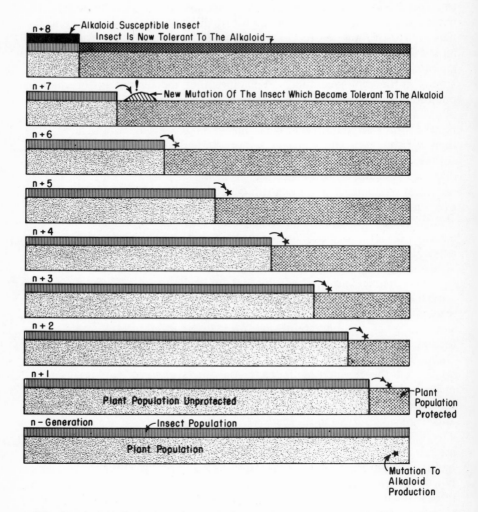

Figure 5.10. Changes in the population of plants under stress of predation by a specialized insect. Generation n—existing balanced situation; the plants are in the light field and the insects are at the top. A mutant which is poisonous to the insect is represented by the asterisk. Generation n + 1—the resistant form has already formed a part of the population (10%) and is resisting the insect. Generations n + 2 through n + 6—the mutant is increasing its share of population by its resistance to the insect and therefore has a higher number of progeny. Generation n + 7—a mutation of an insect capable of living on the previously poisonous mutant arrives on the scene. Generation n + 8—the mutated plant is preyed upon only by the new forms of the insect; thus the original plant is under stress from both types of insect.

Caterpillars feeding on poisonous milkweed *(Asclepias)* developed into poisonous monarch butterflies, but when raised on a nonpoisonous species of *Gonolobus,* they developed into nonpoisonous butterflies. The poisonous principle of milkweeds (not an alkaloid, but a steroid glucoside) turned out, in the course of adaptive evolution, to be a *disadvantage* to the plant because it no longer protected the plant from caterpillars, but it still protected the insect from predation by birds.

An interesting specialization has been found in an aphid genus, *Acyrthosiphon.* Two closely related species of *Acyrthosiphon* feed on a number of leguminous plants. *A. pisii* develops well when placed on alkaloid-free plants such as peas *(Pisum sativum)* or alfalfa *(Medicago sativa)*, but it does not develop and does not suck the juices from bitter plants (Wegorek and Jasienska-Obrebska, 1964; Krzymanska, 1967). Comparatively, the species *A. spartii* develops only on plants rich in alkaloids, such as *Sarothamnus scoparius,* and is actually attracted by sparteine (Wegorek and Czaplicki, 1966; Smith, 1966). The alkaloid-poor mutation in *Lupinus luteus* also became a host to the *A. pisii.* In Poland, where the sweet yellow lupine is one of the more important fodder plants, the invasion of the aphid became a serious problem not only because the aphid enfeebles the plants by sucking the juices, but also because it transfers a virus disease. The disease, known as lupine narrow leafness, decreases the seed production of infected plants, and the infection takes place early, that is, prior to the plants' blossoming. Thus, a mixed population of sweet-bitter lupines can, after a few generations, lose all the sweet forms. The infestation by the aphid and the following infection of the virus accelerate the elimination of the alkaloid-poor plants, which, even without the infections, are already inferior in seed production (Nowacki and Kazimierski, 1971; Kazimierski and Nowacki, 1971).

5.7. Alkaloids and Herbivorous Vertebrates

While most insects which feed on plants are monophagous or oligophagous, i.e., they feed only on one or a few species of plants, the vertebrates are usually polyphagous. There are, of course, some exceptions, such as the Australian koala, but the majority of herbivorous mammals feed on a broad spectrum of plants. For 65 million years, the only truly herbivorous vertebrates have been the mammals. The Mesozoic reptiles disappeared following the mesophytic flora. Birds, while they feed on seeds and berries, seldom do so on leaves, and they frequently use insects, in addition to plant parts, as a food source (Emlen, 1971). While a single plant can be a host for hundreds of insect larvae, hundreds of plants comprise a daily menu for a mammal. This difference is important in any discussion of the protective role of alkaloids. A poisonous compound in a plant will protect the plant against insects even if it has no smell or taste; it is sufficient that the compound kills the insect after it eats a single piece of the plant. Thus, the plant, while suffering a minor injury, escapes a serious hardship. In the case of mammals, the defense must be different. A single swipe of the tongue of a cow can eradicate a plant, and a single poisonous plant diluted by hundreds of nontoxic ones will cause little or no effect. Therefore, a principle which can protect a species must render it unpleasant, which means it is either bitter, pungent, bad smelling, or in some other way repellent. As Browers *et al.* (1968) have shown, the blue jay learned to avoid emetic monarch butterflies after ingesting a single specimen. Compared with mammals, birds have a less well-developed olfactory sense, inferior sense of taste, and a brain developed little above that of the reptiles. Since a bird can learn to avoid an untasty food after one unpleasant experience, the speed of learning in mammals should not be slower. In Przebedowo, Poland a simple experiment was performed. For a few years, an acreage of significant size was sown with sweet blue lupine. The European hare (*Lagopus lagopus*) preferred to feed on the lupine plants, and some isolated plots, especially between grasses or grains, were completely grazed if unfenced. An isolated plot of *Lupinus angustifolius* was planted in a field of ryegrass. During the germination, the plot was protected by a fence, but when the plants were about 8 in. high, the fence was removed. The plants used were a mixture of bitter and sweet forms, actually sibling lines from a single cross. The day after the screen was removed, a number of plants were bitten off by hares, and while the sweet were eaten, the bitter were left. All cut but uneaten plants were analyzed, and all without exception were bitter. At first glance, it seemed of no advantage to the plants to have alkaloids, because for the cutoff harvested plant it did not matter what happened to it. Farmers from different areas where predominantly bitter lupines were planted (for green manure), however, reported that they had not observed hares biting the plants and leaving them. No hares at all were feeding on the bitter lupine fields. The further development of the experiment disclosed the reasons. A few hares were attending the plot and biting the plants at random, devouring the sweet ones and leaving the bitter; after a few days, they were disappointed by the "surprises" and switched to other plants. Only one animal was seen feeding later, but this one practically never committed a mistake during the following 6 weeks, for all the sweet plants were completely grazed, while the bitter ones grew undisturbed. On alpine pastures in Karkonosze, all plants were heavily grazed by sheep, while only poisonous *Aconitum* and the naturalized Washington lupine, *L. polyphyllus,* were intact. Similarly, along the Odra river no pastures were observed with all plants grazed, and again the lupine and some *Ranunculus* spp. were intact. There were signs that the predators were actually searching for edible plants among the lupine swarms.

In North America in the western ranges the situation is similar. Keeler (1969, 1975) describes the hazards of alkaloid toxicoses (see p. 110). He estimates that perhaps tens of thousands of livestock deaths occur yearly in the western United States from toxic plant ingestion and that five or six times that many cases of moderate toxicoses occur.

It is noteworthy that domesticated animals much more frequently commit mistakes than wild ones. In unpublished experiments of ours, white laboratory mice and freshly captured meadow voles (*Microtus arvalis*) were given a choice of cakes, some of which had added gramine. The white mice preferred the alkaloid-free cakes but from time to time also ate these with the alkaloid. The meadow voles never touched the poisoned food.

The origin of the most common herbivorous mammals, the *Rodentia, Lagomorpha,* and *Artiodactylia,* is in the Eocene era. Thus, these groups of animals came into existence circa 40 million years ago, while the plants had been in existence for at least 100 million years. The older orders of herbivores are either extinct or very restricted in number of species, e.g., the *Perissodactyles.* Comparing the susceptibility of *Perissodactylia* with more modern plant eaters reveals a remarkable difference. A horse (*Perissodactylia:Equus caballus*) is much more susceptible to alkaloids than a sheep, goat, or rabbit (sheep and goat *Artiodactylia*) (rabbit *Lagomorpha*). Horses (*Equus*), donkeys (*Asinus*), and zebras (*Hippotigris*) are animals adapted for life in a dry steppe, and it is known from analysis of plants of the steppe that the number of species producing unpalatable compounds is quite high. The horse-like animals seem to have played a part in decimating the palatable species, thus allowing the dispersal of unpalatable forms. In turn, as the steppe became less suitable for the horselike animals, the more resistant *Artiodactylia,* especially the antelopes, replaced the Equuideae (Figure 5.11).

The introduction of some new species of grasses into North America and Australia caused losses in animals. The grass species involved were *Phalaris tuberosa* and *P. arundinacea,* both accumulating indole derivatives. *P. tuberosa* contains predominantly *N,N*-dimethyltryptamine, 5-methoxy-*N,N*-dimethyltryptamine, and bufotenine, while *P. arundinacea* contains mainly gramine and hordenine. The tall fescue, *Festuca arundinacea,* introduced mostly into Missouri, the Dakotas, and Minnesota has other types of alkaloids, loline and perloline. Although the level of these

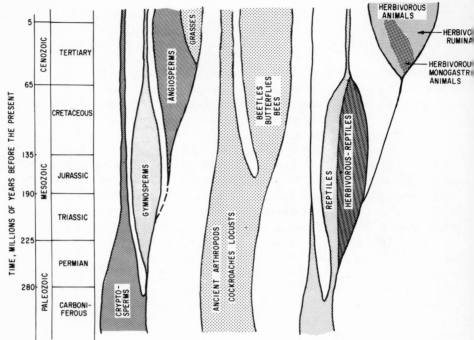

Figure 5.11. Geological time scale showing the approximate evolution of higher plants and animals.

alkaloids is too low to cause acute poisoning in herbivorous animals (Nowacki *et al.*, 1975), it is sufficient to affect the intake of dry matter by animals (Marten *et al.*, 1973; Simons and Marten, 1971; Arnold and Hill, 1971; Gentry *et al.*, 1969) (Figure 5.12a). The toxic effects are caused most probably by infection of the plants by some parasitic fungi (Farnell *et al.*, 1975; Boling *et al.*, 1973; Robbins *et al.*, 1973; Grove *et al.*, 1973). Although the level of the fungal alkaloids is usually very low, it can rise to toxic concentrations depending on the environmental factors and particular genotype used. Therefore, after analysis of the toxic effects of tall fescue and reed canarygrass, and the palatability of these plants, it is possible to demonstrate that the alkaloids causing decreased palatability of the plant are protective devices, but that the toxic effects are caused otherwise (Hagman *et al.*, 1975; Hovin and Marten, 1975).

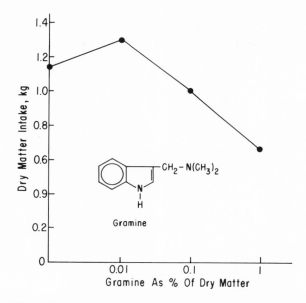

Figure 5.12a. Effect of gramine on the dry matter intake of *Phalaris tuberosa* by sheep (Arnold and Hill, 1971). Courtesy of Academic Press.

Simons and Marten (1971) evaluated the alkaloid content of 411 diverse genotypes of reed canarygrass (*Phalaris arundinacea*) for palatability to sheep.[1] Figure 5.12b shows the linear relationship between alkaloid concentration and palatability rating when 18 classes of experiments A and B were grazed together in September. The correlation coefficient between the ratings and alkaloid concentrations was +0.95. The point that stands out is the strong influence of genotype on alkaloid concentration and palatability in reed canarygrass. Palatability is affected by many plant, animal, and environmental factors; thus it was most interesting and useful that Kendall and Sherwood (1975) successfully developed the use of meadow voles (*Microtus pennsylvanicus*) in a bioassay for estimating palatability of reed canarygrass and tall fescue to large animals. Scientists in all parts of the world will recognize the use of a small laboratory animal for palatability studies as being an approach of inestimable value.

[1]The alkaloids were gramine, 5-methoxy-*N*,*N*-dimethyltryptamine and *N*,*N*-dimethyltryptamine in all genotypes. Two groups of nine genotypes were clonally propagated, and they ranged in alkaloid content from 0.18% to 1.21% of the dry matter.

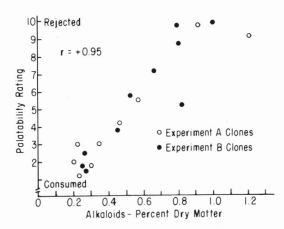

Figure 5.12b. Relationship between palatability rating and alkaloid concentration of 18 reed canary grass clones grazed as one experiment in September, 1970 (Simons and Marten, 1971). Courtesy of *Agronomy Journal.*

The investigation of enzymes in the livers of a number of animals tolerating plants with high alkaloid content revealed that, unlike the susceptible animals, the tolerant ones are provided with enzymes capable of detoxifying a number of alkaloids (Lang and Keuer, 1957; Nowacki and Wezyk, 1960; Tsukamoto *et al.,* 1964). The impact of the ingested alkaloid is usually on the central nervous system. Very often, a certain level of the alkaloid must be accumulated in the blood to produce poisoning; thus any means of slowing down the absorption of the alkaloid from the intestines helps to preserve the animal (Figure 5.13). Furthermore, alkaloids must resemble normal hormones like acetylcholine to interfere with the active centers, and to be eliminated they must be converted into forms readily excretable by the kidneys. Hydroxylation reactions, N-methylation, and conversion into glucuronates are the most common routes used. Only a small number of alkaloids can be converted into compounds that can be degraded to CO_2, urea, and water. It is hard to avoid the teleological notion that enzymes which render alkaloids harmless are produced because they may be needed, but it may be that those that convert alkaloids are actually broad-spectrum enzymes. These usually perform a different function, but upon encountering an alkaloid they can transform it. In most cases, the product of an enzymatic reaction is less toxic; thus strychinine is converted into 2-hydroxystrychnine, which has only 1% the toxicity of strychinine (Tsukamoto *et al.,* 1964). Occasionally, however, a product of the conver-

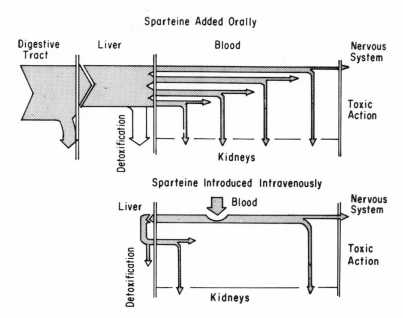

Figure 5.13. Explanation of different toxicity of an alkaloid introduced orally and intrave-
nously. The alkaloid sparteine, when injected, is 20 times as toxic as when given orally. The
alkaloid is absorbed slowly from the intestine and is converted in the liver to a detoxification
product and then excreted by the kidney, which prevents accumulation in the blood over a
dangerous level. When injected the barriers are avoided and the rapid action on the nervous
system does not allow the kidney enough time to excrete the compound (Nowacki and Wezyk,
1960). Courtesy of the journal.

sion of an alkaloid can be more toxic, e.g., the conversion of codeine to
morphine (a more toxic compound in man than in the rat) (see Figure 6.13)
or the conversion of the senecio alkaloids to esters of crotanecine
(Figure 5.14) (Mattocks, 1968a, b). Animal species exposed for many
generations to an alkaloid usually develop a system of increased efficiency
for eliminating the alkaloid from the body. The animal most thoroughly
investigated for possessing enzymes capable of deactivating alkaloids is the
common rabbit (*Oryctolagus cunniculus*). The original area of distribution
of this species was the Iberian peninsula and some other countries in the
western Mediterranean. In this semiarid area, the percentage of alkaloid-
rich plants is high, so natural selection has favored the forms which survive
after accidental ingestion of alkaloid-rich plants. One of the animals most
susceptible to alkaloid poisoning is the cat, a compulsory carnivore which

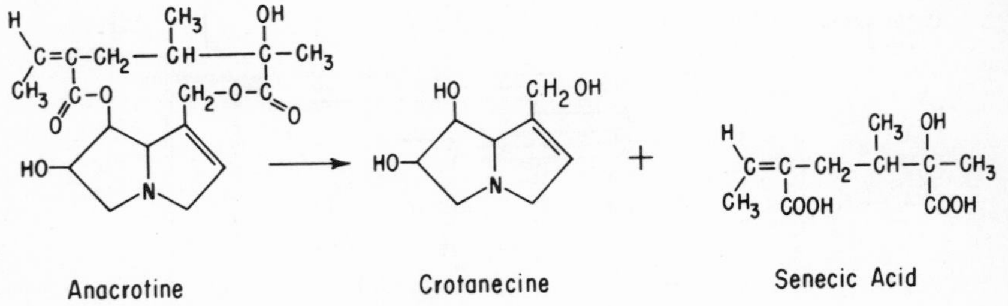

Figure 5.14. Aqueous alkaline hydrolysis of the alkaloid anacrotine to give the alkaloid crotanecine and senecic acid (Mattocks, 1968a).

would seldom come into close contact with alkaloids in nature. Since there has been no selective pressure, no resistance has been developed.

In summary, the insects probably have had a great influence on the establishment of the alkaloid character in some plant families, and the herbivorous mammals are responsible for preserving this character; furthermore, even if the alkaloid, due to simultaneous evolution of insects, can no longer defend the plant from *specialized* vermin, it is still preventing predation by unadapted herbivores.

5.8. Other Roles of Alkaloids

The colored alkaloids of Centrospermae are pigments, replacing the anthocyanins in the pigmentation of flowers in this group of plants (Mabry and Dreidings, 1968; Mabry *et al.*, 1972; Mabry and Difeo (1973); Piattelli and Minale, 1964; Reznik, 1955). How this character developed is hard to judge; a plausible hypothesis would be that as this group of plants lost the ability to synthesize anthocyanins, selective pressure favored plants with colored flowers as more attractive to pollinating insects, and thus a new pigment arose. It could be the reverse, because the Centrospermae developed a system capable of synthesizing betacyanins and the anthocyanin character was lost when it was no longer indispensable.

5.9. Conclusion

A favorite topic for discussion is the metabolic status of alkaloids. The most common notion is that alkaloids are waste products which play an unimportant role as plant-protective compounds and accumulate in the plant because the plant lacks an effective excretory organ comparable to animal kidneys. However, plants do have a surprisingly effective system—the aging leaf; and it is curious that this system is not widely recognized (Nowacki and Waller, 1972). Each plant, without exception, produces new leaves and drops old ones while growing. Before the leaf is ready to drop, all metabolites which can be useful are translocated and only true waste remains. If alkaloids were waste, they would be concentrated in the dead leaves or be converted into another form (see Chapter 6), as happens with introduced foreign alkaloids and unusual biosynthetic compounds. It seems more convincing that, since the alkaloid character is well established in some plants, it is necessary for those species to produce and save some alkaloids in, for example, the seeds, as well as traces in other parts. An inefficient system would long since have been eradicated by selection. The question of the metabolic status of alkaloids arises because only a few of the plants accumulate alkaloids. There is no adequate answer to this question. The role of singular alkaloids may be different in different plants, and in alkaloid-free plants it may be fulfilled by other compounds, such as unusual amino acids, other metabolites, etc., but the protective function, both present and past, may be the *raison d'être* for most alkaloids.

6 | Metabolic (Catabolic) Modifications of Alkaloids by Plants*

6.1. Introduction

The study of alkaloid biosynthesis has been a major effort of natural products chemists, biochemists, and pharmaceutical chemists during the past several decades. Alkaloid catabolism studies have been a minor effort of a small group of scientists, some of whom are oriented primarily toward botany or plant physiology. As Robinson (1974) states, "Alkaloids appear to be active metabolites, but their usefulness to plants remains obscure." It is incorrect to conclude that the problem of metabolism has been solved merely because the routes of biosynthesis have been described; the other half of the problem is what happens to the alkaloid after it is synthesized. Catabolic reactions of some alkaloids have been studied using microorganisms (especially soil microorganisms), and biochemical pharmacology has yielded data on modifications of alkaloids which may be interpreted as detoxification processes in animals. Altschul (1970, 1973) recently completed a study of 2.5 million specimens of flowering[1] plants that have been described as restorative of health, according to medicinal folklore. Altschul (1970) and Schultes (1975a,b) indicate that many species offer, somewhat vaguely, hope for relief of cancer, heart disease, mental illness, and diverse

[1]They might have maximum alkaloid activity, but flowering is not always correlated with an increase in alkaloid production.

---

*This chapter was written by George R. Waller, L. Skursky of the University of J. E. Purkijne, Brno, Czechoslovakia, and T. M. Heinze of the Medical Research Foundation, Oklahoma City, Oklahoma.

metabolic disorders, but these authors have not yet explored the plants influencing the function and diseases of the reproductive tract of the human female.

Interest in alkaloid catabolism in plants that produce them has increased during the last decade, especially when isotopically labeled alkaloidal compounds became available; however, the difficulty in synthesizing alkaloids of high specific radioactivity remains a major drawback to research in this area. Perhaps the widely accepted view that alkaloids are the final waste products of nitrogen metabolism has hindered larger efforts in this direction. The older literature about degradation of alkaloids, probably most completely reviewed by Neumann and Tschoepe (1966), contains scattered data on the metabolism of secondary substances such as the alkaloids, but it is difficult to find any unifying concept. The earliest evidence indicating that decomposition of an alkaloid could occur in the living plant was obtained in 1892 by Clautriau in studies of the alkaloids found in the poppy. In spite of an increasing number of papers on this subject, one of the reviews about alkaloid biogenesis (Ramstad and Agurell, 1964) concludes "and we shall still have to wonder about alkaloid catabolism."

Catabolic or degradative reactions could be restrictive and exclude important transformations of alkaloids which have been detected recently and which might help us obtain a greater understanding of the role of alkaloids in plants, i.e., the interconversion of "free" and "bound" forms (Fairbairn and Ali, 1968a,b). The various biological modifications of alkaloids in organisms can be classified according to the types of chemical reactions the alkaloids undergo, whether the reactions occur in bacteria, fungi, animals, fowl, humans, or in plant tissues. This chapter is restricted to reactions that occur in alkaloid-producing plants (James, 1953), since these metabolic changes may shed some light on alkaloid function. The catabolism (or metabolism) of alkaloids by microorganisms, animals, fowl, and humans will be covered in a later volume.

Hydroxylation and epoxidation reactions, although they can represent catabolic modifications of alkaloids, have not been included here for two reasons. One is that these processes are traditionally included in the biosynthetic pathways of alkaloids that contain the corresponding groups; the other is that only in some alkaloids, such as scopolamine and 6-hydroxyhyoscyamine, are such groups probably introduced by modification of a preformed alkaloid skeleton. In other alkaloids, the hydroxyl groups are already present in the precursors, e.g., phenolic groups in materials from which the isoquinoline esters (Jindra et al., 1964) have not been included. Since this review is mainly about the catabolism of the heterocyclic rings, these have been left out.

Dawson and Osdene's (1972) report gives a speculative view of tobacco alkaloid metabolism and deals primarily with alkaloid biosynthesis. Sadykov, Aslanov, and Kushmuradov (1975) provide a limited view of alkaloid catabolism but cover the biosynthesis of the quinolizidines adequately.

Some known facts and perhaps some future findings of alkaloid catabolism might help explain why some plant species are alkaloid-free, some have a low alkaloid level, and some have a high alkaloid level. Most scientists depend upon modern instrumental-computer methods to detect and determine alkaloids and their metabolic products. This field needs more scientists with training in biochemistry, natural products chemistry, agronomy, botany, plant physiology, and pharmaceutical chemistry. Only after much research may it be possible for us to understand metabolic pathways and their regulation. Low levels of alkaloids might be caused by the rapidity of catabolic reactions. In fact, the term "alkaloid-free plant" may even become questionable; the plant may only degrade an alkaloid as fast as it can synthesize it.

6.2. Methods Used to Study Secondary Metabolism of Alkaloids

Isotopic labeling of substrates is the single most fruitful technique, and it is now used almost exclusively. Nevertheless, in the older literature there are some pertinent studies which, using suitable sampling procedure, followed the changes in alkaloids with time and detected the occurrence of catabolism or reversible metabolic translocation processes (see, for example, Weevers, 1932).

Feeding an experimental plant or an excised part of it with an alkaloid is not without objection, since any compound introduced into the organism may behave differently, due to membrane barriers and the compartmentalization of cells and tissues, from the same compound formed inside. So results gained by feeding may reflect not a completely natural process, but rather some sort of "aberrant" one (Leete, 1969). Generally the concentrations of alkaloids in plant tissue are low. Therefore, any overloading of such tissue by administered compound(s) would be unnatural. In feeding labeled experimental compounds, one should choose those with high specific radioactivity so (a) dilution by the endogenous compound does not invalidate the methods used, and (b) the plant tissue is not overloaded. To minimize dilution of a labeled compound by the endogenous one, several authors (Kisaki and Tamaki, 1964b; Stepka and Dewey, 1961, 1963; Mothes, 1972) have used leaves or shoots of *Nicotiana tabacum* and *Nicotiana glutinosa* plants that had been grafted onto some compatible alkaloid-free stock (see Chapter 4, p. 000). This technique utilized the fact that in several cases alkaloids are produced mainly in the root and transported into the shoot.

The mass spectrometer has been used in determining the structure of alkaloids and their metabolites (Sastry, 1972; Burlingame *et al.*, 1974). Rapidly developing quantitative analytical techniques use the mass spectrometer, and mass fragmentography, where a given fragment ion is monitored by the mass spectrometer for a specific, sensitive, and quantitative detector in gas–liquid chromatography. This has been used for quantitative analysis for drugs including alkaloids at very low concentration in animal tissues or fluids (Ebbighouser *et al.*, 1975; Lee and Millard, 1975; Aldercreutz *et al.*, 1974). It is expected that application of this technique employing a multiple-peak monitoring system coupled with a computer system will contribute much, in coming years, to the study of alkaloid metabolism in plants.

Recently Scott (1974, 1976) described several systems where incorporation of 5% or more might be achieved (usually in fungi or bacteria) through using [^{13}C] as the biosynthetic label; it offers unique advantages and can be combined successfully with the results of others such as radioactive labeling experiments. When the incorporation efficiencies are high, as for ricinine biosynthesis (Johnson and Waller, 1974), the opportunity to study [^{13}C] enrichment becomes really exciting; thus we will experience a rapid expansion, since the method removes the tedious carbon-by-carbon degradation. To quote Scott, "For the next few years, however, the prognosis would seem to favor parallel studies of [^{13}C] and [^{2}H], and of [^{14}C/^{3}H]ratio techniques since the last-mentioned method provides more information concerning the stereoselectivity of labeling processes on the microgram scale."

6.3. Evidence for Metabolism or Catabolism

It is clear that alkaloids are not inert end products but rather active participants in the dynamic metabolic reactions. They fluctuate in total concentration and in rate of turnover. There is an optimum time for harvest of plants containing alkaloids to be used for drugs; this phenomenon has been recognized by "primitive" as well as modern societies. The optimum time frequently occurs while the plant is in its prime growth stage and rarely when senescence begins. This is a clear indication that alkaloid content must increase and then decline with respect to physiologic state of plants. Scientific references about fluctuations in caffeine in the vegetative part of the tea plant established that the caffeine contents were lowest (average 0.49%) in February and highest (average of 1.43%) in August (Mudzhiri, 1955).

James (1950) cites a number of examples of seeds containing no alkaloids *(Nicotiana tabacum),* scant alkaloids *(Atropa belladona),* and abundant alkaloids *(Strychnos nox-vomica).* Thus it becomes very important to have a method capable of distinguishing between the static and dynamic constituents of a plant such as *Vinca rosea,* which at maturity contains about 100 alkaloids, whereas the seeds of *V. rosea* are essentially devoid of material (Scott, 1970). Figure 6.1 (Scott *et al.,* 1973; Scott, 1974) shows the incorporation of DL-tryptophan by the seedlings. From the figure

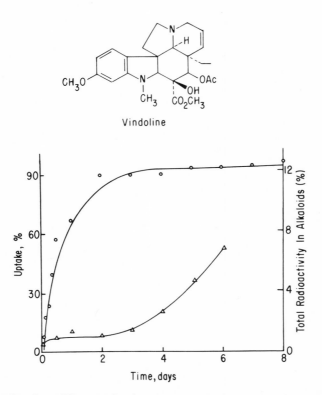

Vindoline

Figure 6.1. Uptake of DL-[^{14}C]tryptophan by *Vinca rosea* seedlings and the radioactivity in vindoline after administration of DL-[^{14}C]tryptophan during 6 days. (Scott, *et al.,* 1972, 1973; Scott, 1974). Courtesy of the authors, Academic Press and the American Association for the Advancement of Science; copyright 1974.

one can see that immediately upon imbibition of the liquid surrounding them, the seeds and the DL-tryptophan absorbed by the seeds undergo a change in dynamic state for the metabolism of tryptophan, which gives rise to the unknown alkaloids. Thus the seeds were converted from a static state to a dynamic state in terms of alkaloid metabolism. If the feeding is allowed to continue for 8 days, vindoline (see Figure 6.1) (a major alkaloid of *V. rosea*) is labeled to the extent of 7% of all the radioactivity. Figure 6.2 shows the behavior of geissoschizine (A), which shows a rapid rise to approximately 4% of the total radioactivity in the alkaloids over the first

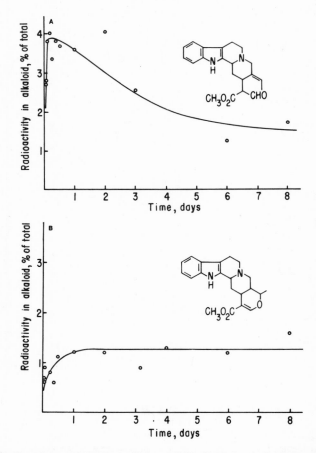

Figure 6.2. Radioactivity of (A) geissoschizine and (B) ajmalicine during 8 days after administration of DL-[^{14}C]tryptophan (Scott *et al.,* 1973; Scott, 1974; courtesy of Academic Press).

60–90 min, followed by a gradual decline during the 8-day experiment. This result may be contrasted with that of ajmalicine (B), which is isomeric with geissoschizine, for which the fraction of total radioactivity shows a rather slow climb to a plateau which is maintained for the 8-day incubation period. Mothes *et al.* (1965) and Coscia (1976) report another example, the periwinkle *(Catharanthus roseus),* from which 60–200 alkaloids have been isolated. Again, no detectable amounts of alkaloids are in the seed. They appear throughout germination, and by 3 weeks they are present throughout the plant. Then the alkaloids disappear almost completely and finally reappear in about 8 weeks.

The position of a leaf on the plant may affect its alkaloid content, independently of the age of the leaf. Cromwell (1933) observed that young bayberry leaves contained no berberine until the bush carrying them had reached an advanced age. Therefore, a young leaf on an old parent plant is not an accurate index of alkaloid content for a leaf borne when the parent plant was young.

The leaves of *Atropa belladonna* show a steady decline from the beginning of the experiment in alkaloid content (atropine mostly) by James (1950) (Figure 6.3). The total alkaloids per leaf increases to a maximum point of accumulation (approximately May 20), which corresponds to the fully expanded leaf, and then drops to an insignificant value before senescence occurs (June 17). Yellowed leaves show a still lower content of alkaloids.

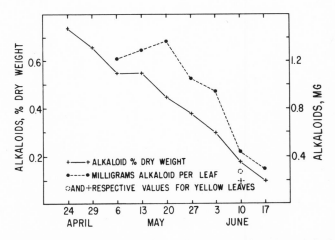

Figure 6.3. Total alkaloids in the basal leaves of belladonna throughout their development (James, 1950). Courtesy of Academic Press.

6.4. Diurnal and Developmental Variations

Another way to show that plants are actively metabolizing alkaloids is by determining the fluctuations that occur during a single day. In essence, this is catabolism of alkaloids with metabolites unknown.

In the late 1950s and early 1960s several papers appeared which documented the decrease of alkaloid content during plant development. Sometimes periodic variations (increases or decreases) were also observed. Modern separation and assay methods made these experiments possible.

Some of the papers to be mentioned presumed that alkaloid translocation partially explains the content changes, but usually catabolic changes have been assumed. In certain examples the relationship between biotransformation and translocation has been suggested. Before Fairbairn's studies (see later in this chapter), bound forms of alkaloids had not been shown present, although Robinson (1971) and Waller and Lee (1969) had obtained some preliminary evidence for that of ricinine. Fluctuations have been observed by Fairbairn and Wassel (1967) for plant concentrations of atropine (1967), the hemlock alkaloids (1964) (Figure 6.4a and 6.4b), and opium alkaloids (Figure 6.5). The hemlock coniine and γ-coniceine vary in concentration in the hemlock fruit in ways that are complementary to each other (diurnal variation). Since the coniine content is much higher than that of γ-coniceine, all the coniine represented by the maxima in both 1958 and

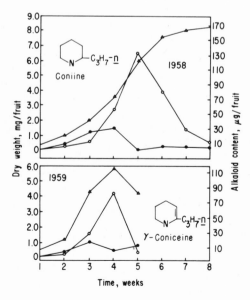

Figure 6.4a. Alkaloidal content and dry weight for weekly samples of whole *Conium maculatum* plants. Dry weight—Δ; coniine—○; γ-coniceine— • (Fairbairn and Suwal, 1962). Courtesy of the authors and Pergamon Press, Ltd., copyright 1962.

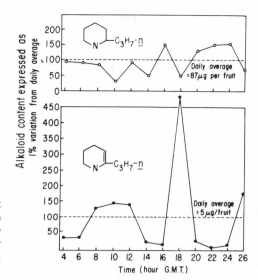

Figure 6.4b. Diurnal changes in content of alkaloids of *Conium maculatum* in week 4. Coniine—○; γ-coniceine—● (Fairbairn and Suwal, 1962). Courtesy of the authors and Pergamon Press, Ltd., copyright 1962.

1959 cannot be produced from the γ-coniceine that disappears concurrently, unless it appears that additional γ-coniceine is synthesized at a very rapid rate and then converted into coniine (see Section 6.9.1).

Morphine, which is mostly found in the roots of young poppy plants, gradually increases in concentration in leaves. At the stage of fruit formation, morphine disappears from leaves and accumulates in high levels in the fruit capsule (Sárkány, 1962; Sárkány and Dános, 1958; Fairbairn, 1965; Fairbairn and Wassel, 1964, 1965). Diurnal variation of alkaloid concentration in the plant (Figure 6.5) shows a suggestion of complementarity between thebaine and codeine, but both drop just before the morning rise in morphine. This is consistent with the known biosynthetic pathway: thebaine → codeine → morphine. Morphine is always predominant (like coniine in the hemlock alkaloids), and more morphine is made than can be accounted for by simple conversion of thebaine and codeine, both of which disappeared. A reasonable hypothesis is rapid biosynthesis of thebaine and conversion to codeine and codeine to morphine followed by rapid conversion of morphine to nonradioactive substances. Such a process would result in the observed daily concentrations of morphine.

Figure 6.6a shows the diurnal variation in atropine content of *Atropa belladonna* (Fairbairn and Wassel, 1967) and there are two- to threefold changes in both the large leaf and the fruit.

On the other hand, the role of perloline in *Festuca arundinacea* (tall fescue) is quite different, because it shows almost a fourfold increase during

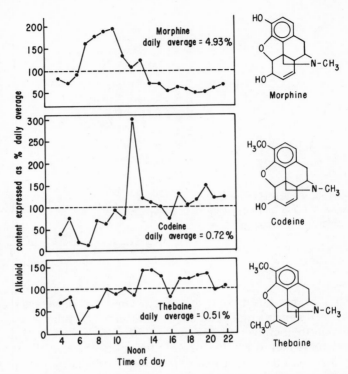

Figure 6.5. Diurnal changes in content of alkaloids of whole *Papaver somniferum* plants. (Fairbairn and Wassel, 1964). Courtesy of the authors and Pergamon Press, Ltd., copyright 1964.

July and August and a decrease to normal when it is stabilized (Figure 6.6b) (Gentry, 1968).

Birecka and Zebrowska (1960) studied the diurnal variations of lupine alkaloids in two species of *Lupinus*. In *Lupinus albus* (white lupine), the alkaloid content increased in the day and decreased at night. The largest changes were in lupanine and another alkaloid (see formula on page 220), thought to be hydroxylupanine. In *Lupinus luteus* (yellow lupine), the changes of total alkaloids were different, with sparteine increasing during the night. Remarkable variations have been observed with steroid alkaloids. Willuhn (1966) studied the disappearance of steroid glycoalkaloids by assaying for their aglycone tomatidenol and solanidine (Figure 6.7) during the maturing process of *Solanum dulcamara* fruits. Mature fruits do not contain more than 1–2% of the maximal value, which occurs at the early stages of fruit development.

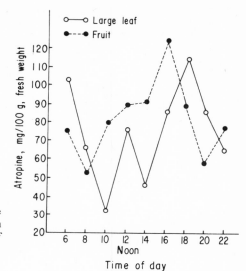

Figure 6.6a. Diurnal changes in atropine content of *Atropa belladonna* (Fairbairn and Wassel, 1967). Courtesy of *Journal of Chemistry of the United Arab Republic,* Cairo.

α-Tomatine (Figure 6.7), another steroidal glycoalkaloid produced by the tomato plant, has also been found to disappear from the maturing fruits (Sander, 1956). Some suggestions were made that the aglycone tomatidine might be metabolized into carotenoids and that tomatidine was converted into allopregnenolone in the ripe tomato (see page 194).

Change of an alkaloid spectrum in a plant organ during its development often indicates at least one phenomenon: *de novo* synthesis, translocation, and degradation (e.g., tabersonine and the pair dehydroaspidospermidine-vincadifformine on page 217).

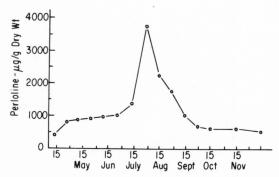

Figure 6.6b. Seasonal distribution of perloline in the tall fescue (Gentry, 1968). Reproduced from Crop Science Society of America Special Publication No. 4, "Tall Fescue Toxicity," 1972, p. 106, by permission of the Crop Science Society of America.

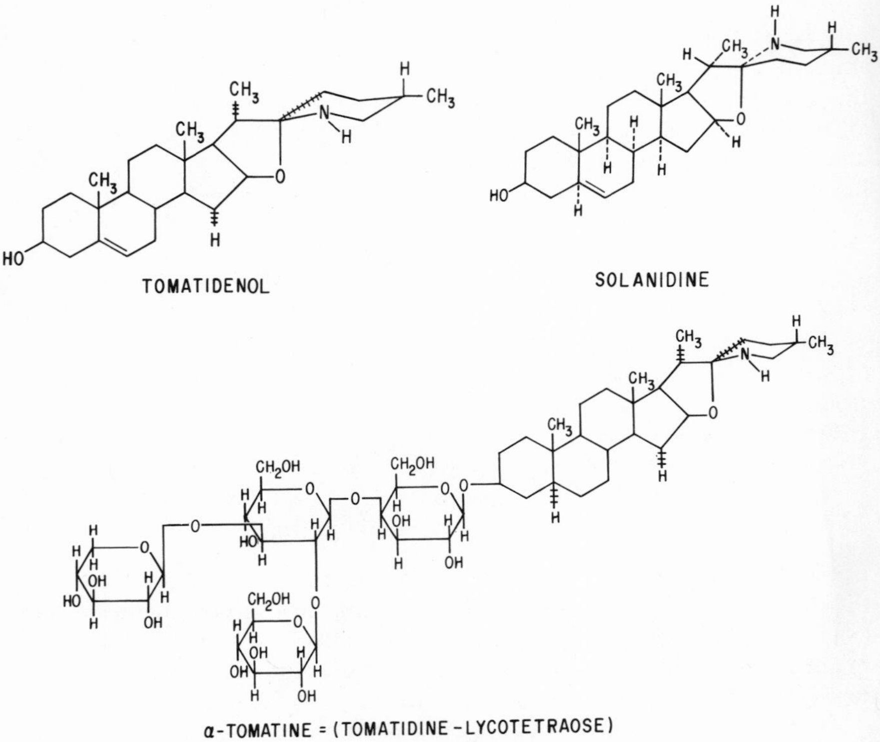

Figure 6.7. Structures of the steroid alkaloids, tomatidine and solanidine, and the steroid alkaloid glycoside, α-tomatine.

A very interesting fact is that caffeine is found in all parts of the coffee plant, from roots to fruits (Herndlhofer, 1933), although the highest contents are found in the leaves and fruits, where they vary depending upon the season of the year. In fact the coffee plant of 30 to 60 years produces the same amount of caffeine in its leaves or fruit that it was capable of producing in its younger stages (5 years to 15–20 years, which is the life expectancy of the average coffee plant). The coffee bean requires about 7 to 10 months to develop from flower into ripened fruit and the amount of caffeine varies during that time (Beaudin-Dufour and Muller, 1971; Keller *et al.*, 1972).

Alkaloid-bearing plants, and indeed all plants used by "primitive" societies for maintaining and restoring health, were generally collected at a specific time of day or night. Robinson (1974) tells us that according to

Theophrastos the herb gatherers of his time (about the 4th century B.C.) prescribed that "some roots should be gathered at night, others by day, and some before the sun strike on them." Comparing Figures 6.4, 6.5, and 6.6, one can see that yields would vary widely with specific alkaloids. Alkaloid catabolism (or metabolism) might provide insight into the *materia media* and nutritional patterns of peoples who are now extinct or whose cultures have been absorbed by Western civilization.

6.5. Turnover Rates

The intensity of alkaloid metabolism can be measured through the use of isotopically labeled molecules. With this technique, the turnover rate of a compound can be determined even if the total amount remains constant. Disregarding fluctuations, Robinson (1974) assumed first-order reactions and complete mixing of pools to estimate half-times for the disappearance of several alkaloids. The results are shown in Table 6.1.

Robinson (1974) calculated that about 35 mmol of net CO_2 is fixed per plant per day in the 60-day-old tobacco plant, which contains 250 mg of

Table 6.1. Turnover Rates of Some Alkaloids[a]

Alkaloid	Half-life	Reference
Tomatine (in tomato fruit)	6 days	Sander (1956)
Hordenine (in several species of barley	42 hr	Frank and Marion (1956)
Morphine (in several species of poppies)	7.5 hr	Fairbairn and Paterson (1966)
Nicotine	22 hr	Leete and Bell (1959) Ill'in and Lovkova (1966)
Ricinine	4 hr[b] or 6.7 days[c]	Waller *et al.* (1965, 1969)
γ-Coniceine-coniine-*N*-methylconiine[e]	1–4 days[d]	Dietrich and Martin (1969)

[a]Patterned after Robinson (1974).
[b]By extrapolation of a very rapid decline.
[c]By a similar extrapolation but over a longer period.
[d]These two values suggest two nonequilibrating pools.
[e]The *total alkaloid concentration* bases of both varieties of *Conium maculatum* since they are a reflection of both the rate of synthesis as well as that of turnover. The large pool of γ -coniceine in var. California compared with var. Minnesota would appear to be due to a combination of a faster rate of synthesis (35 mμmol/hr/g) and a slower turnover (4 days) compared with the slower rate of synthesis (4.5 mμmol/hr/g) and a faster turnover (1 day) for var. Minnesota. In either case the methylation of coniine does not appear to be limiting since it did not occur at a measurable rate in var. Minnesota which had total activity and pool ratios consistent with a rapid conversion of γ-coniceine into coniine.

nicotine. From the turnover data, 92 mg must be catabolized and biosynthesized in a 10-hr day; this is equivalent to 0.6 mmol. Since there are 10 carbon atoms in nicotine, 6 mmol of fixed carbon dioxide is required to replace that amount of degraded nicotine. If 30% of the gross CO_2 is lost by photorespiration (a reasonable estimate for tobacco), about 50 mmol of nicotine is metabolized. This is a startling result, particularly for anyone who believes that alkaloid metabolism (or more correctly catabolism) is a trivial part of a plant's total metabolic activity. The results are summarized in Figure 6.8a. According to Robinson, the foregoing considerations are actually ambiguous, since the 0.6 mmol of nicotine merely replaces another 0.6 mmol that disappears. The disappearing nicotine may contribute to the expired CO_2 and to the biosynthesis of other products by the plant. The major point is that the rate of nicotine turnover is quite appreciable in comparison to the main business of the plant—fixing carbon dioxide. Daddona and Hutchinson (1976) recently observed a dramatic difference in the turnover rates of vindoline and catharanthine between intact *Catharanthus roseus (Vinca rosea)* plants and their apical cuttings, which shows that the indole-dihydroindole alkaloids can be actively utilized in the plants' biochemical processes (Table 6.2). The catabolism of vindoline and catharanthine occurs more rapidly in apical cuttings than in intact plants as shown by feeding $^{14}CO_2$ under single pulse and steady state conditions and by examining the variation of the percentage of the total incorporation versus time of radioactivity into vindoline and catharanthine. It is possible to make

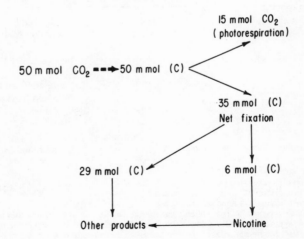

Figure 6.8a. Quantitative aspects of carbon dioxide and nicotine metabolism in tobacco plants: carbon dioxide assimilation by one tobacco plant per day. Courtesy of Robinson, 1974; copyright 1974 by the American Association for the Advancement of Science.

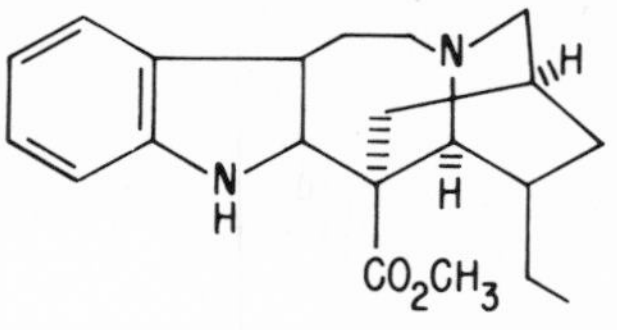

Figure 6.8b. Structure of catharanthine. Catharanthine

some tentative conclusions about the biosynthetic and catabolic relation-ships between plants and their apical cuttings. It appears that catharanthine (Figure 6.8b) is biosynthesized and catabolized faster than vindoline in the intact plants [see Scott (1974) for agreement]. However, the reverse rela-tionship between catharanthine and vindoline appears to be present in the apical cuttings. These observations with *C. roseus* appear to be unique in the literature dealing with plant alkaloid biochemistry. Further research in this area of alkaloid metabolism offers great promise.

Table 6.2. Variation in the Percentage Total Incorporation of Radioactivity from $^{14}CO_2$ into Vindoline (1) and Catharanthine (2)[a]

	Metabolism period, days	Total incorporation[b]	
		1	2
Apical cuttings	0.25[c]	0.0013	0.0008
(3-hr pulse	0.5	0.0031	0.0024
feeding)	1	0.0095	0.0084
	2	0.004	0.003
Intact plants	1	0.0016	0.010
(4-hr pulse	5	0.023	0.065
feeding)	11	0.046	0.028
	20	0.057	0.024
(7-hr steady	0.12	0.004	0.007
state feeding)	0.25	0.010	0.010
	0.30	0.018	0.029
	1	0.035	0.043
	30	0.20	0.17
	60	0.13	0.09

[a]From Daddona and Hutchinson, 1976. Courtesy of the authors and Pergamon Press, Ltd., copyright 1976.
[b]Total dpm of [^{14}C] in 1 or 2 divided by the total dpm of $^{14}CO_2$ taken up by the plant material times 100.
[c]For periods longer than the $^{14}CO_2$ feeding period, the plants were grown in $^{12}CO_2$ in a growth chamber.

6.6. Demethylation

6.6.1 Relation between Methylation and Demethylation and Their Biological Significances

Cyclization reactions form the basic skeletons of alkaloids. The introduction of methyl groups constitutes one of the most important reaction types in alkaloid biosynthesis. The opposite reaction, demethylation, is the most frequently described degradative step in alkaloid catabolism. In several cases, both processes appear to be formally reversible in the sense that both methylated and demethylated alkaloids exist in the same plant. However, the mechanisms for introduction and removal of methyl groups appear to be of different nature and are generally not reversible.

The appropriate time in the biosynthesis for the methylation reaction to occur and, of course, the correct position of the methyl group in the molecule, have important influences on modifying the skeleton. This is particularly true for the isoquinoline skeleton, especially in alkaloids of the morphine and aporphine groups (see p. 204 for structures). Methylation is considered to be a type of detoxication reaction. Thus Pöhm (1966) has observed that in seedlings of *Cytisus laburnum* with removed cotyledons, cultivated in appropriate feeding solution, addition of small amounts of cytisine was inhibitory, whereas 50 times as much methylcytisine had no effect. Moreover, it seemed that methylcytisine acted as a type of antagonist toward cytisine. Generally methylation is considered to diminish the reactivity of a compound (Mothes, 1965, 1966).

N-Demethylation has been found to be connected with senescence of the relevant plants in the cases of nicotine (Mothes *et al.*, 1957), hyoscyamine (Romeike, 1964), and ricinine (Skursky *et al.*, 1969). Demethylation seems to have some connection with the transport of the alkaloid within the plant body (Skursky and Waller, 1972; Nowacki and Waller, 1973). In several alkaloids, e.g., nicotine (Kisaki and Tamaki, 1966), demethylation probably initiates more extensive processes of degradation of the molecule.

The opium poppy has been shown by Miller *et al.* (1973) to effect a demethylation, rapidly forming normorphine from morphine (Figure 6.9a). Normorphine was not converted back to morphine but rather disappeared, forming unknown products that had a turnover time of approximately 4 hr. It was suggested by Miller *et al.* that morphine is functionally important as a methylating agent, but the transfer of intact methyl groups to other compounds was not shown.

Kirby *et al.* (1972) showed that *Papaver somniferum* was capable of converting "unnatural" precursors to unnatural alkaloids. Various *O*-demethylated codeine derivatives were converted to "unnatural" derivatives of morphine. One "unnatural" conversion which was more efficient

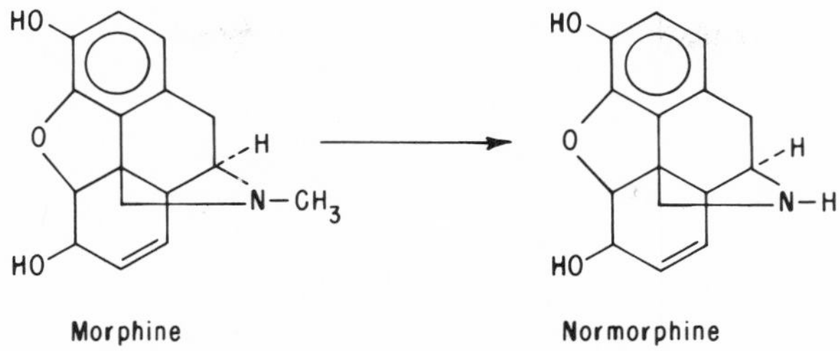

Morphine   Normorphine

Figure 6.9a. The conversion of morphine to normorphine by *Papaver somniferum* (Miller *et al.*, 1973). Courtesy of Pergamon Press, Ltd., copyright 1973.

than the natural one was the conversion of dihydrodeoxycodeine to dihydro-deoxymorphine (Figure 6.9b). The authors concluded that neither the 6-hydroxy group nor the 7,8-double bond in codeine was important for binding to the enzyme responsible for demethylation of the 3-methoxy group of codeine. Such methyl ethers as reticuline and its isomers are not demethylated by the poppy plant (Barton *et al.*, 1965; Battersby *et al.*, 1965) and only one enzyme is responsible for the *O*-demethylation of codeine and its derivatives.

No discussion has appeared in the literature about carbon demethylation (*C*-demethylation) as plants degrade alkaloids. This subject seems to have some interest, since certain amino acids, nucleic acids, sugars, lignin, etc., may be formed before the alkaloid is metabolized completely to CO_2.

6.6.2. *N*-Demethylation of Nicotine

Plant growth and development are complicated biological phenomena, dependent upon genetic and environmental variables. In tobacco products,

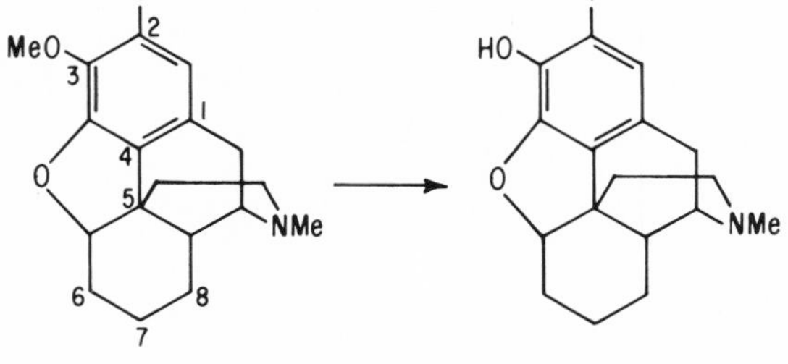

Figure 6.9b. The conversion of an "unnatural" precursor to an "unnatural" alkaloid (Kirby *et al.*, 1972). Courtesy of the Chemical Society.

these processes are not the final ones. They are complicated by postharvest handling such as airing, aging, and fermentation, processes which alter the leaves. There are many psychological reasons why people smoke, but the physiological basis for smoking is the stimulation derived from the nicotine absorbed from tobacco smoke. Leaf composition among some species is known to vary from crop to crop (Weybrew, 1975). Even in the same plant, leaves from various positions on the stalk vary considerably in their alkaloid composition; however, these plants carry on active alkaloid metabolism.

N-Demethylation of nicotine to nornicotine has been one of the most extensively studied biodegradative reactions in plants. Mothes and Schuette (1969) reviewed numerous papers dealing with this reaction in several species and varieties of tobacco (Figure 6.10). It is difficult to present a unified view based on the described experimental results. The reason is that different techniques were used (isotopic in more recent and nonisotopic in the older studies), and the plant materials were of different origins and were subjected to different treatments before and during the experiments. The site of demethylation of nicotine has been found mostly to be the leaves (Schuette, 1969). Dawson (1948, 1952), working with *Nicotiana glutinosa,* in which nornicotine is the major alkaloid, observed intense demethylation of nicotine in excised leaves fed with nicotine via the transpiration flow. From the time lag of nornicotine production, the author concluded that the chemical process occurs in the leaf blade (laminar) tissue; the delay is believed to occur during transport through the vein system, since disks cut from the leaf were infiltrated with nicotine under reduced pressure which exhibited no time lag.

N-Demethylation has been observed also in leaves of *Nicotiana alata,* a species with a very low nicotine content in its leaves. Schroeter (1958) and Griffith and Griffith (1964) experimented with *Nicotiana rustica,* a species which normally contains very little nornicotine, and found that nicotine demethylation occurs in both the root and, more rapidly, in the shoot.

Demethylation and consequent loss of nicotine (Yoshida, 1971) in tobacco leaves during the postharvest period is important commercially, because nornicotine is less toxic than nicotine. These postharvest pro-

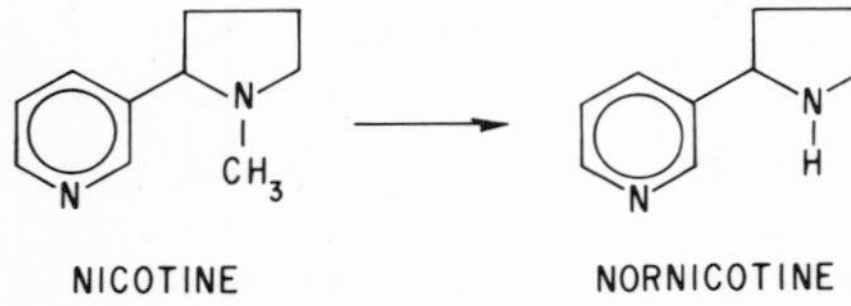

NICOTINE   NORNICOTINE

Figure 6.10. N-Demethylation of nicotine to yield nornicotine.

cesses (curing and fermentation) of tobacco leaves are not dealt with in this chapter; they cannot be considered as "normal," due to the almost certain participation of microorganisms (see Merker and Pyriki, 1966).

6.6.2.1. Physiological Role or Significance of Nicotine N-Demethylation. The physiological role or significance of nicotine N-demethylation is unclear. Leete and Bell (1959) administered [*methyl*-^{14}C]nicotine to *Nicotiana tabacum* and after one week isolated 90% of the radioactivity in the form of choline. They concluded that nicotine acts as a methyl donor in the tobacco plant. Later papers on the mechanism of the demethylation reaction do not agree with these results (see below).

Mothes (1928) related the active demethylating process of nicotine in tobacco plants (*Nicotiana tabacum* L.) to leaf senescence. Similarly, Mothes *et al.* (1957) and Schroeter (1966) observed in crude homogenates that older leaves demethylated nicotine much more extensively than did the young leaves.

Dawson (1952) did not find any difference in the demethylation activity of normal fresh leaves of *Nicotiana glutinosa* and leaves which had been starved for 30 hr (by being cultured in water). He concludes that the demethylation process does not serve as some sort of mobilization of reserves.

6.6.2.2. Reversibility of the Reaction Nicotine ⇌ Nornicotine. Early hypotheses about the biosynthesis of tobacco alkaloids assumed that nornicotine is the immediate precursor of nicotine or at least that the process is reversible. The literature on this subject is controversial. The disagreements could be caused by use of different experimental techniques. Convincing data are presented against regarding nornicotine as a nicotine precursor by Rapoport (1966), Leete (1973), Leete and Chedekel (1974). Rapoport fed $^{14}CO_2$ to *Nicotiana glutinosa* and found, beyond any doubt, that the specific radioactivity of the nicotine isolated afterwards was always higher than that of nornicotine. He does not report on the distribution of radioactivity between the skeleton and the methyl group of nicotine (to exclude the possibility of an overwhelming contribution of radioactivity from the methyl group to the total radioactivity of nicotine); however, this has been done by Zielke *et al.* (1968). After feeding $^{14}CO_2$ to *Nicotiana glutinosa* for 6 hr, the [^{14}C] distribution in nicotine was as follows: pyridine ring, 77.8%; C-2', 4.3%; C-3', 4.4%; C-4', 4.7%; C-5', 4.1%; and N-methyl, 5.0%.

The results by Tso and Jeffrey (1957) about a separate origin of part of the nornicotine molecule does not seem too appropriate in view of the present situation; however, their findings generally support either that separate pathways can exist or that nicotine can produce nornicotine. So it seems well established that the methylation of nornicotine does not play a significant role (if any) in nicotine formation. Nevertheless, the problem is very complicated and the results which present conflicting data are obtained from several *Nicotiana* species under varying conditions (discussed by Alworth and Rapoport, 1965).

Another type of "demethylation" of nicotine which involves the insertion of the methyl group into the six-membered ring of anabasine will be described in Section 6.7.1.

6.6.2.3. Mechanism of the Reaction.

Ill'in and Serebrovskaja (1958) found that in the varieties of *Nicotiana* which bear nornicotine, demethylation of administered nicotine occurs parallel with the uptake of oxygen.

As a possible intermediate in the oxidative process between nicotine and nornicotine, nicotine 1'-oxide was suggested by Wenkert (1954). For some time this possibility appeared to have some experimental support, since Egri (1957) found that nicotine 1'-oxide (nonradioactive) infiltrated into tobacco leaves gave an increase in nornicotine. But in experiments in which radioactive nicotine 1'-oxide was infiltrated into nicotine-free leaves of *Nicotiana tabacum* previously grafted on tomato stocks, the nicotine 1'-oxide was very inefficiently demethylated to yield nornicotine, whereas it was extensively reduced to nicotine (Kisaki and Tamaki, 1964a). This grafting technique has proved to be very helpful, especially in locations of the site where alkaloids are being synthesized (see Chapter 4 on The Sites of Alkaloid Formation). For instance, tobacco shoots grafted on tomato stock remain nicotine-free, since normally the alkaloid is formed only in the root of *Nicotiana*. Nicotine 1'-oxide has been proven by Alworth *et al.* (1969) not to be the intermediate in nicotine demethylation in *Nicotiana glutinosa*. *N*-Formylnicotine, another compound considered as a possible intermediate, was experimentally excluded, too (Kisaki and Tamaki, 1964a).

The results of Leete and Bell (1959)[2] strongly supported the view that the process was a transmethylation; but later papers failed to prove this. Nevertheless, Kisaki and Tamaki (1964b), who basically showed the involvement of oxidation in the demethylation process, also demonstrated that the transmethylating pathway also exists, especially in young leaves. The discrepancies seem to have been caused by use of different experimental conditions and plant materials.

(−)-Nicotine demethylated by excised *Nicotiana tabacum* leaves yields partially racemized nornicotine (Kisaki and Tamaki, 1961a,b). This has caused speculation that the pyrrolidine ring opens during the demethylation, which would cause the racemization. But this problem remained unsolved (Kisaki and Tamaki, 1964c) until Leete and Chedekel (1974) performed some new and clever experiments.

---

[2]These authors also state that the reaction was recognized by Ill'in in 1948–49.

The results (Leete and Chedekel, 1974) indicate that 48% of [2-³H](−)-nornicotine and 52% [2-¹⁴C](+)-nornicotine is incorporated from the labeled nicotine. Thus if (+)-nornicotine is formed from (−)-nicotine, the transformation must involve the loss of the hydrogen from C-2'; however, almost the same [³H/¹⁴C]ratio occurs as in the administered mixture of [2'-³H](−)-nicotine and [2'2-¹⁴C](+)-nicotine, which indicates that in *N. glauca* (+)- and (−)-nicotine are demethylated at similar rates. If the demethylation had been stereospecific for (−)-nicotine, the resultant nornicotine would have its [³H/¹⁴C]ratio doubled. This led the authors to propose a scheme for the formation of (+)-nicotine and (−)-nicotine (Figure 6.11). This mechanism accounts for the partial racemization of the nornicotine derived from (−)-nicotine; a Cope elimination of nicotine N'-oxide (a) would involve one of the hydrogens at C-3', which would provide the unsaturated compound (b). Elimination of water from this hydroxyamine yields the Schiff base (c), which upon hydrolysis yields formaldehyde or other C_1 metabolite and the primary amine (d). Cyclization of this intermediate yields (+)- and (−)-nornicotine. Leete and Chedekel presume that these steps are enzyme mediated, and it is to be expected that the final cyclization would yield a preferential amount of (+)- or (−)-nornicotine; however, the mechanism which yields (+)-nornicotine from (−)-nicotine remains unknown.

Some authors referring to nicotine demethylation call the reaction nonspecific. Dawson (1948, 1951) has found that not only nicotine but also N-ethylnornicotine, N-methylanabasine, and N-ethylanabasine (Figure 6.12)

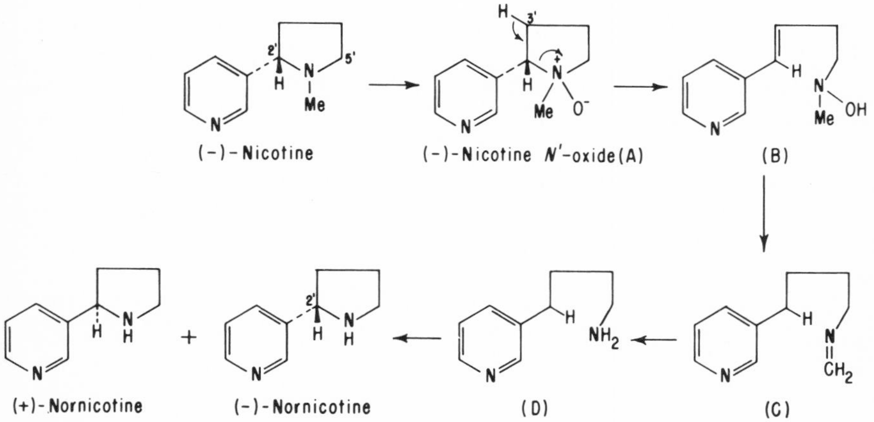

Figure 6.11. Hypothesis for the formation of (+)- and (−)-nornicotine from (−)-nicotine (Leete and Chedekel, 1974). Courtesy of the authors and Pergamon Press, Ltd., copyright 1974.

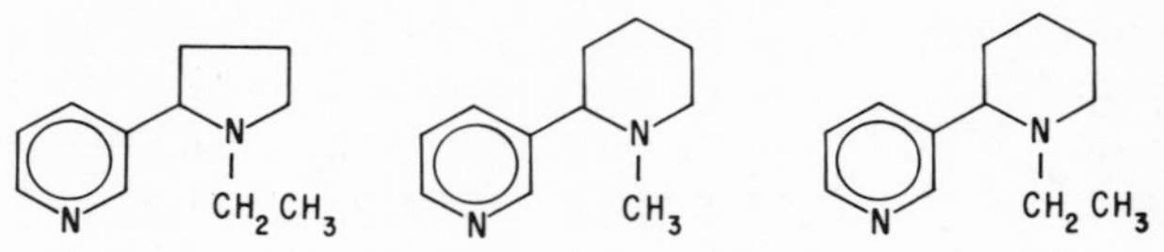

ETHYLNORNICOTINE   METHYLANABASINE   ETHYLANABASINE

Figure 6.12. Structures of some compounds related to nicotine or anabasine.

were dealkylated by excised leaves of *Nicotiana glutinosa* at comparable rates. Slight but definite demethylation of administered nicotine was also observed in the tomato plant (Tso and Jeffrey, 1959), which normally is completely free of nicotine. Neumann and Tschoepe (1966) have shown that not only *Nicotiana* species but also several other plants of the Solanaceae demethylated administered nicotine, although this alkaloid is usually not present in their tissues.

In a discussion of the nonspecific character of nicotine demethylation, it is appropriate to mention a curious finding of Gruetzmann and Schroeter (1966). They were looking for a system to accomplish the biological production of codeine and morphine from thebaine, which does not have as much pharmacological use. They succeeded with a tissue culture of *Papaver somniferum* and surprisingly also with a tissue culture from *Nicotiana allata,* which also *O*-demethylated and reduced thebaine to yield codeine and morphine (Figure 6.13). No details have been published, probably because of the practical value of the procedure.

6.6.3. *N*-Demethylation of Ricinine

The origins of the two methyl groups of ricinine have been extensively studied, but the order of their introduction into the molecule has not yet

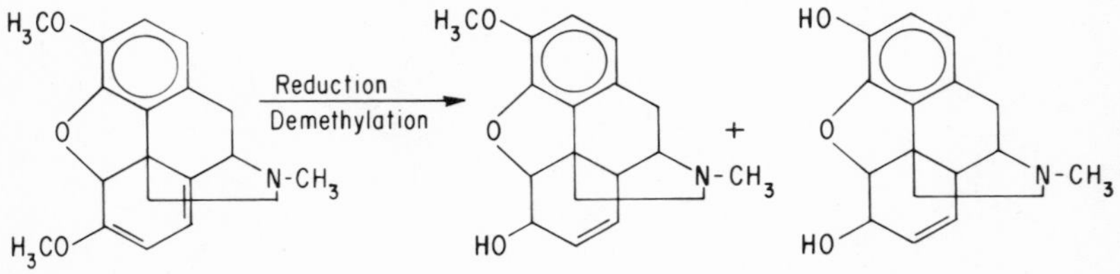

THEBAINE   CODEINE   MORPHINE

Figure 6.13. Formation of morphine and codeine from thebaine (Gruetzmann and Schroeter, 1966).

been determined. Excised senescent yellow leaves rapidly demethylated administered ricinine to *N*-demethylricinine (Figure 6.14) (Skursky *et al.*, 1969). This was the first report (a) describing a specific alkaloid catabolic reaction occurring in the plant that produced it, (b) showing that the catabolic product in yellow leaves can also serve as a precursor of the alkaloid in green leaves, and (c) showing that alkaloid catabolism may be related to the aging process. *N*-Demethylricinine, as well as ricinine, is normally absent in old yellow leaves, but the two are present in leaves in early stages of senescence; it seems probable that demethylation may play some role in the excretory function of ricinine in the plant (Waller and Skursky, 1972). The interconversion of these compounds can be correlated with the age of the plant. *N*-Demethylricinine was found to be a normal (but minor) constituent of mature seeds and to disappear during germination (see Figure 5.2 for a fuller explanation) (Skursky and Waller, 1972). The physiological significance of these processes is not yet known.

6.6.4. O-Demethylation of Ricinine

A new ricinine metabolite was conclusively identified as *O*-demethyl-ricinine (see Figure 6.14) (Lee and Waller, 1972a) (*N*-methyl-3-cyano-4-hydroxy-2-pyridone), and this interconversion with ricinine in senescent and green castor plant leaves was demonstrated. [3,5-^{14}C]Ricinine adminis-

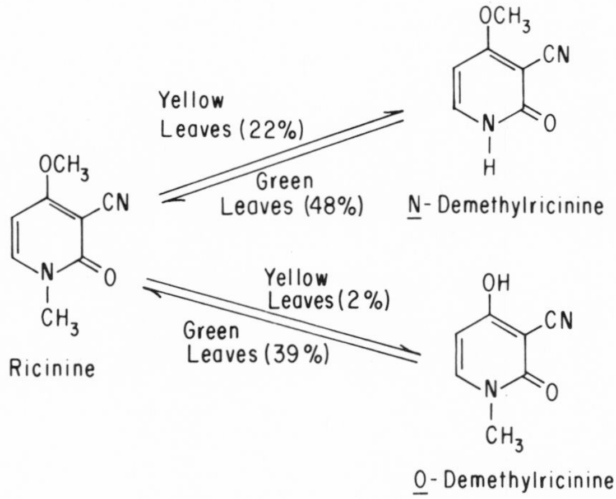

Figure 6.14. Methylation and demethylation reactions involving ricinine in castor bean plant leaves. Conversion percentages are based on the results obtained when 1 mg of precursor was administered to one leaf weighing about 5 g (Skursky *et al.*, 1969). Courtesy of the American Association of Biological Chemists.

tered in the yellow leaves was translocated to healthy parts of the plant, especially the growing apex. The results indicated that translocation of ricinine may be performed to neutralize chemically the alkaloid in the leaves which are being prepared for abscission. The facts that both ricinine and its demethylated forms (N- and O-demethylricinine) are present and that the ratios among them in stems and yellow leaves are not strikingly different indicate that there is no preferred form for translocation.

The formation of N-demethylricinine, O-demethylricinine, and carbon dioxide from different amounts of administered ricinine in excised senescent castor bean plant leaves is summarized in Table 6.3. Three levels of ricinine, 34.5, 84, and 143 µg/g fresh weight, were administered to senescent leaves. These results suggest that ricinine inhibited the O- and N-demethylating reactions. They also indicated that senescent leaves have a certain limitation of the rate of demethylation activity. The low extent of N-demethylating activity obtained after the administration of the high level of ricinine agreed with previous work (Skursky et al., 1969). The results showed that majority of the administered ricinine could be converted to its demethylated forms at low dosage levels comparable to the normal physiological levels. The approximate disappearance rate of ricinine obtained by extrapolating the ricinine curve is 3.8 µg/g fresh weight/day.

6.6.5. Possible Physiologic or Metabolic Significance of Ricinine, N-Demethylricinine, and O-Demethylricinine

Export of ricinine from senescent leaf tissues has been unequivocally established (Table 6.4) in a time course study about the accumulation of ricinine and its demethylated forms in the stems and yellow and green leaves following the administration of [3,5-^{14}C]ricinine to a yellow leaf. In

Table 6.3. Formation of N-Demethylricinine, O-Demethylricinine, and Carbon Dioxide from Ricinine in Excised Senescent Leaves of *Ricinus communis*[a,b]

Isolated compound	Expt. I		Expt. II		Expt. III	
	dpm × 10^{-2}	Percent	dpm × 10^{-2}	Percent	dpm × 10^{-2}	Percent
Ricinine	13.51	11.5	16.4	62.83	12.6	38.2	143.20	10.1	43.5
N-Demethylricinine	32.20	10.7	39.1	47.31	11.4	28.7	71.81	8.4	21.8
O-Demethylricinine	4.94	9.9	6.1	4.12	11.5	2.5	4.94	8.5	1.5
Carbon dioxide	0.18		0.2	0.29		0.2	1.32		0.4
Total	50.83		61.8	114.55		69.6	221.27		67.2

[a]Lee and Waller (1972). Courtesy of Pergamon Press, Ltd., copyright 1972.
[b]Percentage of incorporation was determined by dividing total radioactivity administered by the total amount recovered: Expt. I—8.23 × 10^3 dpm (0.5 mg), 34.5 µg ricine/g fresh weight; Expt. II—1.646 × 10^4 dpm (1 mg), 83.0 µg ricinine/g fresh weight; Expt. III—3.292 × 10^4 dpm (2 mg), 143.0 µg ricinine/g fresh weight.

Table 6.4. Demethylation and Translocation of [3,5-^{14}C]Ricinine from Yellow Leaf to Green Leaf of Castor Bean Plant Cuttings[a,b]

Experiment number	Isolated compound	Green leaf	Yellow leaf	Stem	Total	Fraction of radio-activity accounted for, %
		Radioactivity of plant part, dis/min × 10^{-2}				
I	Ricinine	2.6	129.3	47.3	179.2	54.4
(12 hour)	N-Demethylricinine	1.3	14.8	12.3	28.6	8.7
	O-Demethylricinine	—	0.1	0.8	0.9	0.3
	Total	3.9	174.2	60.4	208.7	63.4
II	Ricinine	4.5	90.8	56.9	152.2	46.2
(24 hour)	N-Demethylricine	1.5	2.8	28.5	62.8	19.1
	O-Demethylricinine	0.4	2.0	1.1	3.5	1.1
	Total	6.4	95.6	86.5	218.5	66.4
III	Ricinine	8.6	78.4	36.8	123.8	37.6
(36 hour)	N-Demethylricinine	7.3	52.1	8.5	67.9	20.6
	O-Demethylricinine	1.5	2.0	0.3	3.8	1.2
	Total	17.4	132.5	45.6	195.5	59.4

[a] Lee and Waller (1972). Courtesy of Pergamon Press, Ltd., copyright 1972.
[b] The plants used were grown in the greenhouse. [3,5-^{14}C]ricinine (2 mg) with total radioactivity 3.292 × 10^4 dis/min was administered. Percentage of incorporation was determined by dividing the activity recovered by the amount administered.

addition, it was demonstrated that the [8-^{14}C]ricinine administered in the stem could be transported to the seeds since a radioactive alkaloid administered to a young nonflowering castor bean plant was isolated from the seeds of matured plants (Waller et al., 1965). It is of interest that the demethylated form of ricinine is practically absent in the green leaves and that very small amounts of ricinine and N-demethylricinine are present in the yellow leaves. The very small amounts of ricinine and N-demethylricinine found in the yellow leaves might be due to the incomplete process of senescence since naturally detached leaves were found to be void of both compounds. The facts that (a) both ricinine and its metabolites are absent from the yellow leaves, (b) the methylation and demethylation reactions occur in the green leaves and yellow leaves, respectively, and (c) the α-pyridone ring of ricinine is not intensively degraded in the yellow leaves, supported the conclusion that ricinine and/or N-demethylricinine or O-demethylricinine in the yellow leaves is translocated from the senescent tissue to other parts of the plant, especially to the growing apex.

Figure 6.15 shows the demethylation and translocation of [3,5-^{14}C]ricinine administered to a yellow leaf attached to the lower part of the stem of a mature castor bean plant. The radioactivity of ricinine and the

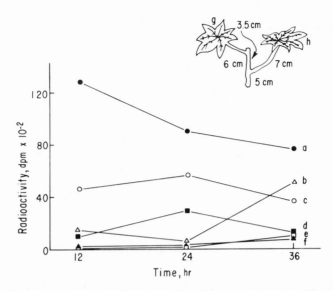

Figure 6.15. Demethylation and translocation of [3,5-¹⁴C]ricinine from the yellow leaf to the adjacent green leaf of the castor bean plant cutting with one green and one yellow leaf. *a*— Ricinine in yellow leaves; *b*—demethylated ricinines in yellow leaves; *c*—ricinine in stems; *d*—demethylated ricinines in stems; *e*—demethylated ricinines in green leaves; *f*—ricinine in green leaves; *g*—green leaf; *h*—yellow leaf (Lee and Waller, 1972). Courtesy of Pergamon Press, Ltd., copyright 1972.

demethylated forms of it were found to be highest in the yellow leaf where ricinine was injected. Radioactive *N*-demethylricinine was found in all parts of the plant except in the root. The second highest recovery of radioactive ricinine and *N*-demethylricinine was in the stems. However, the recoveries of radioactive ricinine per gram of fresh plant weight and of *O*-demethylricinine were high in the growing apex and stem, respectively (Lee and Waller, 1972a).

Interpretations of the above two results were complicated by the fact that the radioactivity ratio between ricinine and its demethylated forms did not necessarily indicate the ratio of translocated forms since the demethylated form of ricinine upon arrival to the green leaf from the yellow leaf may undergo methylation. The result showed that ricinine administered in the yellow leaves was translocated to healthy parts of the plant, especially the growing apex where ricinine is actively synthesized. The result supports the idea that a ricinine translocation process might be a salvage operation performed by the plant in order to reutilize ricinine from the leaves which are being prepared for abscission. This is a broad concept of the traditional source–sink relationship (Crafts, 1961); the translocation of ricinine is a reutilization or saving-type process. Another possible speculation about the phenomenon is that the demethylation reactions, which one generally believes make a compound more metabolically active, may maintain the vital precursors for the compounds in the pyridine nucleotide cycle within the yellowing leaves. This process might inhibit the progression of senescence.

Ricinine is obviously translocated via the phloem tissue, since its movement via xylem would result in accumulation in the large older leaves (Crafts, 1961). There is a possibility that ricinine passively accompanies other metabolites from leaves to sinks or sites of utilization moving along by "mass flow." However, the reason for its intense export from leaves prior to abscission remains unclear. Perhaps this alkaloid is eventually utilized for the synthesis of, or controls the synthesis of, compounds in or closely related to the pyridine nucleotide cycle (Waller *et al.*, 1966).

The observation that ricinine is translocated out of large leaves more readily than *N*-demethylricinine and *O*-demethylricinine suggests that the formation of these demethylated ricinines in senescent leaves is a protective mechanism to prevent such leaves from losing all their ricinine, i.e., *N*-demethylricinine and *O*-demethylricinine would tend to stay in the leaf tissue and not be translocated to growing areas as readily as ricinine. In contrast, it appears that ricinine exported from the leaves to the roots of the nutrient-deficient plant (and ricinine synthesized *in situ* by roots) must be converted to *N*-demethylricinine and *O*-demethylricinine for upward transport in the xylem transpiration stream. Since demethylation increases water solubility, it seems likely the *N*-demethylricinine and *O*-demethylricinine are transported in the xylem tissue.

The facts that the plant translocates ricinine away from senescent leaves to young, developing tissues and that it is abundant in young tissues suggest that this compound has a specific physiologic or metabolic function in growth. A mere waste product of metabolism would hardly be transported into seeds and other vitally important tissues from a leaf which was to be lost by abscission.

6.6.6. *N*-Demethylation of Hyoscyamine

Demethylation of hyoscyamine has been studied extensively by Romeike (1962, 1964) (Figure 6.16). Norhyoscyamine has been found in low amounts in older plants of different *Datura* species. The *N*-demethylation reaction has been observed in the shoots of *Datura* which had been made alkaloid-free by previous grafting on *Cyphomandra betacea* (see page 126, which deals with such grafting). These alkaloid-free shoots were fed with a

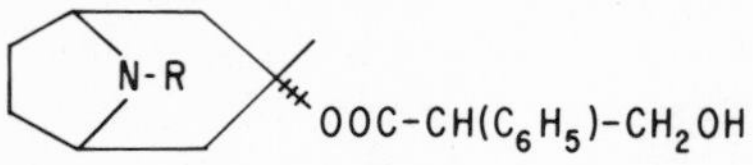

HYOSCYAMINE : R = CH₃

HYOSCYAMINE : R = CH$_3$
NORHYOSCYAMINE : R = -H

Figure 6.16. Structures of hyoscyamine and norhyoscyamine.

hyoscyamine solution and later analyzed for the demethyl derivative. The extent of the demethylation was different in various species, and it was established that the process takes place in the leaves. Addition of a wide variety of metabolic inhibitors into the feeding solution increased the extent of demethylation 16 to 40 times. Since the inhibitors used (azide, fluoride, iodoacetate, cysteine, cystine, chloromercuribenzoate, semicarbazide, hydroxylamine) were very different, their influence is probably nonspecific. High demethylating activity was always accompanied by marked damage of the experimental plants. Thus the process is believed to be part of the general degradation during the course of death of the tissues.

Fodor (1976) has prepared a review of biosynthesis of tropane alkaloids for publication in *Specialist Periodical Reports* of the Chemical Society of London and, although it is too early to say for certain, some of the biosynthetic intermediates may be similar to the biodegradation intermediates.

6.6.7. Demethylation of Hordenine

Hordenine occurs mainly in young barley plants, but has been found in millet or proso *(Panicum miliaceum)* too. It is found for a short time during plant development; this is supposedly due to catabolism (Brady and Tyler, 1958; Neumann and Tschoepe, 1966). The course of the catabolic reactions is not known. In 1956, Frank and Marion studied the *N*-demethylation of hordenine in barley. Radioactivity from D-[¹⁴C]hordenine fed to barley seedlings was found in *N*-methyltyramine. Since no tyramine (Figure 6.17) was found at all, it appears likely that the demethylation affects only one of the two methyl groups of hordenine. The mechanism of the reaction was not discussed in the cited paper, but the isolated choline was found nonradioactive, so transamination is not likely.

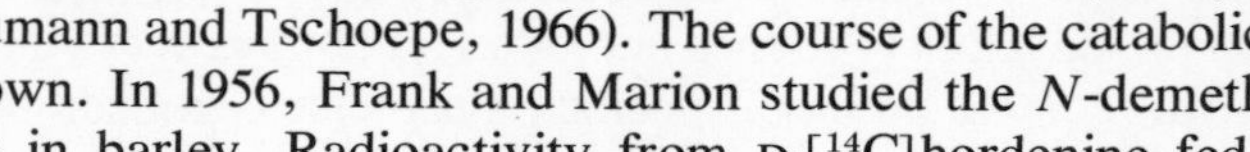

HO—⟨○⟩—CH$_2$-CH$_2$-N(CH$_3$)$_2$ ⟶ HO—⟨○⟩—CH$_2$-CH$_2$-NHCH$_3$

HORDENINE   N-METHYLTYRAMINE

Figure 6.17. Formation of *N*-methyltyramine from hordenine.

6.6.8. Demethylation of Caffeine

Wanner and Kalberer (1966) studied the extensive degradation of caffeine (formula on page 224) in *Coffea arabica,* and reported the release of $^{14}CO_2$ by excised older leaves after feeding [*methyl*-^{14}C]-labeled alkaloid. No radioactivity was found in amino acids connected with C-1 metabolism, e.g., serine, citrulline, arginine, and methionine (Kalberer, 1964). Therefore, nothing definitive could be stated about the mechanism of the demethylating reaction; however, it may be possible that the demethylation enzyme systems present in either tea *(Thea sinensis)* or coffee have been developed simultaneously as the tissue develops its capability of synthesizing the caffeine. The methylation of the purines (e.g., caffeine or theobromine from tea) makes them more hydrophilic and therefore makes their excretion easier[3] (Ogutuga and Northcote, 1970).

6.7. More Extensive Degradation of Alkaloids and Introduction of Their Catabolites into Primary Metabolic Pathways

6.7.1. Nicotine and Other Tobacco Alkaloids

Literature about the extensive degradation of nicotine is richer than that on the catabolism of any other alkaloid. In spite of that, it is not yet possible to give a clear-cut and concise review of definite, established pathways.

Tso and Jeffrey (1959) were the first to show convincingly that nicotine undergoes more extensive degradation than mere *N*-demethylation. Administration of [^{15}N]-labeled anabasine, nornicotine, nicotine, and doubly [^{15}N,^{14}C]-labeled nicotine to *Nicotiana rustica, N. glauca,* and *N. glutinosa,* showed that interconversion of all the alkaloids occurred. It was suggested, therefore, that this interconversion probably occurred more directly than through the general metabolic pool. However, significant portions of the isotopes from the administered alkaloids were found in other unidentified organic compounds, particularly in the insoluble residue of the plants.

Yoshida (1962) injected [^{15}N]-labeled nicotine into *Nicotiana tabacum* at various stages of development and found that in young plants, up to 80% of the labeled nicotine was degraded during the 2 weeks after feeding. In older plants such degradation amounted to only 50% and in topped plants only 20% during the same period of time. The author did not study the

[3]For a recent article on the biosynthesis of caffeine see Looser *et al.* (1974) and the references therein.

degradation products but demonstrated convincingly that nicotine can be translocated from the leaves to other parts of the plant. High degradation rates observed in young plants appear to be correlated with comparatively low levels of nicotine at the earlier stages of development.

During the air-curing of tobacco leaves, loss of nicotine decreased (27–52%) with leaf maturity, and the loss was accompanied by the familiar browning of the leaves during curing (Yoshida, 1971).

6.7.2. Specific Metabolites of Nicotine

A substantial contribution to the problem of nicotine degradation was the isolation of radioactive nicotinic acid from *Nicotiana rustica* after feeding radioactive nicotine (Griffith *et al.*, 1960). Isolated nicotinic acid contained a significant amount of the isotope with dilution ranging from 18 to 88. Thus, nicotinic acid was the first known specific metabolite from the extensive degradation of nicotine.

After a longer feeding of *Nicotiana rustica* with uniformly labeled [^{14}C]nicotine, Tso and Jeffrey (1961) showed that radioactivity was detectable in amino acids, pigments, organic acids, and sugars; large portions of the isotope from uniformly labeled [^{15}N]nicotine administered simultaneously were found in the amino acid fraction (both free and after hydrolysis). Although none of the labeled specific metabolites were characterized, the authors demonstrated that nicotine in the plant is in a dynamic state that is connected with the primary metabolic pathways.

Lovkova (1964) administered uniformly labeled [^{14}C]nicotine to *N. tabacum* and after 4 days recovered only 3.5% of the administered radioactivity in nicotine. About the same percentage of total radioactivity was recovered in the free amino acid fraction, which seems to have been convincingly purified from any contamination by nicotine. No separation of this amino acid mixture was undertaken, and no data were presented on the fate of the remaining majority (about 90%) of the administered radioactivity. Ill'in, from the same laboratory (Ill'in, 1965, 1966), reviewed his concept that the appearance of nicotine, especially in germinating seeds, and its disappearance during the maturation process have some relation to the metabolism of proteins and their constituents and that nicotine is not just a metabolic waste product. To support these hypotheses, he administered uniformly labeled [^{14}C]nicotine to maturing fruits of *Nicotiana* plants. After 14 days, a radioactive protein fraction was isolated, but, because of the minute amount of nicotine in seeds and a high protein concentration, the total specific radioactivity of the isolated protein remained very low. Such incorporation of radioactivity into proteins was prevented by chloramphenicol. In later papers, more detailed experiments were described

(Ill'in *et al.*, 1968; Ill'in and Lovkova, 1972); the radioactive nicotine used for feeding was biosynthesized from different precursors ([¹⁴C]glutamate, [¹⁴C]aspartate, [¹⁴C]acetate), so that the label in the nicotine molecule was localized either predominantly in the pyrrolidine ring or in both the pyridine and pyrrolidine rings (the position of the labels was not exactly determined). Differences in relative labeling of the protein-bound amino acids after feeding the three types of radioactive nicotine led to some very preliminary guesses about the pathways of nicotine degradation in maturing tobacco seeds.

Leete (1968) reported the very extensive degradation of [2'-¹⁴C]nicotine in *N. glauca* and announced his search for nonalkaloidal metabolites. Anabasine was not found as a metabolite of nicotine; however, 5.7% of the radioactivity was found in nornicotine. The difference between Lovkova's and Leete's findings may be because nicotine may not be a direct precursor of anabasine in excised *N. glauca* shoots. Perhaps [2'-¹⁴C]nicotine could be degraded to [7-¹⁴C]nicotinic acid. If this were reincorporated into anabasine, no activity would appear in the anabasine. The [2'-¹⁴C]nicotine underwent extensive metabolism, as shown by the fact the only 0.41% was recovered after 5 days.

Kisaki and Tamaki (1966) found myosmine (1',2'-dihydronornicotine) as a metabolite of nornicotine in *N. tabacum*. The results (together with earlier papers of the same authors) show a rather complicated pattern of optical rotations of all alkaloids in the tobacco plant. To explain this, the authors assume that the nornicotine might be synthesized as a racemate and later optically be activated by stereospecific dehydrogenation to form myosmine. This assumption is supported by the fact that the rate of (−)-nornicotine degradation in the plants is several times faster than that of the (+)-isomer. The authors proposed such a tentative pathway for nicotine degradation (Figure 6.18a).

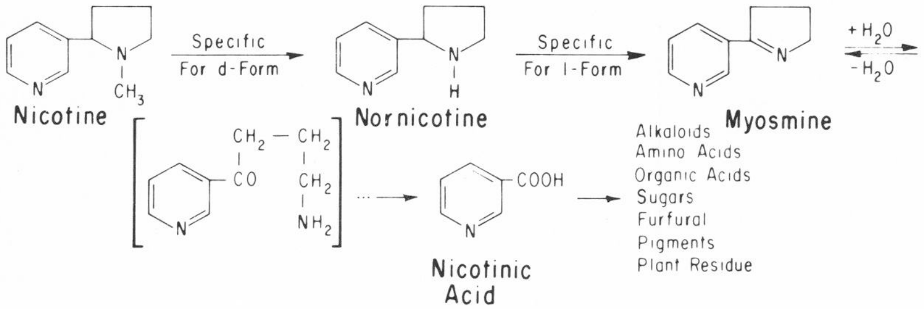

Figure 6.18a. A postulated metabolic pathway of nicotine (Kisaki and Tamaki, 1966). Courtesy of the authors and Pergamon Press, Ltd., copyright 1966.

   Leete (1973) and Leete and Chedekel (1974), following the administration of a mixture of [2'-³H](−)-nicotine and [2'-¹⁴C](+)-nicotine to *N. glauca,* obtained a series of observations similar to those of Kisaki and Tamaki (1966), but they were able to go farther than the Japanese scientists because of their starting materials and their degradation procedures. They developed a method by which they could specifically degrade [2'-¹⁴C,³H]nornicotine. The major pathway is nornicotine (50%), myosmine (2.05%), cotinine (0.30%), and nicotinic acid (0.14%) (Figure 6.18b). Myosmine (a component of tobacco smoke) is presumably formed from nicotine via nornicotine; however it was shown by feeding [2'-¹⁴C]myosmine to *N. glauca* that the dehydrogenation was not reversible. Nicotinic acid was presumed to be a metabolite of myosmine since all of its activity was located in the carboxyl carbon. They reported negligible activity in anabasine (which is different from Lovkova *et al.,* 1973, Figure 6.23) and 3-acetylpyridine, a metabolite of cured tobacco. Cotinine, a metabolite of nicotine in humans and other animals, is also labeled. It presumably arises by a different pathway directly from nicotine.

6.7.3. Ricinine

   Ricinine has also been shown to be in a dynamic state in *Ricinus communis,* since its α-pyridone ring is degraded to CO_2 (Figure 6.19a).

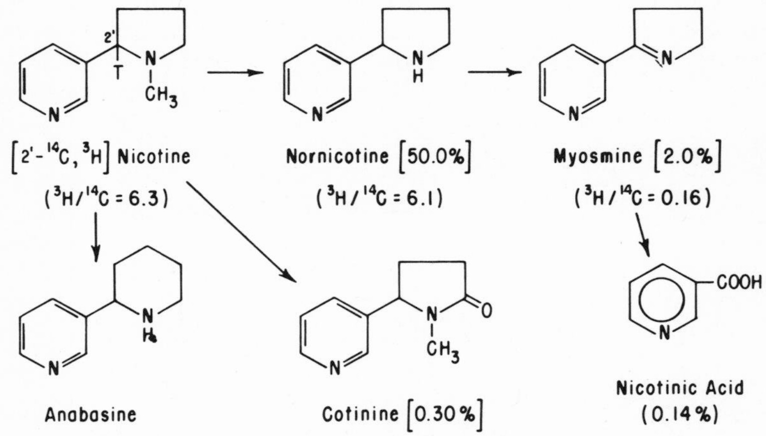

Figure 6.18b. Partial metabolism of *Nicotiana glauca* (Leete, 1973; Leete and Chedekel, 1974). Courtesy of the authors and Pergamon Press, Ltd., copyright 1974.

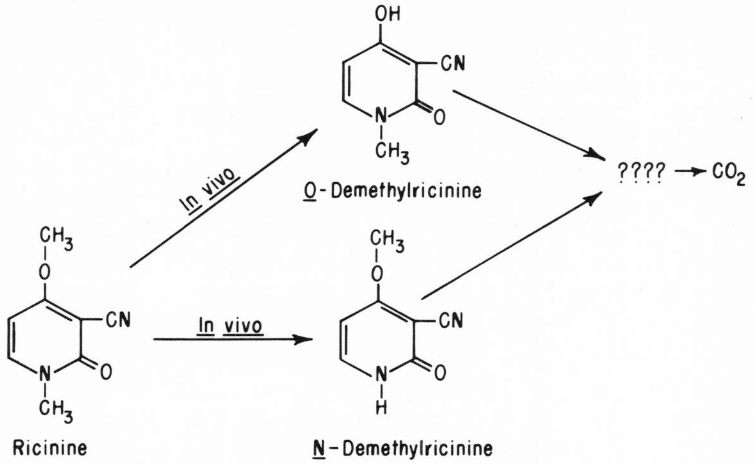

Figure 6.19a. Proposed route of metabolism of ricinine in aging castor leaves.

Chromatography of an extract from an experimental plant fed with labeled [3,5-^{14}C]ricinine (Figure 6.19b) revealed that at least 15 radioactive compounds were formed, none of which have yet been identified (Waller and Lee, 1969). They also showed that yellow leaves which contained a trace amount of alkaloid could metabolize [3-5-^{14}C]ricinine to CO_2. Synthesis of ricinine occurred continually during these experiments. Figure 6.20 shows the rate of disappearance (Waller *et al.,* 1965) of [^{3}H]ricinine, [8-^{14}C]ricinine and total ricinine found in castor plants in 1961. From these results, it was clear that rapid destruction of ricinine occurred while synthesis progressed. For example, the [^{14}C]ricinine curve shows that nicotinic acid was converted to ricinine and that the ricinine thus formed was 90% degraded in 11.5 weeks. The [^{14}C]ricinine was about 95% degraded during the 10-week experiment. The highest rate of metabolism occurred during the first 3 weeks, when the plants started to flower; the rate of disappearance of the ricinine fraction was considerably slower thereafter.

6.7.4. Degradation of Alkaloids to Other Alkaloidal Compounds

Biosynthesis of some alkaloids of rather complicated structures might involve some steps in which one alkaloid is being transformed to another by degradative reactions. It is a formal question (to be solved by some sort of convention) whether such a metabolic transformation should belong to the

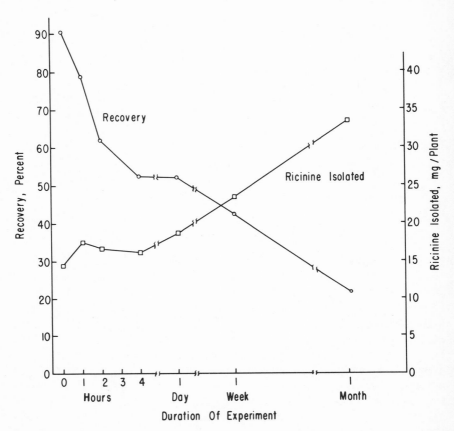

Figure 6.19b. Recovery of [¹⁴C]ricinine from 6-week-old castor plants administered 155 mμC [3,5-¹⁴C]ricinine (specific activity 360 μc/mmol) (Waller and Lee, 1969). Courtesy of the American Society of Plant Physiology.

class of either biosynthesis or biodegradative reactions. The classical reaction sequence of this type is the one-directional pathway thebaine → codeine → morphine (see Figure 6.13). "Biodegradative" steps in biosynthesis could be comparatively more frequent within the groups of structurally complicated alkaloids which are biogenetically linked with the isoprenoids. To illustrate, several examples are given.

During the ripening of the seeds of *Amsonia tabernaemontana* one of the alkaloids, (+)-vincadifformine, loses a C_2 unit and undergoes a rearrangement to (+)-1,2-dehydroaspidospermidine (Figure 6.21) (Zsadon *et al.*, 1975). The same process has also been found in the leaves during the period of flowering and development of the crop (Zsadon and co-workers, 1970,

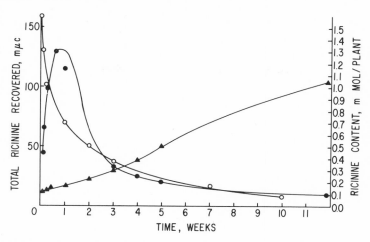

Figure 6.20. Rate of disappearance of [³H]ricinine and [8-¹⁴C]ricinine in castor plants. They were 26 days of age and were at the fifth to sixth nodal stage when the labeled compounds were injected. [³H]ricinine, 9.7 μc/mmol, was dissolved in 200 μl of distilled water and injected in a single plant. [7-¹⁴C]Nicotinic acid, 0.108 μmol, with a specific activity of 9.2 mc/mmol was dissolved in 10 μl of distilled water and injected in a different single plant, ○—[³H]ricinine, ●—[¹⁴C]ricinine, ▲—ricinine content per plant. The entire plant was used for ricinine isolation (Waller *et al.,* 1965). Courtesy of the American Society of Plant Physiology.

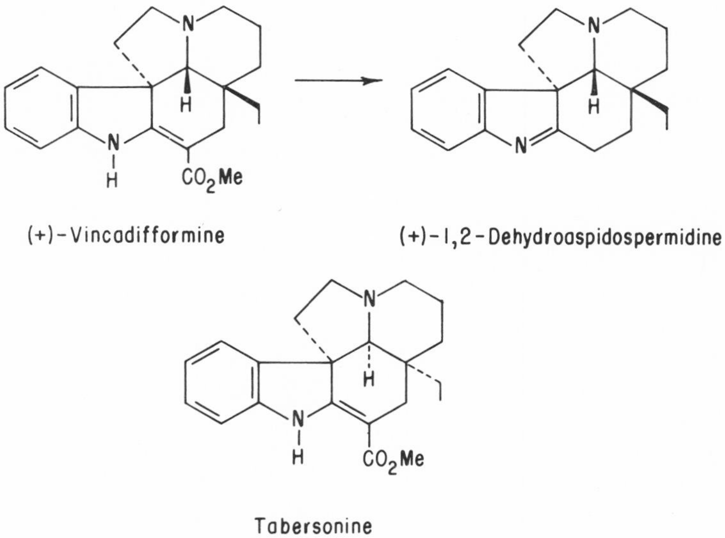

(+)-Vincadifformine   (+)-1,2-Dehydroaspidospermidine

Tabersonine

Figure 6.21. Transformation of (+)-vincadifformine to (+)-1,2-dehydroaspidospermidine (courtesy of Zsadon *et al.,* 1975).

1971, 1972). Both of these bases are only minor components of the plant, and it is interesting that tabersomine, the main alkaloid which accumulates in seeds during the course of ripening, is a compound isomeric with (+)-vincadifformine, having just the opposite configuration of the three centers of asymmetry (Figure 6.20).

Genista alkaloids have a pattern which leads Steinegger and Bernasconi (1964) to the observation that as the level of one alkaloid goes down, that of the other alkaloids increases. It does not necessarily follow that there is a precursor–product relationship. These proposed pathways are shown in Figure 6.22; the concentrations involved are compatible with the conversions that are proposed.

A novel metabolic conversion of an alkaloid involves the expansion of the methylpyrrolidine ring of nicotine into the piperidine ring of anabasine (Figure 6.23) and possibly to anatabine. This possibility was first given

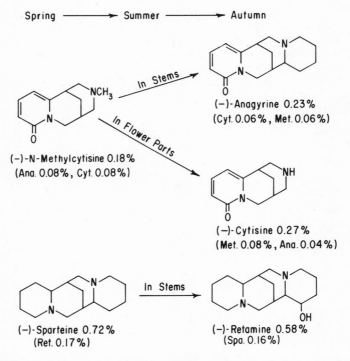

Figure 6.22. Alkaloid transformations in *Genista aethnensis* during a growing season (Steinegger and Bernasconi, 1964). Courtesy of the authors and *Pharmaceutica Acta Helvetiae.*

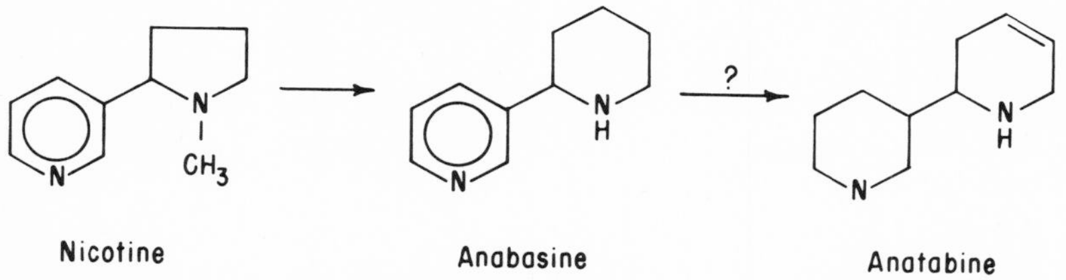

Figure 6.23. Proposed metabolic relationship of nicotine to anatabine in *Nicotiana glauca* (courtesy of Lovkova *et al.*, 1973).

some support by Alworth and Rapoport (1965), who fed *Nicotiana gluti-nosa* with $^{14}CO_2$ and studied the rate and the extent of isotope incorporation into the alkaloids. More recently Lovkova *et al.* (1973) have proved the conversion of nicotine into anabasine by feeding *Nicotiana glauca* shoots with randomly labeled [^{14}C]nicotine. They found that 83% was metabolized during the three days immediately following feeding. All of the adminis-tered radioactivity (see also Section 6.7) was found in the group of alka-loids: nicotine 4%, anabasine 1.5%, anatabine 0.1%, and nornicotine 0.02%. Thus anatabine is not likely to be the precursor; it seems more likely to be a metabolite of anabasine.

Nowacki and Waller (1975) studied the metabolism of L-[^{14}C]sparteine, DL-[^{14}C]lupanine, and L-[^{14}C]thermopsine in species of Leguminoseae and observed that they were converted to make highly oxidized alkaloids. Certain species of Leguminosae produce, besides the common lupine alkaloids, alkaloids with a dehydrogenated ring such as thermopsine, anagyrine, baptifoline, cytisine, and methylcytisine which are found in species of *Baptisia, Thermopsis,* and in *Lupinus nanus.* Figure 6.24 shows the structural formulas of the alkaloids investigated. The radioisotopic tracer study indicated that in *Thermopsis* and *Baptisia* plants the saturated tetracyclic lupine alkaloids were converted into thermopsine and cytisine, and the specific activity of the isolated compounds suggests that there is a direct pathway between the two. When an alkaloid with a dehydrogenated ring A, e.g., thermopsine, was introduced, only 12% of the radioactivity was found in alkaloids containing hydrogenated rings (Figure 6.24, Table 6.5) (sparteine, lupanine, hydroxysparteine, and angustifoline), and the remainder was in thermopsine, cytisine, methylcytisine, baptifo-line, and about one-fourth of the radioactive labeled compounds were not identified. The experiments involving the feeding of lupine seeds with radioactive alkaloids seemed to indicate that the alkaloids were utilized during germination and then converted into other compounds, e.g., amino acids.

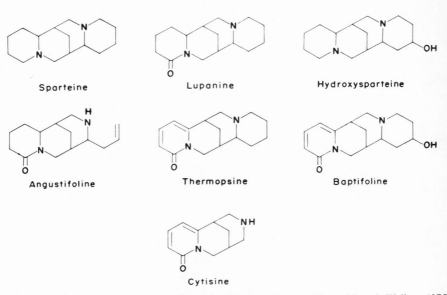

Figure 6.24. Structural formulas of alkaloids investigated (Nowacki and Waller, 1975). Courtesy of Pergamon Press, Ltd., copyright 1975.

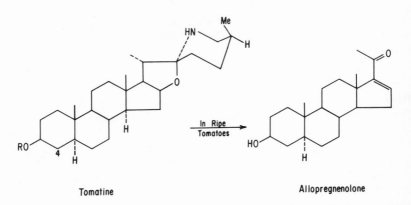

Figure 6.25. The conversion of tomatine into allopregnenolone (Heftman and Schwimmer, 1972). Courtesy of the authors and Pergamon Press, Ltd., copyright 1972.

Table 6.5. Metabolism of DL-[^{14}C]Thermopsine in *Thermopsis* and *Baptisia*[a]

	Hydroxy-sparteine	Sparteine	Lupanine	Thermo-psine	Cytisine	Methyl-cytisine	Bapti-foline	Not identified
Thermopsis macrophylla								
Distribution of isolated alkaloids (%) (410 mg total)	—	10.1	26.4	27.2	8.3	18.7	—	9.3
Distribution of radioactivity (%) (1157 dpm total)	—	—	11.8	51.1	10.3	12.6	—	14.2
Specific activity (dpm/mmol)	—	—	129	530	274	160	—	—
Baptisia leucophaea								
Distribution of isolated alkaloids (%) (702 mg total)	1.5	2.0	2.6	10.5	35	31	3.7	14.3
Distribution of radioactivity (%) (1375 dpm total)	1.0	1.1	1.7	30.7	12.7	7.9	9.9	34.9
Specific activity (dpm/mmol)	23	18	23	981	95	72	897	—

[a]Nowacki and Waller (1975a). Courtesy of Pergamon Press, Ltd., copyright 1975.

6.7.5. Degradation of Steroid Alkaloids

A very interesting biological transformation is the degradation of the toxic alkaloid, tomatine, which produces allopregnenolone (Heftman and Schwimmer, 1972) (Figure 6.25). Radioactive allopregnenolone was formed from ripe tomatoes which had been fed [4-^{14}C]tomatine. The reaction might have some economic importance. Green (unripe) tomatoes do not degrade the alkaloid (Sander, 1956).

6.7.6. Hordenine

In a study on the metabolism of hordenine administered to barley seedlings Frank and Marion (1956) reported that they found some [^{14}C]radioactivity from the α-[^{14}C]atom incorporated into lignin. However, they were unsure that they had a pure lignin preparation because of the severity of their procedure. It is significant that hordenine was metabolized by the barley seedling.

6.7.7. Conversion of Gramine into Tryptophan

Gramine has been long known to be a tryptophan metabolite in germinating barley (Figure 6.26). After the 50th day of plant age, it disappears from the shoot completely. Digenis *et al.* (1966) fed [^{14}C]-labeled gramine to young barley plants and after 7 days detected 0.84% of the radioactivity

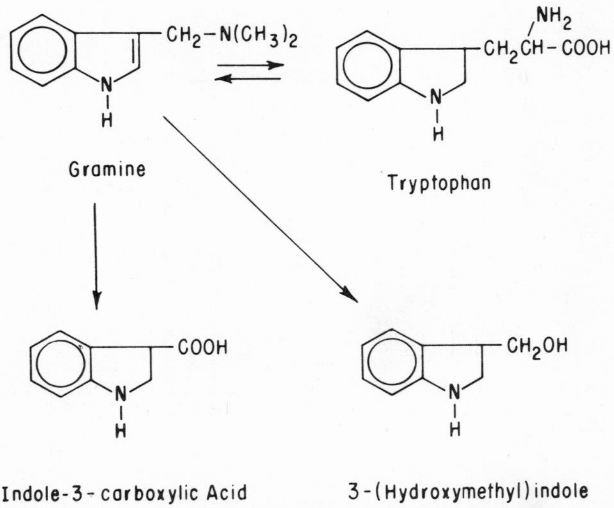

Figure 6.26. The conversion of gramine into tryptophan, 3-(hydroxymethyl)indole, and indole-3-carboxylic acid (Digenis *et al.*, 1966). Courtesy of the *J. Pharmaceutical Science.*

in total (free and bound) tryptophan. Localization of the label in tryptophan has not been determined. But since no radioactivity was detected in serine (nor in any other amino acid), it was concluded that the pathway from gramine to tryptophan is different from the known formation of tryptophan by molecularly fusing indole and serine. A possible route, speculative so far, is proposed in the article.

In a later paper, Digenis (1969) showed that [*methylene*-$^{14}C^3H$]gramine, [^3H:^{14}C] = 0.5, was converted to 3-(hydroxymethyl)indole (10.1% incorporation), indole-3-carboxylic acid (6.2% incorporation), and tryptophan (0.84% incorporation). It was indeed surprising that 7 days after feeding the gramine he isolated tryptophan with the same [^3H:^{14}C]ratio. Thus gramine becomes the second example of an alkaloid (after nicotine) that is transformed into an amino acid through catabolic reactions in the plant that produces it.

Radioactive indole-3-carbaldehyde labeled specifically at its aldehyde carbon was isolated from 8-week-old seedlings of *Lupinus hartwegii* which had been administered L-[3-^{14}C]tryptophan. The *L. hartwegii* which is capable of forming gramine used the same biosynthetic route as barley. However, it was an important observation that no significant amount of indole-3-carbaldehyde was found in 7-week-old seedlings. This is consistent with the hypothesis that the indole-3-carbaldehyde arises by the metabolism of gramine in the maturing plant (Leete, 1974). It is well-known that tryptophan can serve as a precursor of gramine (O'Donovan and Leete, 1963). Yet the significance of these processes for the plant's general metabolism cannot be judged, since, as with the case of nicotine (Lovkova, 1964), only a small fraction of the administered isotope has been accounted for and no data are given about most of the administered material. Certainly this is an expensive (energywise) way for amino acids to be produced.

6.7.8.   Caffeine

Degradation of caffeine occurs in older leaves of *Coffea arabica*. The roots do not contain caffeine except in the young seedlings (Wanner, 1963), and little is known about the role of roots in catabolism. The degradation process has been studied by Swiss workers (Kalberer, 1964, 1965), and a review of this and other papers as well as some facts are given by Wanner and Kalberer (1966). Experiments have been done with [*methyl*-^{14}C]-labeled as well as with ring-labeled caffeine.

The $^{14}CO_2$ arises from the methyl carbon and also from the ring-labeled caffeine, proving that the entire molecule can be degraded. Allantoin, allantoic acid, and at least three unidentified compounds have been detected as intermediates (Figures 6.27a and 6.27b). Recently Wanner's group (1975) identified N^3- and N^7-methylxanthine as breakdown products,

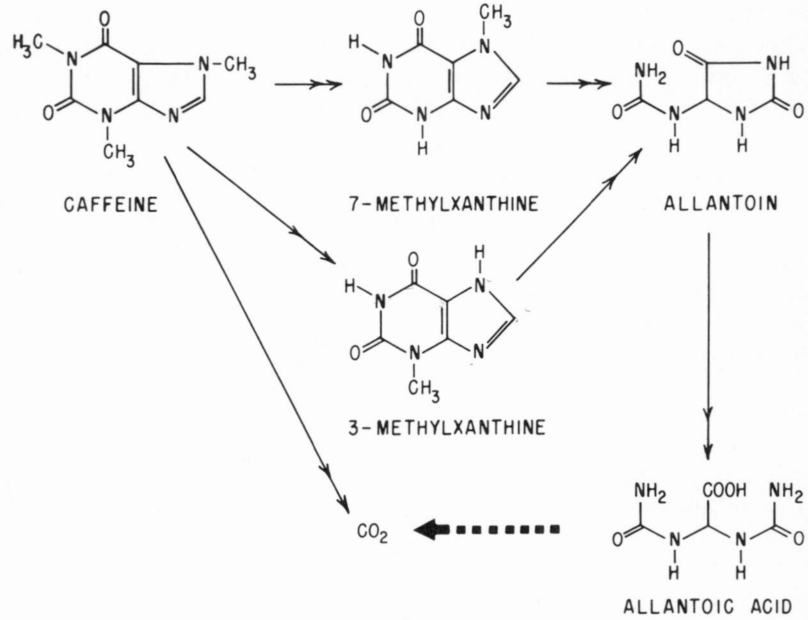

Figure 6.27a. Partial metabolic route for caffeine in *Coffea arabica* (Kalberer, 1964, 1965; Wanner and Kalberer, 1966; Wanner *et al.*, 1975). Courtesy of the authors and journals.

whereas no radioactivity was detected in the mono- and dimethyluric acids. This corresponds to the biosynthesis route leading to caffeine from xanthine which is methylated according to the pathway

$$
\text{xanthine}
\begin{array}{l}
\nearrow N^3\text{-methylxanthine} \searrow \\
\searrow N^7\text{-methylxanthine} \nearrow
\end{array}
\text{theobromine} \rightarrow \text{caffeine}^4
$$

(Luckner, 1972).

Thus it would seem that the coffee plant might be capable of degrading caffeine by merely reversing the biosynthetic pathway, a truly remarkable feature! The distribution of radioactivity between the identified and uniden-

[4]Alternative pathways have been proposed but 7-methylxanthine and theobromine are generally regarded to be the most immediate precursors of caffeine. In a recent study on the origin of the purine ring of caffeine, [8-^{14}C]adenine was a highly effective precursor, where in contrast [8-^{14}C]guanine was a poorer precursor. This led Suzuki and Takahashi (1976) to propose that caffeine was synthesized from purine nucleotides from the nucleotide pool rather than from the nucleic acids.

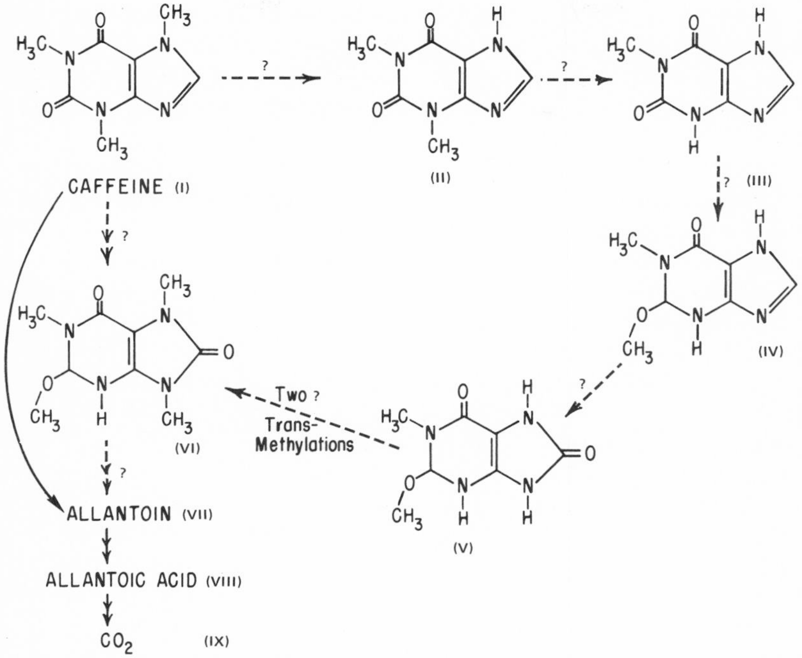

Figure 6.27b. Proposed metabolic route of caffeine in *Coffea* species (the solid line, Wanner *et al.*, 1975), courtesy of the author and Pergamon Press, Ltd., copyright 1975; (the arrows shown on the dotted lines, Waller, 1976).

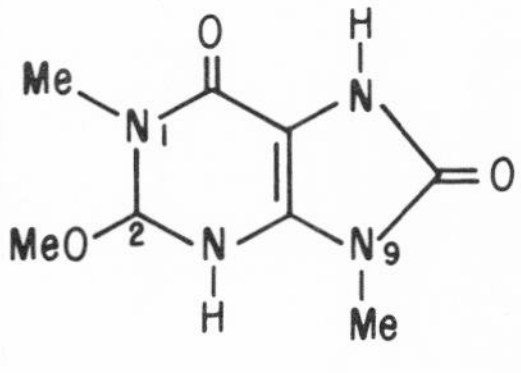

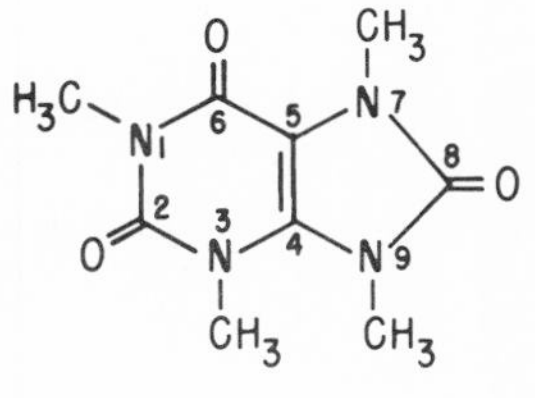

Figure 6.27c. Compounds isolated from several *Coffea* species. *Top:* 2-methoxy-1,9-dimethyl-7,9-dihydro-1*H*-purine-6,8-dione [O^2,1,9-trimethyluric acid]; *bottom:* 1,3,7,9-tetramethyl-7,9-dihydro-1*H*-purine-2,6,8(3H)-trione (1,3,7,9-tetramethyluric acid) (Wanner *et al.*, 1975). Courtesy of the authors and Pergamon Press, Ltd., copyright 1975.

tified metabolites was slightly different when leaves that were still attached to the plant were fed; however, the same compounds were isolated. Apparently the transport from the leaves plays a role. It is probably not caffeine itself but allantoin that is translocated from the old leaves to the young ones.

In Fig. 6.27b a hypothetical pathway is proposed for the route of caffeine (I) in *Coffea* species. This scheme shows two alternative pathways: (1) an N^1, N^3-methylxanthine (II) which is subsequently demethylated at the N^3-position (III) and remethylated on C-2 to form the methoxy derivative (IV) and oxidized on C-8 (V) which can be converted to compound VI, which is one of the compounds recently isolated (see Fig. 6.27c, Wanner *et al.*, 1975); (2) the direct conversion to compound VI which can proceed by several alternate pathways. Compound VI can be oxidized to allantoin (VII) and allantoic acid (VIII) and then to CO_2 (IX). Caffeine may be degraded by an unknown pathway into allantoin, allantoic acid, and finally, to CO_2.

Caffeine may be transported in small amounts from leaf to leaf, as was demonstrated by the application of $[N^1\text{-}^{14}C]$caffeine and $[^3H]$caffeine (Baumann and Wanner, 1972). There was translocation from the subtending leaves into the fruits, but considerable translocation occurred from the pericarp into the seed tissue. The results were confirmed by the analysis of grafts between caffeine-free and caffeine-containing species. Apparently the shoot always governs the amount of caffeine in the bean (Melo *et al.*, 1975). During fruit development the relative caffeine content *(Coffea arabica)* of the pericarp decreases from approximately 1.68% to 0.24% (dry-weight basis), but the seeds remain constant at about 1.25%. Calculated on an absolute basis, the pericarp has twice as much caffeine and the seed 20 times the amount of caffeine at maturity as at the beginning of fruit development (Keller *et al.*, 1972).

El-Hamidi and Wanner (1964) reported on the occurrence of caffeine and chlorogenic acid in the coffee plant and found that the variable quantitative relation existed during the development of the leaves and fruit. It led the authors to propose "in view of this evidence we consider it improbable that the complex could act as a metabolic link between caffeine and chlorogenic acid." It is possible that the complex is formed *in vitro* by chemical-physical interaction of chlorogenic acid and caffeine. The metabolic link could then serve as a sink or a trap for the caffeine (in the vacuole?). The concentrations of the two substances were *not* found to change together, and they clearly have different metabolic pathways.

There are two methylated uric acids which are considered by Wanner *et al.* (1975) to be the first products of caffeine degradation. They are testing that hypothesis. This is a particularly attractive hypothesis since the two isolated methylated uric acids were not detected before in any *Coffea* species. These were 2-methoxy-1,9-dimethyl-7,9-dihydro-1*H*-purine-6,8-dione [O^2,1,9-trimethyluric acid] and 1,3,7,9-tetramethyl-7,9-dihydro-1H-purine-2,6,8(3H)-trione (1,3,7,9-tetramethyluric acid) (Figure 6-27c). They were found together in young leaves of *Coffea liberica, C. arnoldiana, C. dewevrei,* and *C. aruwimiensis.* Many *Coffea* species have not been investigated chemically, hence they examined 26 species collected in Africa, Madagascar, and Indonesia before they found the two uric acids. Shortly before the process of decaffeination was developed, Bertrand (1905 and earlier references) discovered the existence of a coffee tree which produced seeds without caffeine, *Coffea humblotiana,* a species from the Grande Comore Island (Chevalier, 1937). Since that time a number of caffeine-free species have been found in Madagascar and the Mascarene Islands (Sylvain, 1976). The possible use of these types of coffee was pointed out by Trigg (1922) who stated that: "Just why the coffee men have not taken advantage of naturally caffeine-free coffee, or of the possibility of obtaining coffee low in caffeine content by chemical selection from the lines (varieties) now used is a difficult question to answer." The idea of using naturally caffeine-free coffee has been considered by roasters of decaffeinated coffees. Crossbreeding of standard varieties with caffeine-free varieties has been the subject of experimentation; however, the coffee brews produced from the experimental varieties with caffeine-free varieties have so far been "undrinkable." So more than fifty years later, little significant progress has been made (Sylvain, 1976).

The possibility that caffeine is harmful to humans is highly controversial. Caffeine is listed as a noncarcinogen and a nonmutagen based on the *Salmonella*–microsome test that uses over 300 compounds (McCann *et al.,* 1975). The literature is abundant with controversial articles like "Coffee and Coronary Disease," "Coffee and Myocardial Infarctions," "Coffee Drinking Prior to Acute Myocardial Infarction," "Cigarettes, Alcohol, Coffee, and Peptic Ulcer," "Coffee Drinking and Death Due to Coronary Heart Disease," and "The Truth About Coffee and Your Health" (respectively, Kannel and Dawber, 1973; Jick *et al.,* 1973, Klatsky *et al.,* 1973, Friedman *et al.,* 1974; Hennekins *et al.,* 1976; Dempewolff, 1975). The most emphatic coffee clearance came from Dawber, Kannel, and Gordon (1974) (see also U.S. Dept. of Health, Education, and Welfare, 1966), the directors of the famed Framingham Study which was the most definitive epidemiological study on cardiovascular disease ever conducted. After following the health histories and social habits of over 5000 people for a period of 12–25 years, the researchers concluded that coffee drinking, as engaged in by the general population, is *not* a factor in the development of heart disease. One reason is that caffeine is rapidly metabolized in the body of humans: it has an average metabolic half-life of 3.5 hr; by the following day the caffeine has virtually disappeared from the blood tissues (Axelrod and Reichenthal, 1953; Burg, 1976). Ames (1975) has calculated that the

Figure 6.27d. Scientists disagree sharply as to whether caffeine is a hazard to humans. Caffeine is consumed in the largest amounts in coffee. Charles J. Grebinger (left) and John Adinolfi (right) taste several coffee samples at the former Coffee Brewing Center in New York City.

caffeine content is approximately 1×10^{-5} M concentration in the body of an average person weighing 150 pounds ($\sim$73 kg) that consumes from two to four cups of coffee per day. No statistically significant relationship between coffee consumption and cardiovascular diseases could be found in either men or women (Dawber, 1976). Nevertheless, we should be concerned about it as humans, since we daily consume relatively large quantities of coffee (Figure 6.27d), tea, and other caffeine-containing drinks and drugs (Sanders, 1969; Boecklin, 1975; Stagg and Millin, 1975).

Little is known about the physiological significance of caffeine formation or the role of caffeine in plants. Perhaps some of these answers will be forthcoming during the next few decades as researchers begin to probe into the metabolism (biosynthesis and catabolism) of caffeine and relate it to purine nucleotide, nucleic acid, and other types of metabolism in coffee plants.

6.7.9. Morphine

Fairbairn and El-Masry (1967), studying the fate of [^{14}C]-labeled morphine (formula on page 192) in *Papaver somniferum,* found that part of the radioactivity could be detected among sugars and amino acids. None of these radioactive metabolites were identified (see Fairbairn and Wassel, 1964; Fairbairn and El-Masry, 1968; Fairbairn *et al.,* 1968a,b); however,

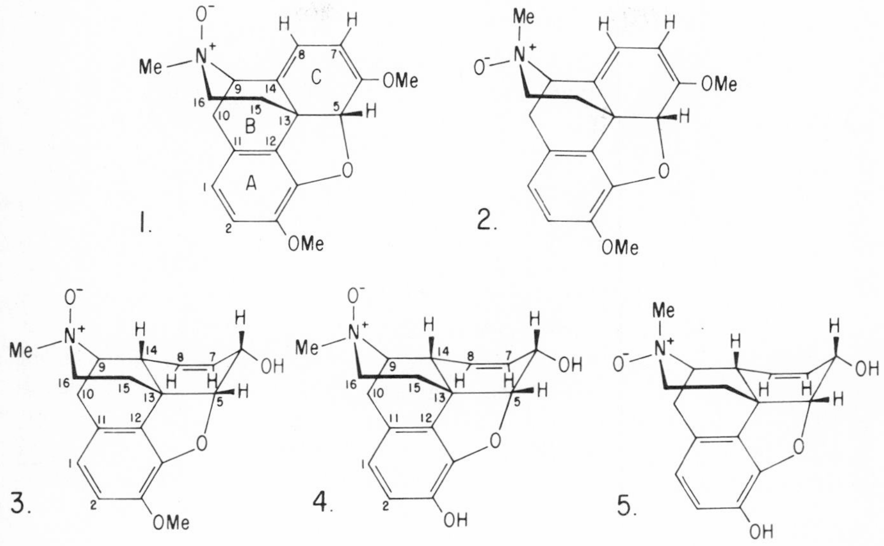

Figure 6.28. *N*-oxides of *Papaver* alkaloids. Thebaine *N*-oxide, major isomer (1); minor isomer (2); codeine *N*-oxide, major isomer (3); minor isomer (4); morphine *N*-oxide, major isomer (5). 1 and 2 were isolated from *P. bracteatum* (thebaine-rich strain), 3, 4, and 5 were isolated from *P. somniferum* (Halle strain), courtesy of J. D. Phillipson and Pergamon Press, Ltd., copyright 1976.

extracts of *P. somniferum* showed the presence of several polar unidentified Dragendorff positive spots. The high rate of morphine turnover led the University of London group to conclude that the alkaloid has an active role in metabolism; one of its roles is to serve as a methylating agent (see normorphine, page 192).

The recent finding of *N*-oxides of thebaine, a relatively minor alkaloid, codeine and morphine (Figure 6-28) by Phillipson *et al.* (1976) in *P. bracteatum* and *P. somniferum* (Halle strain) permits the speculation that they might be involved in active metabolism. Phillipson *et al.* found low yields of the three alkaloid *N*-oxides and suggested that they do not accumulate but are either transformed into other metabolites or returned to the corresponding bases. *P. somniferum* seeds contain "bound" forms of alkaloids (Sec. 6.9.2) which on hydrolysis yields tertiary bases (Fairbairn and El-Masry, 1968) and their finding of *N*-oxides offers one of the best plausible explanations at this time.

6.8. Subcellular Localization of Alkaloids

In 1973, Fairbairn *et al.* showed that the pellet obtained by centrifuging the latex of *Papaver somniferum* was capable of performing certain biosynthetic reactions leading to morphine. The pellet consists of vacuolar particles, and the name proposed for these particles was "alkaloid vesicle." The alkaloid vesicles were suggested to have the following functions: (1) storage (95–99%) of alkaloids, (2) biosynthesis and catabolism of alkaloids, and (3) translocation of alkaloids. It is indeed a remarkable finding that a specific organelle in the plant has been credited with these three functions. They found that the alkaloids appear to be stored in the vacuolar sap of the vesicles rather than membrane-bound; the evidence indicated that the stem latex and vesicles are translocated into the capsule during its rapid expansion after the petal fall. At this time these authors think that morphine is capable of being biosynthesized and metabolized in the vesicles and that the metabolites pass out of the latex into the ovules and the pericarp, which indicates that the vesicles are therefore not passive accumulators of alkaloids (Fairbairn *et al.*, 1974). In the same laboratory Antoun and Roberts (1975a,b,c) could not find in the vesicles those particular enzymes which were involved in the biosynthesis of the morphine-type alkaloids; therefore, more detailed investigation is necessary.

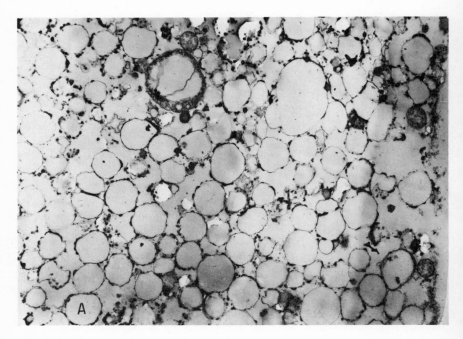

Figure 6.29a. A longitudinal ultrathin section of part of a latex vessel in a *Papaver somniferum* capsule showing smooth and granulated alkaloidal vesicles. Glutaraldehyde/osmium tetroxide fixation with lead citrate poststaining and Epon embedding × 10,000 reduced 25% for reproduction.

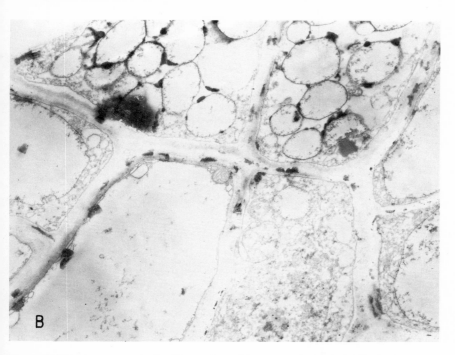

Figure 6.29b. A transverse ultrathin section of latex vessels in the pedicel of *Papaver somniferum* showing smooth and granulated alkaloidal vesicles. Glutaraldehyde/osmium tetroxide fixation with lead citrate poststaining and Araldite embedding × 18,000 reduced 20% for reproduction.

Dickenson and Fairbairn (1975), working on the ultrastructure of the alkaloidal vesicles of *Papaver somniferum,* found that these could be obtained in two forms: the first with a smooth but progressively granulated outer membrane (Figures 6.29a and 6.29b) and the second, probably derived from the first, with adherent caplike structures which in the heavier centrifuged fractions possessed a zonally ordered interior (Figure 6.29c).

Of considerable interest is the work of Madyastha *et al.* (1976), who reported the subcellular localization of a cytochrome P-450 dependent monooxygenase (see page 245) in vesicles of *Catharanthus roseus.* By differential and sucrose density gradient centrifugation, the vesicles could be distinguished from the endoplasmic reticulum, Golgi bodies, mitochondria, and plasma membranes, monitored by marker enzymes and electron microscopy. Examination by electron microscopy (Ridgway and Coscia, 1976) revealed vesicles of varying diameters (Figure 6.30), which contained all of the geraniol hydroxylase activity. Examination of the band obtained by sucrose gradient centrifugation of the microsomal (1,000,000g) pellet show vesicles varying in both size and membrane profiles, Figure 6.30(A).

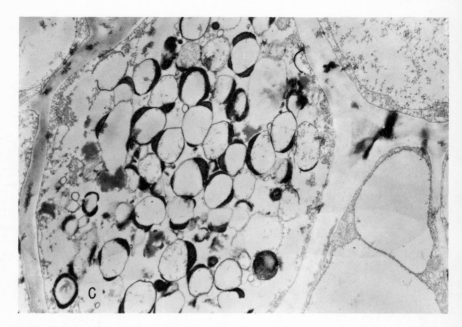

Figure 6.29c. A transverse ultrathin section of a latex vessel in the pedicel of *Papaver somniferum* showing alkaloidal vesicles with adherent caps. Glutaradehyde/osmium tetroxide fixation with lead citrate poststaining and Araldite embedding ×18,000 reduced 25% for reproduction. Courtesy of Dickenson and Fairbairn (1975).

--→

Figure 6.30. (A) Electron micrograph of the major protein band obtained by sucrose gradient centrifugation of the microsomal pellet. Note the presence of membrane profiles and vesicles measuring 50 nm in thickness, × 66,000, reduced 25% for reproduction. (B) Electron micrograph of the pelleted yellow band obtained by sucrose gradient centrifugation. The vesicles which contained 90% of the total hydroxylase activity were 0.2–2 μm in diameter and 40–60 nm in membrane thickness, × 26,000, reduced 25% for reproduction. (C) Electron micrograph of intact mesophyll cell from 5-day-old etiolated cotyledons. Note the large central vacuole (V) with dense inclusions, ×17,000, reduced 25% for reproduction. (D) Electron micrograph of intact mesophyll cell from 5-day-old etiolated cotyledons. Note the presence of provacuoles (pv) undergoing fusion with each other and coalescing with the central vacuole (V) (arrow), × 17,000, reduced 25% for reproduction. The micrographs are from the *Catharanthus roseus (Vinca rosea)*. Courtesy of John E. Ridgway and Carmine I. Coscia, St. Louis University.

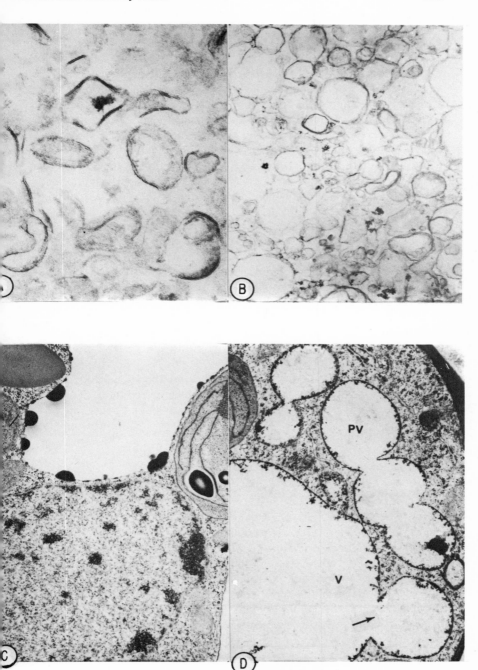

The membranes comprising this fraction measure 40–90 nm. The thicker, pleomorphic, membranes which are in greatest abundance in this fraction appear to be derived from the plasmalemma. In contrast, electron micrographs, Figure 6.30(B), of the yellow hydroxylase-containing band reveal vesicles 0.2–2 in diameter, with only 40–60 nm in membrane thickness. Vesicles of a similar size containing a yellow pigment can be observed by light microscopy of free-hand sections of cotyledons. Examination by electron microscopy of fixed tissue of 5-day-old etiolated cotyledons shows a large central vacuole with dense inclusions, Figure 6.30(C). In Figure 6.30(C) it can be seen that the large central vacuole is derived by coalescence of provacuoles which resemble the vesicles in membrane thickness and size range and which were found in the yellow hydroxylase-containing band, see Figure 6.30(B). At this time they have no evidence that these are "alkaloid vesicles." It is not known if these vesicles carry out metabolism of the indole alkaloids, although the evidence for their role in biosynthesis of the alkaloids seems to have been established unequivocally.

Roddick (1976) isolated pericarp tissue from green tomato fruits *(Lycopersicon esculentum)* homogenized and separated into organelle fractions by differential centrifugation. He found that α-tomatine (Figure 6.7) was mainly located in the 105,000g supernatant, with small amounts in the microsomes. He suggested that tomatine accumulates in the vacuole and/or soluble phase of the cytoplasm and is possibly synthesized in microsomal organelles.

Thus this new body of evidence implicating the vacuole containing the alkaloids and possessing some of the enzymes required for alkaloid biosynthesis offers intriguing possibilities concerning alkaloid degradation.

6.9. Bound Forms of Alkaloids

6.9.1. Coniine and γ-Coniceine

Evidence for a type of "bound alkaloid" emerged in the course of Fairbairn's studies on the distribution and variation in content of alkaloids in hemlock *(Conium maculatum)*. The "bound form" is supposed to yield an alkaloid in the plant tissue upon hydrolysis. It may be possible that nucleotide-like compounds exist (see Section 6.4).

The first paper of the series (Fairbairn and Challen, 1959) documented marked changes of alkaloid content during the period from the development of flowers up to the production of the mature fruits. Most striking were the variations in content of γ-coniceine and coniine, which provided a hypothesis of actual interconversion of these two alkaloids (Fairbairn and

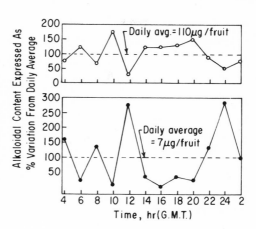

Figure 6.31. Two-hourly changes in the coniine and γ-coniceine content of 2-week-old fruits of *Conium maculatum* during a period of 24 hr. Variation expressed as percentage of the daily average; coniine daily average content (24 hr) = 110 μg/fruit—open circles; γ-coniceine daily average content (24 hr) = 7 μg/fruit—filled circles (Fairbairn and Suwal, 1961). Courtesy of the authors and Pergamon Press, Ltd., copyright 1961.

Suwal, 1961) (Figure 6.31). The alkaloid contents studied at 2-hr intervals (Fairbairn and Suwal, 1961) showed surprisingly rapid changes during the day, and the complementary nature of coniine and γ-coniceine contents was even more surprising (Figure 6.31). The authors note that the change from γ-coniceine to coniine and vice versa might be related to an oxidation–reduction process. But the two alkaloids do not appear to be quantitatively interconvertible. The amount of coniine that disappears during any 2-hr interval is several times as high as that which could be accounted for with γ-coniceine (or as any other of the minor alkaloids which also have been determined). The authors come to the possibility that coniine could reversibly combine with another molecule to form a substance behaving as a nonalkaloid.

The third paper of Fairbairn's series brings proof of existence of such substances in developing fruit (Fairbairn and Ali, 1968a). Three types of unidentified compounds have been separated on the basis of solubility in water, in 70% ethanol, and in pure ethanol. Upon hydrolysis by acid or alkali, all three groups of compounds yielded Dragendorff-positive substances. Some were not identified, but coniine and γ-coniceine were positively detected.

Radioactive γ-coniceine fed to plants bearing maturing fruit was rapidly deposited into the fruit, and the radioactivity could be recovered as bound and free forms of not only γ-coniceine itself but also of coniine. The same occurred after administration of radioactive coniine (Fairbairn and Ali, 1968b). Specific radioactivities of bound forms of alkaloids fluctuated. This is regarded as indicating the involvement of bound alkaloids in important metabolic processes. On the contrary, free coniine and γ-coniceine retained their specific radioactivity essentially unchanged even after 19 days, a fact which might be explained by the existence of some recycling systems. It was speculated that the bound forms might represent types of nucleotide-like compounds with a function similar to that of the couple $NAD^+ \rightleftarrows NADH + H^+$.

The kinetics of [^{14}C] incorporation into the alkaloidal fraction of *Conium maculatum* (10–12 weeks old), followed for periods of 0–7 days after 1-hr photosynthesis in an atmosphere of $^{14}CO_2$, were studied by Dietrich and Martin (1969). The specific activity time curves, crossover points, and specific activity to pool ratios were generally consistent with the sequence $CO_2 \rightarrow \gamma$-coniceine $\rightarrow$ coniine $\rightarrow$ N-methylconiine (Figure 6.32). If γ-coniceine were synthesized by the "bound" pools, its specific activities would be expected to fluctuate accordingly (Fairbairn and Challen 1959; Fairbairn and Suwal, 1961; Fairbairn and Ali, 1968a, b) when the ripening fruits have hour-to-hour and day-by-day fluctuations (see Figures 6.4

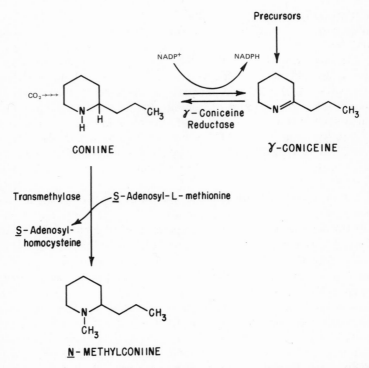

Figure 6.32. Interconversion of the hemlock alkaloids coniine and γ-coniceine; the pathway of N-methylconiine formation (Roberts, 1971, 1974, 1975). Courtesy of the author and Pergamon Press, Ltd., copyright 1971, 1974, 1975. (Dietrich and Martin, 1969). Courtesy of the authors and the journal.

and 6.31). However, since the specific activities were independent of pool fluctuations, the free (extractable) pool of γ-coniceine most likely represents the newly synthesized material and its removal to bound pools would not affect its specific activity. A similar consideration would apply to coniine and N-methylconiine.

A comparison of the *total activity* ratios (since all three compounds were recovered to the same extent) is indicative of the relative turnover of these three compounds; the total activity ratios γ-coniceine:coniine:N-methylconiine were 1770:1:1 at 4 hr, 165:26:1 at 11 hr, and 80:40:1 at 35 hr. These ratios indicate that γ-coniceine is not turning over faster than coniine, nor the latter faster than its N-methyl derivative, and that γ-coniceine is not being converted into other compounds at rates which compare with its conversion into coniine.

Recently Roberts (1975) isolated a γ-coniceine reductase from leaves and fruit of a number of *Conium maculatum* cultivars which are NADH-dependent (Figure 6.32). The ready interconversion of γ-coniceine to (+)-coniine may have an important role in the regulation of alkaloid metabolism. The reduction of γ-coniceine requires NADPH rather than NADH (the reaction proceeds approximately seven times faster with NADPH), and experiments indicate that the stereospecific removal of hydride is from the B(pros) side of NADPH. The availability of precursors and products may have an effect on the substrate concentration in the *in vivo* systems.

6.9.2. Morphine and Related Alkaloids

On the basis of experiences with fluctuation of the alkaloid content in hemlock, Fairbairn turned to *Papaver somniferum*. His first such study (Fairbairn and Wassel, 1964) brings clear evidence for rapid variations of the content of major alkaloids in the latex of developing fruit. This proved that thebaine, codeine, and morphine (formulas in Figure 6.12) might play some dynamic role in the plant. Especially surprising was a decrease in morphine concentration.

Since it has been unequivocally proven by Stermitz and Rapoport (1961) that the biosynthetic sequence thebaine → codeine → morphine is irreversible, and since the transport of morphine had been made, very probably, the subject of other pertinent papers, Fairbairn and Wassel (1964) suggested the presence of some sort of bound morphine. This conclusion was strengthened by the study with [14C]-labeled tyrosine (Fairbairn *et al.*, 1964) as a precursor. Biosynthesis of alkaloids in the isolated latex provided evidence for the variation in content of all alkaloids and the disappearance of morphine. This indicated that morphine is transformed into some nonalkaloidal compounds in the latex itself rather than elsewhere in the plant.

When [^{14}C]-labeled morphine was administered (Fairbairn and El-Masry, 1967), it was rapidly metabolized into nonalkaloidal molecules that disappeared from the latex. Recently, it was shown that the reaction proceeds through alkaloidal substances (Fairbairn and Steele, 1977). Most of the new radioactive substances appeared in the pericarp and the ovules.

Finally Fairbairn and El-Masry (1968) succeeded in proving the existence of bound alkaloids in poppy seeds. Poppy seeds normally contain only traces of alkaloids. The authors discovered "alkaloidlike" substances, codeine among them, after acid hydrolysis or pepsin digestion of ground seeds. These compounds were radioactive if the maturing capsule had been fed radioactive morphine. They also reported formation of these alkaloidlike compounds as a result of the action of a crude enzyme preparation from the plant upon morphine.

A possible biological role of the bound alkaloids in poppy seeds was established by the following experiment: During the maturation process, the capsules were deprived of most of their latex by its bleeding out. Mature seeds did not show any visible difference in comparison with controls, but during the first 10 days of germination the development of the deprived seeds significantly lagged behind that of the normal seeds. Later the differences disappeared, and normal, healthy plants developed; possibly *de novo* synthesis of alkaloids provided what the seeds had been missing.

6.10. Regulation of Metabolic Pathways

It has been demonstrated in this book (Chapter 2) that alkaloid formation is genetically controlled and is also affected by the plant's environment. It should be remembered that, according to Mothes (1965), the presence of an alkaloid in a plant means that the plant has the ability both to biosynthesize and catabolize the alkaloid. More importantly, it means that the alkaloid is nontoxic to the plant and that the rate of catabolism is slow enough, or the rate of biosynthesis is fast enough, to permit some alkaloids to accumulate.

Floss *et al.* (1974) have recently published a review entitled "Regulatory Control Mechanisms in Alkaloid Biosynthesis." While the article is not all-inclusive on the subject of alkaloid biology and metabolism, we will use some of its arguments and examples, since it represents the first extensive treatment of the subject of regulation. The pyridine alkaloids, e.g., nicotine, anabasine, and ricinine, have a special relationship to primary metabolic reactions that occur in *Nicotiana* spp. and *Ricinus communis*. The control mechanism appears to involve the regulation of the pyri-

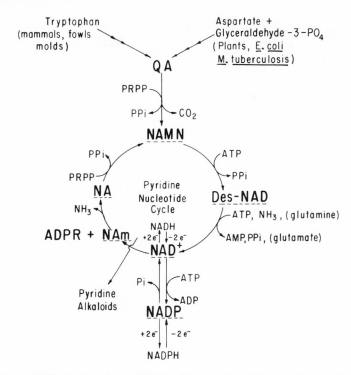

Tryptophan
(mammals, fowls
molds)

Aspartate +
Glyceraldehyde $-3-PO_4$
(Plants, E. coli
M. tuberculosis)

Figure 6.33. Pyridine ring metabolism (for meaning of abbreviations see text).

dine nucleotide cycle and the metabolic pathway leading to the alkaloids. Once quinolinic acid is formed, NAD and nicotinic acid biosynthesis appear to proceed by the same pathway in almost all organisms (Figure 6.33) (Waller *et al.*, 1966).

Quinolinic acid[5] is the common intermediate in the forms of life that produce it by the oxidative degradation of the tryptophan pathway (e.g., animals, molds, and fungi) and those that use condensation of a C_3

[5]Gholson *et al.* (1976) have studied quinolinic acid (QA) biosynthesis, in a cell-free system prepared from *E. coli* mutants. In this system QA is synthesized by the condensation of aspartate and a [3-^{14}C]dihydroxyacetone phosphate which is incorporated into the C-4 of QA. An FAD-requiring reaction is catalyzed by two partially purified proteins which they call quinolinate synthetase. Quinolinate synthetase is composed of protein A (MW about 35,000) and protein B(MW about 85,000). Preincubation of A and B proteins leads to inactivation of at least the A protein and DHAP prevents this inactivation reaction. Neither the A nor the B protein binds aspartate-^{14}C but in the presence of both proteins aspartate is bound to an entity with an apparent MW greater than the B protein.

compound with a C_4 compound (e.g., plants and bacteria). The existence of these two pathways represents a major deviation from the "unity principle," which has become a widely accepted tenet in modern biochemistry. The chemical nature of the diversity in the sequence of reactions is well established, but many questions remain to be answered.

The first step in conversion of quinolinic acid is the reaction catalyzed by quinolinic acid phosphoribosyl transferase to give nicotinic acid mononucleotide. Nicotinic acid mononucleotide which reacts with ATP in the presence of phosphoribosyladenyl transferase gives desamidonicotinamide adenine dinucleotide (desNAD). NAD is formed by amidation of desNAD in the presence of ammonia or glutamine, ATP and NAD synthetase (Preiss and Handler, 1958; Imsande, 1961; Spencer and Preiss, 1967). NAD and NADP are interconvertible in the cell by the action of NAD kinase and NADP phosphatase (Gholson, 1966). NAD synthesis from nicotinic acid was first observed by Preiss and Handler in 1958 in erythrocytes and is usually referred to as the Preiss–Handler pathway. NAD may be cleaved to form nicotinamide and ADPR by NAD glycohydrolase (Kaplan, 1961). The reutilization of nicotinamide is initiated by its hydrolysis to nicotinic acid by nicotinamidase, an enzyme found in plants (Joshi and Handler, 1960; Newell and Waller, 1965), mammalian liver (Petrack *et al.*, 1963), and bacteria and yeast (Joshi and Handler, 1960, 1962). The cycle is completed by the synthesis of NAMN from free nicotinic acid and PRPP by nicotinic acid phosphoribosyl transferase (Gholson and Kori, 1964). Alternatively, nicotinamide is converted directly to nicotinamide mononucleotide, which is converted directly to NAD, which is converted back to nicotinamide in rat liver (Keller *et al.*, 1971). It has been postulated that this shunt can operate in plants (Ryrie and Scott, 1969), but definitive evidence is lacking.

Loss of the nicotinyl moiety from the cycle may occur from any compound; however, the predominant evidence suggests that such loss occurs as nicotinic acid or nicotinamide, which may be excreted either as the free compounds or as derivatives, depending on the species involved. Methylation of nicotinamide occurs in liver and provides an important excretion product of NAD-related compounds in many mammalian species (Kaplan, 1961). In plants, methylation of nicotinic acid to give trigonelline is widespread, and the trigonelline can be reutilized to provide free nicotinic acid for NAD biosynthesis. The recent finding in this laboratory (Lee and Waller, 1972b) of the accumulation of large amounts of N-methylnicotinamide from [6-^{14}C]NAD and [6-^{14}C]nicotinic acid in *Tripterygium wilfordii* makes it necessary to reexamine the widely accepted view that N-methylnicotinic acid is the most common storage form of the vitamin in plants.

The reaction sequence leading to the condensation of a C_3 and a C_4 compound to form quinolinic acid(QA), NAMN, desNAD, and NAD makes up the *de novo* pathway of NAD biosynthesis. The formation of NAMN from nicotinic acid serves as the regular "salvage" or recycling pathway. The reaction involving the direct conversion of nicotinamide to NAMA constitutes a "shunt" route to NAD formation. The biosyntheses of ricinine, nicotine, *N*-methyl-2-pyridone-5-carboxamide, nudiflorine, wilfordic acid, hydroxywilfordic acid, and other plant-produced nicotinic acid derivatives represent metabolic branches from an unknown point in the cycle. Recent evidence from inhibition studies in this laboratory (Johnson and Waller, 1974) indicates that the branch point for ricinine biosynthesis probably occurs between NAMN and desamido NAD; also, the possibility still exists that the biosynthesis of ricinine from quinolinic acid occurs by a separate pathway (Hiles and Byerrum, 1969), independent of the pyridine nucleotide cycle. The inhibitor experiments (Johnson and Waller, 1974) indicated an interdependency between ricinine biosynthesis and the pyridine nucleotide cycle, but the order of intermediates in ricinine biosynthesis was not elucidated. These experiments suggest that the biosynthesis of ricinine (in the family Euphorbiaceae) is regulated at the precursor stage.

Evidence for the obligatory role of the pyridine nucleotide cycle in pyridine alkaloid biosynthesis was provided by Byerrum's laboratory (Mann and Byerrum, 1974a,b). Using etiolated seedlings of *Ricinus communis,* these authors found a 12-fold increase in ricinine content over a 4-day period and a sixfold increase in quinolinic acid phosphoribosyl transferase activity that preceded the onset of ricinine biosynthesis by 1 day. The castor bean endosperm was an especially rich source of quinolinic acid phosphoribosyl transferase; this material was found to contain 150 times as much as was found in most plants. The enzyme catalyzes the formation of nicotinic acid mononucleotide from quinolinic acid and phosphoribosyl pyrophosphate, and it appears to be at a point at which regulation may take place in the pyridine nucleotide cycle. Along the same line of reasoning, Gholson *et al.* (1964) showed that a crude beef liver homogenate was capable of converting quinolinic acid to nicotinic acid mononucleotide which was inhibited by 10^{-4}M concentration of NAD: the finding led them to speculate that the NAD in liver may serve to regulate its biosynthesis from quinolinic acid. In a cell-free extract from the alga *Astosia longa,* nicotinic acid inhibited quinolinic acid phosphoribosyl transferase activity (Kahn and Blum, 1968). Nicotinic acid also represses the synthesis of the enzyme in *Bacillus subtilis* (Gholson and Kori, 1964). However, for ricinine biosynthesis the point of regulation appears not to be in the conversion of quinolinic acid to nicotinic acid mononucleotide, since Mann and Byerrum (1974a, b) found no inhibiting action by ricinine of quinolinic acid phosphoribosyltransferase from etiolated castor bean endosperm which had been purified approximately 500-fold.

At this time, the evidence indicates that the formation of the pyridine moiety of the tobacco alkaloids (family Solanaceae), nicotine and anabasine; the *Tripterygium wilfordii* alkaloids, wilfordic acid and hydroxywilfordic acid (family Celastraceae); and the alkaloids of *Trewia nudiflora* (family Euphorbiaceae), *N*-methyl-2-pyridone-5-carboxamide and nudiflorine (*N*-methyl-5-cyano-2-pyridone) closely parallels the biosynthesis of ricinine. It was demonstrated that nicotinic acid can serve as a precursor to nicotine (Dawson *et al.*, 1960), anabasine (Solt *et al.*, 1960), *T. wilfordii* alkaloids (Lee and Waller, 1972b), and those of *T. nudiflora* (Sastry and Waller, 1972). For *T. wilfordii* alkaloids and nicotine (Yang *et al.*, 1965; Frost *et al.*, 1967), each of the compounds present in the PN cycle is incorporated with an efficiency similar to that of quinolinic acid and nicotinic acid. Chandler and Gholson (1972) found an O_2-dependent release of $^{14}CO_2$ from [^{14}C]nicotinic acid in tobacco roots which did not occur in the leaves or the stem. It was not found in extracts of cotyledon of *Ricinus communis* seedlings. Mann and Byerrum (1974a,b) discovered elevated levels of quinolinic acid phosphoribosyl transferase in *Nicotiana rustica* roots that synthesized nicotine. If, as it now appears, the biosynthesis of ricinine is regulated at the level of precursor formation, Floss *et al.* (1974) speculate that the formation of nicotine and anabasine (and we include the alkaloids from *T. wilfordii* and *T. nudiflora*) may also be regulated in a similar way because of its close similarity to the biosynthesis of the pyridine moiety of the PN cycle compounds. This is summarized in Figure 6.34.

The three enzymes that catalyze the first three reactions leading to nicotine from ornithine were isolated from *Nicotiana tabacum* by Mizusaki *et al.* (1968, 1971, 1972, 1973). The 4-week-old plants were decapitated, and the three enzyme activities increased two- to tenfold, reaching maximum activity 24 hours after the decapitation, and then decreased again during the following 1–2-day period. The nicotine that accumulated in the roots followed a similar pattern. The three enzyme activities that were measured were ornithine decarboxylase, putrescine *N*-methyl transferase, and *N*-methyl putrescine oxidase. *In vitro* experiments were conducted with nicotine as an inhibitor; the results were negative. Indoleacetic acid (IAA, 2.5–5.0 μm) significantly increased these enzyme activities in roots of decapitated plants, whereas higher concentrations of IAA prevented the rise in enzyme activities promoted by decapitation. Nicotine strongly inhibited the rise in enzyme activities in roots of decapitated plants in all cases. Thus it appears that the regulation of the synthesis of nicotine from ornithine (see Figure 6.32) results in the three enzymes being under the

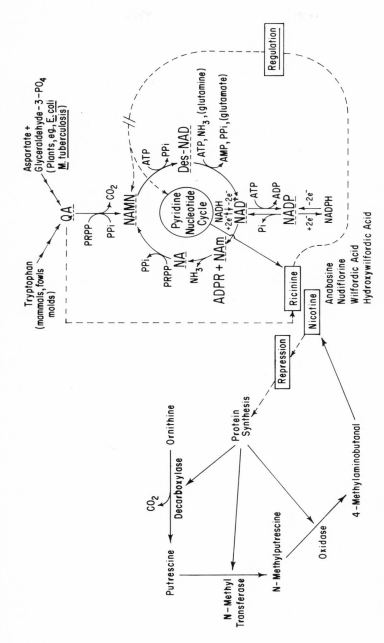

Figure 6.34. Proposed regulation of the pyridine moiety of the pyridine alkaloids and of the *N*-methylpyrrolidine moiety of nicotine.

control of a common regulatory system in which auxin and nicotine are important components. An increase in enzyme concentration in the roots 0.02–0.5 mm is high enough to repress the formation of the three enzymes at least 50%. This would be an example of a feedback regulation of alkaloid biosynthesis.

Another system which Floss *et al.* (1974) describe regulates alkaloid precursor synthesis, which had been studied to some extent in the ergot fungus *Claviceps purpurea*. The clavine ergot alkaloids are derived from tryptophan, mevalonic acid, and the methyl group of methionine (Weygand and Floss, 1963) in the biosynthetic pathway shown in Figure 6.35. Floss *et al.* (1974) and Arcamone *et al.* (1962) have studied the effect of various

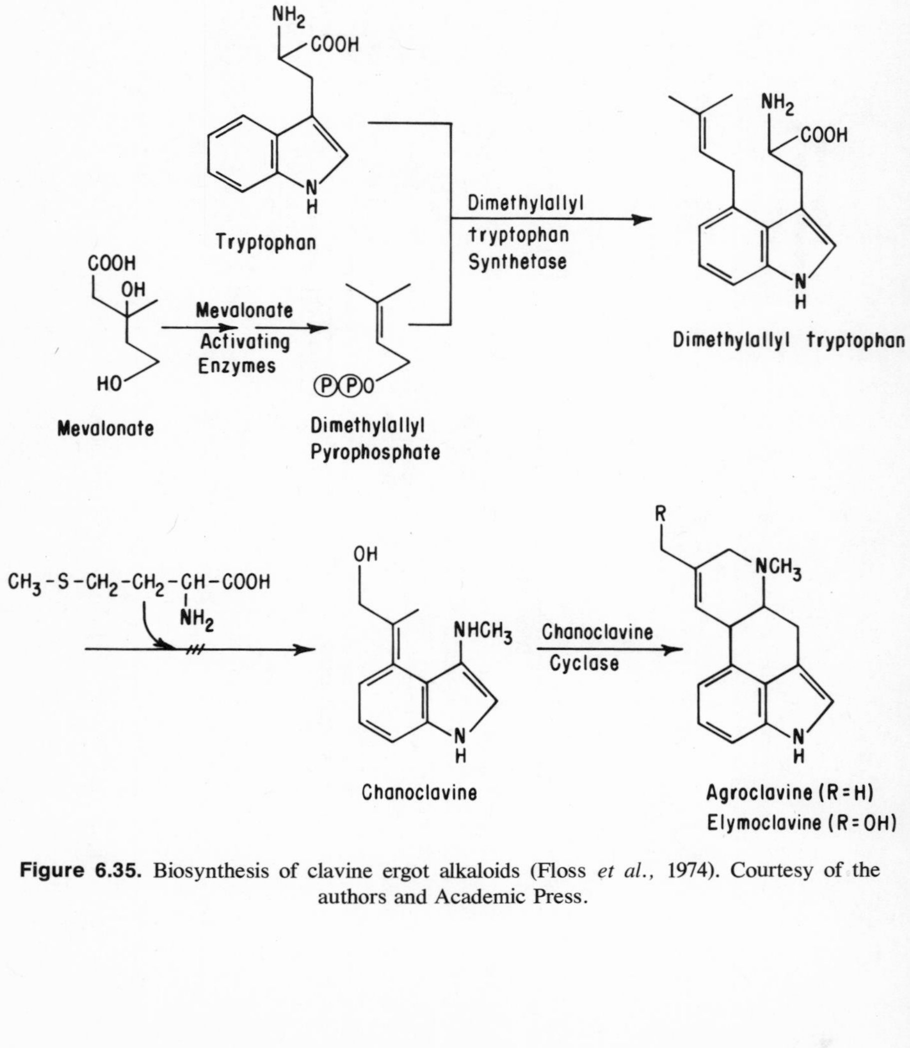

Figure 6.35. Biosynthesis of clavine ergot alkaloids (Floss *et al.*, 1974). Courtesy of the authors and Academic Press.

tryptophan analogues. Some of these analogues increase alkaloid formation and are effective inhibitors of tryptophan synthesis in other microorganisms. These observations provided a stimulus for activity by Lingens *et al.* (1967), Lingens (1972), Eberspaecher (1970), Schumauder and Groeger (1973), Heinstein *et al.* (1971); Mothes *et al.* (1965); ar. 1 of course by Floss *et al.* (1974) to study the regulation of aromatic amino acid biosynthesis in *Claviceps*.

Floss *et al.* (1974) goes ahead to describe, in succession, induction of alkaloid biosynthesis, feedback regulation of alkaloid biosynthesis, and induction of alkaloid degradation (catabolism). Only two references are to be found about the induction of alkaloid catabolism—those of Schuette and his group (Gross *et al.*, 1970) and Digenis (1969). Schuette and his group found that feeding increasing amounts of [*methylene*-^{14}C] gramine produced increasing specific radioactivity of cellular gramine but did not increase the total amount of gramine in the plant. This observation leads one to think that at a certain concentration gramine induces its own degradation. Much work remains to be done; for example, the catabolism of gramine has not been thoroughly investigated. If the catabolic pathway is a reversal of the biosynthetic pathway (Digenis, 1969), the balance of the cellular concentrations of gramine and tryptophan could force the equilibrium either to tryptophan formation from gramine or to the formation of gramine from tryptophan.

Indole alkaloid biosynthesis presents us with another possible control mechanism, although it is in biosynthesis rather than in catabolism. The alkaloid catharanthine is biosynthesized in seedlings of the plant *V. rosea* from three different primary metabolites: tryptophan, methionine, and the monoterpene geraniol. These same three primary metabolites give four different families of indole alkaloids, the iboga, strychnos, aspidosperma, and corynanthe (Waller, 1969; Scott, 1970). Meehan and Coscia (1973), Madyastha *et al.* (1976), and McFarlane *et al.* (1975), from Coscia's group, isolated a cytochrome P-450-dependent monooxygenase which converts both geraniol and its *cis*-isomer, nerol, to their corresponding 10-hydroxy derivatives. Geraniol is also incorporated into triterpenoids, steroids, and other systems; its hydroxylation probably represents one of the first "committed" steps of the alkaloid pathway. Catharanthine was found to inhibit (Table 6.6) the membrane-bound cytochrome P-450 monooxygenase which oxidizes geraniol to 10-hydroxygeraniol. Although vinblastine is a *bis*-indole alkaloid containing monomeric structures having the skeleta of vindoline and vincadine, an iboga alkaloid structurally similar to catharanthine, it does not inhibit the hydroxylase. Since this hydroxylase catalyzes one of the first committed steps in the biosynthesis of indole alkaloids, these authors postulate feedback inhibition of indole alkaloid production

Table 6.6. Effect of Some Indole Alkaloids *(Vinca rosea)* on
Geraniol Hydroxylase Activity[a]

Alkaloid	Concentration, mM	Percent activity
Vinblastine	0.5	79
	1.0	83
Vindoline	0.5	88
	1.0	70
Catharanthine	0.5	68
	1.0	40
Catharanthine[b]	0.5	44
	1.0	23
Catharanthine[c]	0.5	33
	1.0	13

[a] From McFarlane *et al.* (1975). Courtesy of C. I. Coscia, St. Louis University.
[b] Solubilized protein was used in this assay. In the absence of an alkaloid enzyme, activity was 6.1 nmol/hr/mg protein versus 9.65 nmol/hr/mg protein for the 20,000g pellet.
[c] Solubilized protein in these experiments was subjected to ultrafiltration to remove about 90% of the cholate, which had been added to produce emulsification, and enzyme activity was 4.66 nmol/hr/mg protein without alkaloid.

(Figure 6.36) based upon the results with geraniol hydroxylase and loganic acid methyl transferase (Madyastha *et al.,* 1973). Of five alkaloids tested (ajmalicine, perivine, vinblastine, vindoline, and catharanthine), only catharanthine inhibits geraniol hydroxylation appreciably in a 0.25–1.0 mm concentration range. On the other hand, vindoline is an effective inhibitor of the methyl transferase, whereas preliminary results with other alkaloids (ajmalicine, perivine, and catharanthine) show no inhibition. The monoterpene hydroxylase reaches maximal specific activity by the 5th day after germination and this activity decreases thereafter. The mature plant possesses only 10% of the activity of the 5-day-old seedling. Interestingly, Scott *et al.* (1962) has found catharanthine to be synthesized by the fifth day and accumulate after that. Similarly, the methyl transferase activity is maximal in 8 days and its inhibitor vindoline appears in substantial amounts by the 9th day (Scott, 1970). One can speculate that geraniol is converted to hydroxylated derivatives and subsequent intermediates of the indole alkaloid pathway until the catharanthine concentration increases. Then, monoterpene hydroxylation is attenuated by feedback inhibition while intermediates beyond will be methylated rapidly until vindoline inhibits the methyl transferase. This mechanism might be described as a type of metabolite pulse and is consistent with the accumulation of metabolites either nonalkaloidal or alkaloidal between steps. It is suggestive of the possibility of synergistic or cumulative-type of feedback inhibition.

6.11. Use of the Metabolic Grid to Explain the Metabolism of Ricinine in *Ricinus communis* and Quinolizidine Alkaloids in Leguminosae

Although the results presented so far indicate an interdependency between the PN cycle and ricinine biosynthesis, the details concerning the order of intermediates in ricinine biosynthesis have not been elucidated. The conflicting evidence (Waller *et al.*, 1965; Hiles and Byerrum, 1969) in which the PN cycle intermediate is the immediate precursor of ricinine may best be explained by a complicated relationship between the cycle and ricinine biosynthesis. One possible relationship is the metabolic grid concept, which is defined as a series of parallel reactions in which analogous transformations occur, but at different rates; thus a compound may be converted to a product by several different parallel pathways (Bu'Lock, 1965). The proposed metabolic grid, as shown in the structure, Figure 6.37, in which several of the PN cycle intermediates enter into the ricinine biosynthesis pathway, explains the similarity of radioisotopic incorporation results when each of the PN cycle intermediates ([^{14}C]-labeled) were fed (Waller *et al.*, 1965). It is presumed that these reactions are catalyzed by their respective enzymes. The report that a tenfold excess of NAD$^+$ blocks

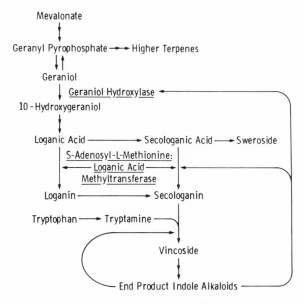

Figure 6.36. Proposed feedback of regulation of genaniol hydroxylase by indole alkaloids *(Vinca rosea)*. Courtesy of Carmine I. Coscia, St. Louis University).

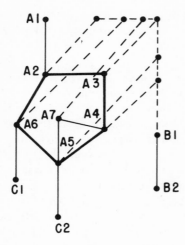

Figure 6.37. A metabolic grid proposed for the biosynthesis and control of metabolism of ricinine. A1—quinolinic acid; A2—nicotinic acid mononucleotide; A3—nicotinic acid adenine dinucleotide; A4—nicotinamide adenine dinucleotide; A5—nicotinamide; A6—nicotinic acid; A7—nicotinamide mononucleotide; B1—*N*-demethylricinine; B2—ricinine; C1—*N*-methylnicotinic acid; and C2—*N*-methylnicotinamide. pyridine nucleotide cycle and postulated reaction sequence (Nowacki and Waller 1975a). Courtesy of Pergamon Press, Ltd., copyright 1975.

the cycle at the step of NAD^+ of radioactivity into ricinine from $[6\text{-}^{14}C]QA$ (Hiles and Byerrum, 1969) does not necessarily mean that ricinine was biosynthesized from QA independently of the PN cycle, but it can be readily explained as a shunting of radioactivity from $[6\text{-}^{14}C]QA$ into ricinine through the first two cycle intermediates, nicotinic acid mononucleotide (NaMN) and desNAD$^+$, since excess NAD^+ blocks the cycle at the step of NAD^+ formation from desNAD$^+$. It has also been reported that QA is a better precursor of ricinine than the PN cycle intermediates (Hiles and Byerrum, 1969); but this conflicts with other results (Waller *et al.*, 1966). The possibility remains that ricinine is biosynthesized from QA by two pathways, one through the cycle and one independent of it. However, this seems less likely, since inhibition of the conversion of QA to the PN cycle intermediates by azaserine and azaleucine also inhibited incorporation into ricinine. If there were a major cycle-independent pathway from QA to ricinine, one would not expect to see an essentially complete block of ricinine formation upon inhibition of the PN cycle.

Nowacki and Waller (1975a,b) have recently employed the metabolic grid to explain the metabolism of quinolizidine alkaloids in the Leguminosae. It was used, together with available data, in speculation on the origin of alkaloids in species of Sophorae, Podalyrieae, and Genisteae (see Chapter 1, p. 33). Cyclic pathways are possible in some parts of the metabolic grid, e.g., for lupanine, hydroxylupanine, and hydroxylupanine esters. In most cases, there is only circumstantial evidence for this, based on concentration changes of alkaloids during plant growth and on evidence from *in vitro* chemical experiments. The possibilities for regulatory metabolic control are only conjecture on our part at this time.

6.12. Summary

In confirmation of medical folklore, alkaloid-containing plants often go through a maximum in total alkaloid content in the course of their lives; this is important for the botanical-pharmaceutical industry in choosing when to harvest plants for collecting drugs. The rates of biosynthesis and catabolism reflected in the pool size of alkaloids in the tissue of plants are varied with respect to physiological states of development, diurnal variation, and functionally different parts of the plants. There are also different variation patterns for different alkaloids in the same plant. Disappearance of an alkaloid for which half-life can sometimes be estimated implies catabolism, which is conveniently studied by feeding the plant with labeled alkaloids. Alkaloids undergo diversified and extensive biotransformation in plants which produce them; their catabolites vary from cellular macromolecules such as protein to expired CO_2.

Demethylation is the simplest and most investigated biodegradation process; it may mobilize an alkaloid for transport and is connected with senescence, but its general physiological function is unclear. A few pathways and specific metabolites of more extensive degradations have been characterized. Regulatory mechanisms have been worked on very little, hardly enough to deserve mention in alkaloid metabolism. It is to be emphasized, however, that the old belief that alkaloids are inert end products of plant metabolism is quite erroneous; they undergo a variety of degradations, some at appreciable rates.

References

References for Chapter 1, Chemotaxonomic Relationships

Ackermann, D., and List, P. H. (1958) *Hoppe-Seyl. Z. Physiol. Chem.* **313**, 30.

Aniol, A., Kazimierski, T., Nowacki, E., and Waller, G. R. (1972) in *Abh. Dtsch. Akad. Wiss. Kl., Chem. Geol. Biol., Berlin; 4. Internationales Symposium Biochemie und Physiologie der Alkaloide,* Halle (Saale), Germany, June, 1969, pp. 481–491.

Arigoni, D. (1972) in *Abh. Dtsch. Akad. Wiss. Kl., Chem. Geol. Biol., Berlin; 4. Internationales Symposium Biochemie und Physiologie der Alkaloide,* Halle (Saale), Germany, June, 1969, p. 317.

Aslanov, K. A., Kushmuradov, Yu. K., and Sadykov, A. S. (1972) in *Abh. Dtsch. Akad. Wiss. Kl., Chem. Geol. Biol., Berlin; 4. Internationales Symposium Biochemie und Physiologie der Alkaloide,* Halle (Saale), Germany, June, 1969, p. 463.

Balcar-Skrzydlewska, E., and Borkowski, B. (1966) in *Abh. Dtsch. Akad. Wiss. Kl., Chem. Geol. Biol., Berlin; 4. Internationales Symposium Biochemie und Physiologie der Alkaloide,* Halle (Saale), Germany, June, 1969, pp. 265–269.

Balcar-Skrzydlewska, E., and Borkowski, B. (1972) in *Abh. Dtsch. Akad. Wiss. Kl., Chem. Geol. Biol., Berlin; 4. Internationales Symposium Biochemie und Physiologie der Alkaloide,* Halle (Saale), Germany, June, 1969, pp. 493–504.

Bandoni, A. L., Rondina, R. V. D., and Coussio, J. D. (1972) *Phytochemistry,* **11**, 3547.

Bandoni, A. L., Stermitz, F. R., Rondina, R. V. D., and Coussio, J. D. (1975) *Phytochemistry* **14**, 1785.

Barton, D. H. R. (1964) *Pure Appl. Chem.* **9**, 35.

Barton, D. H. R., and Widdowson, D. A. (1972) *Adh. Dtsch. Akad. Wiss. Kl., Chem. Geol. Biol., Berlin; 4. Internationales Symposium Biochemie und Physiologie der Alkaloide,* Halle (Saale), Germany, June, 1969, p. 245.

Bate-Smith, E. C., and Swain, T. (1966) *Comparative Phytochemistry* (T. Swain, ed.), Academic Press, New York, p. 168.

Battersby, A. R. (1961) *Quart. Rev. (London)* **15**, 259.

Battersby, A. R. (1966) in *Abh. Dtsch. Akad. Wiss. Kl., Chem. Geol. Biol., Berlin; 3. Internationales Symposium Biochemie und Physiologie der Alkaloide,* Halle (Saale), Germany, June, 1965, pp. 295–308.

Battersby, A. R., and Gregory, B. (1968) *Chem. Commun.* **1968**, 135.

Battersby, A. R., Burnett, A. R., and Parsons, P. G. (1969) *J. Chem. Soc.* **1969**, 1187–1193.

Battersby, A. R., Herbert, R. B., and Santávy, F. (1965a) *Chem. Commun.* **1965**, 415–416.

Battersby, A. R., Herbert, R. B., Pijewska, L., and Santávy, F. (1965b) *Chem. Commun.* **1965**, 228–230.

Battersby, A. R., Kapil, R. S., Martin, J. A., and Mo, L. (1968a) *Chem. Commun.* 1968, 133.
Battersby, A. R., Kapil, R. S., and Southgate, R. (1968b) *Chem. Commun.* 1968, 131.
Bell, E. A. (1961) *Biochem. Biophys. Acta* **47**, 602.
Bell, E. A. (1962) *Biochem. J.* **85**, 91–93.
Bell, E. A. (1971) "Comparative biochemistry of non-protein amino acids," in *Chemotaxonomy of the Leguminosae* (J. B. Harborne, D. Boulter, and B. L. Turner, eds.); Academic Press, New York and London, p. 179.
Bentham, G., and Hooker, J. D. (1876) *Genera Plantarum,* Reeve, London.
Bernasconi, R., Gill, S., and Steinegger, E. (1965a) *Pharm. Acta Helv.* **40**, 246.
Bernasconi, R., Gill, S., and Steinegger, E. (1965b) *Pharm. Acta Helv.* **40**, 275.
Bhakuni, D. S., Tewari, S., and Dahr, M. M. (1972a) *Phytochemistry* **11**, 1819–1822.
Bhakuni, D. S., Satish, S., and Dahr, M. M. (1972b) *Tetrahedron* **28**, 4579–4582.
Birdsong, B. A., Alston, R., and Turner, B. L. (1960) *Can. J. Bot.* **38**, 499.
Bisset, N. G. (1975) *Pharm. Weekblad* **110**, 425.
Bisset, N. G., and Choudhury, A. K. (1974a) *Phytochemistry* **13**, 259–263.
Bisset, N. G., and Choudhury, A. K. (1974b) *Phytochemistry* **13**, 265–269.
Boit, H. G. (1961) *Ergebnisse der Alkaloid-Chemie bis 1960,* Akademie-Verlag, Berlin.
Bratek, M. D., and Wiewiórowski, M. (1959) *Rocz. Chim.* **33**, 1187.
Brechbuhler-Bader, S., Coscia, C. J., Loew, P., Von Szczepanski, Ch., and Arigoni, D. (1968) *Chem. Commun.* 1968, 136–137.
Bu'Lock, J. D. (1965) *The Biosynthesis of Natural Products,* McGraw-Hill, New York, p. 82.
Chang, C., Kimler, L., and Mabry, T. J. (1974) *Phytochemistry* **13**, 2771–2775.
Chao, J. M., and DerMarderosian, A. H. (1973) *Phytochemistry* **12**, 2435–2440.
Cranmer, M. F., and Mabry, T. J. (1966) *Phytochemistry* **5**, 1133.
Cranmer, M. F., and Turner, B. L. (1967) *Evolution* **21**, 508.
Cronquist, A. (1968) *The Evolution and Classification of Flowering Plants,* Houghton Mifflin Co., Boston.
Cronquist, A. (1960) *Bot. Rev.* **26**, 425–482.
Crowley, H. C., and Culvenor, C. C. J. (1962) *Aust. J. Chem.* **15**, 139.
Culvenor, C. C. J., and Smith, L. W. (1967) *Ind. J. Chem.* **5**, 665.
Culvenor, C. C. J., Koretskaya, N. I., Smith, L. W., and Utkin, L. M. (1968a) *Aust. J. Chem.* **21**, 1671.
Culvenor, C. C. J., Sawhney, R. S., and Smith, L. W. (1968b) *Aust. J. Chem.* **21**, 2135.
Engler, A. (1954) *Syllabus der Pflanzenfamilien* (H. Melchior und E. Werdermann, eds.), Bornträger, Berlin, 12 Ed., Vols. I and II.
Essery, J., Juby, P. F., Marion, L., and Trumbull, E. (1962) *J. Am. Chem. Soc.* **84**, 4597–4600.
Evans, W. C., and Treagust, P. G. (1973) *Phytochemistry* **12**, 2505–2507.
Faugeras, G. and Paris, M. (1968) *Ann. Pharm. Fr.* **26**, 265.
Florkin, M., and Stotz, E. H., eds. (1974) *Comprehensive Biochemistry,* Elsevier, Amsterdam, 29A. (See Florkin, M. "Concepts of molecular biosemiotics and of molecular evolution" and T. Swain "Biochemical evolution of plants.")
Gill, St., and Steinegger, E. (1963) *Sci. Pharm.* **31**, 135.
Gill, St., and Steinegger, E. (1964a) *Pharm. Acta Helv.* **39**, 508.
Gill, St., and Steinegger, E. (1964b) *Pharm. Acta Helv.* **39**, 565.
Gmelin, R. (1959) *Hoppe-Seyl. Z. Physiol. Chem.* **316**, 164–169.
Gmelin, R., Strauss, G., and Hasenmayer, G. (1958) *Z. Naturforsch.* **13b**, 252–256.
Gmelin, R., Strauss, G., and Hasenmayer, G. (1959) *Hoppe-Seyl. Z. Physiol. Chem.* **314**, 28–32.
Gohlke, K. (1913) *Die Brauchbarkeit der Serum-Diagnostik für den Nachweis zweifelhafter Verwandschaftsverhältnisse im Pflanzenreiche,* F. Grub, Stuttgart.

Goldberg, S. I., and Moates, R. F. (1967a) *J. Org. Chem.* **32**, 1832.
Goldberg, S. I., and Moates, R. F. (1967b) *Phytochemistry* **6**, 137.
Goldberg, S. I., and Sahli, M. S. (1967) *J. Med. Chem.* **10**, 124.
Gortner, R. A. (1929) *Outlines of Biochemistry*, John Wiley, New York, pp. 541–545.
Gottlieb, O. R. (1972) *Phytochemistry* **11**, 1537–1570.
Goutarel, R., Janot, M-M., Prelog, V., and Taylor, W. I. (1950) *Helv. Chim. Acta* **33**, 150.
Gross, D. (1969) *Fortschr. Bot.* **32**, 93.
Guarnaccia, R., and Coscia, C. J. (1971) *J. Am. Chem. Soc.* **93**, 6320.
Guarnaccia, R., Botta, L., and Coscia, C. J. (1970) *J. Am. Chem. Soc.* **92**, 6098.
Guggenheim, M. (1951) *Die Biogenen Amine*, 4th ed., S. Karger, Basel.
Hadorn, E. (1956) *Cold Spring Harbor Symp. Quant. Biol.* **21**, 363–373.
Hadorn, E. (1961) *Developmental Genetics and Lethal Factors*, J. Wiley and Sons, London.
Haensel, R. (1956) *Arch. Pharm.* **289**, 619–625.
Hasse, K., and Maisack, H. (1955) *Biochem. Z.* **327**, 296.
Hargreaves, R. T., Johnson, R. D., Millington, D. S., Mondal, M. H., Beavers, W., Becker, L., Young, C., and Rinehart, K. L., Jr. (1974) *Lloydia* **37**, 569.
Hawkes, J. G., and Tucker, W. G. (1968) in *Chemotaxonomy and Serotaxonomy* (J. G. Hawkes, ed.), Academic Press, New York.
Hegnauer, R. (1958) *Planta Med.* **6**, 1–64.
Hegnauer, R. (1959) *Pharm. Tijdschr. Belg.* **36**, 35.
Hegnauer, R. (1962–1973) *Chemotaxonomie der Pflanzen*, Vols. 1–6, Birkhauser Verlag, Basel und Stuttgart.
Hegnauer, R. (1965) *Lloydia* **28**, 267.
Hegnauer, R. (1966) "Comparative phytochemistry of alkaloids" in *Comparative Phytochemistry* (T. Swain, ed.), Academic Press, New York, p. 211.
Hegnauer, R. (1967) *Pure Appl. Chem.* **14**, 173.
Hennig, W. (1966) *Phylogenetic Systematics* (D. D. Davis, R. Zangerl, trs.), Univ. of Illinois Press, Urbana.
Hofmann, A., and Tscherter, H. (1960) *Experientia* **16**, 414.
Holloway, J. K. (1957) *Sci. Amer.* **197** (No. 1), 56.
Hughes, D. W., and Genest, K. (1973) "Alkaloids" in *Phytochemistry*, Vol. II (L. P. Miller, . ed.) Van Nostrand Reinhold Corp., New York, p. 118.
Hutchinson, J. (1959) *The Families of Flowering Plants*, 2nd ed., Clarendon Press, Oxford.
Hylin, J. W. (1964) *Phytochemistry* **3**, 161.
Inouye, H., Ueda, S., and Takeda, Y. (1968) *Tetrahedron Lett.* **1968**, 3453.
Jackanicz, T. M., and Byerrum, R U. (1966) *J. Biol. Chem.* **241**, 1296–1299.
James, W. O., and Butt, V. S. (1957) in *Abh. Dtsch. Akad. Wiss. Kl., Chem. Geol. Biol., Berlin; 1. Internationales Symposium Biochemie und Physiologie der Alkaloide*, Halle (Saale), Germany, June, 1956, p. 182.
Johnson, R. D., and Waller, G. R. (1974) *Phytochemistry* **13**, 1493–1500.
Kasprzykowna, Z. (1957) *Hodowla Rosl. Aklim. Nasienn* **1**, 309.
Kustrak, D., and Steinegger, E. (1968) *Pharm. Acta Helv.* **43**, 482.
Leete, E. (1969) *Adv. Enzymol.* **32**, 373–422.
Leete, E., and Adityachaudhury, N. (1967) *Phytochemistry* **6**, 219–223.
Loew, P., and Arigoni, D. (1968) *Chem. Commun.* **1968**, 137–138.
Loew, P., Szezepanski, C. V., Coscïa, C. J., and Arigoni, D. (1968) *Chem. Commun.* **1968**, 1276.
Lörincz, Gy., and Tétényi, P. (1966) *Herba Hung.* **5**, 95.
Mabry, T. (1966) "The betacyanins and betaxanthins" in *Comparative Phytochemistry* (T. Swain, ed.), Academic Press, London, p. 231.

Mabry, T., and Dreiding, A. A. (1968) "The betacyanins and betaxanthines" in *Rec. Adv. Phytochem.* **1**, 145.

Macholan, L. (1965) *Naturwissenschaften* **52**, 186.

Mankinen, C. B., Harding, S., and Elliott, M. (1975), *Taxon* **24**, 415–429.

Marion, L. (1958) *Bull. Soc. Chim. France* **1958**, 109–115.

Mascré, M. (1937) *C. R. Hebd. Seances Acad. Sci.* **204**, 890–891; *Chem. Abstr.* **31**, 4369 (1937).

Matveyev, N. D. (1959) *Osnovy Sortovodno-Semennovo Dela po Lekarstvom Kulturam,* Selkhozgiz, Moscow.

McNair, J. B. (1930) *Am. J. Bot.* **17**, 187.

McNair, J. B. (1965) *Studies in Plant Chemistry Including Chemical Taxonomy, Ontogeny, Phylogeny,* Private publisher, 818 S. Ardmore Ave., Los Angeles, pp. 1–389.

Mears, J. A., and Mabry, T. J. (1971) "Alkaloids in the leguminosae" in *Chemotaxonomy of the Leguminosae* (J. B. Harborne, D. Boutler, and B. L. Turner, eds.), Academic Press, New York, p. 73.

Mettler, L. E., and Gregg, T. G. (1969) *Population Genetics and Evolution.* Prentice Hall, Englewood, N.J.

Mez, C., and Gohlke, K. (1913) *Beitr. Biol. Pflanz.* **17**, 301.

Mez, C., and Ziengenspek, H. (1926) *Bot. Arch.* **13**, 483.

Minale, L., Piattelli, M., Stefano, S. D., and Nicolaus, R. A. (1966) *Phytochemistry* **5**, 1037.

Mirov, N. T. (1963) *Lloydia* **26**, 117.

Molisch, H. (1933) *Pflanzenchemie und Pflanzenverwandschaften,* G. Fischer, Jena.

Mothes, K. (1965) *Naturwissenschaften* **52**, 571.

Mothes, K. (1966a) in *Abh. Dtsch. Akad. Wiss. Kl., Chem. Geol. Biol., Berlin; 3. Internationales Symposium Biochemie und Physiologie der Alkaloide,* Halle (Saale) Germany, June, 1965, p. 27.

Mothes, K. (1966b) *Lloydia* **29**, 156.

Mothes, K. (1969) "Biologie der Alkaloide" in *Biosynthese der Alkaloide* (Mothes, K. and Schuette, H. R., eds.), VEB Deutscher Verlag der Wissenschaften, Berlin, pp. 1–39.

Mothes, K. (1975) *Rec. Adv. Phytochem.* **10**, 385–405.

Mowszowicz, J. (1974) *Zarys Systematyki Roślin,* Państwowe Wydawnictwo Naukowe, Warszawa.

Mukherjee, R., and Chatterjee, A. (1964) *Chem. Ind.* (London), 1524.

Nalborczyk, T. (1968) *Bull. Acad. Pol. Sci. Ser. Sci. Biol.* **16**, 317.

Neumann, D., and Tschoepe, K. H. (1966) *Flora (Jena)* **156**, 521.

Notation, A. D., and Spenser, I. D. (1964) *Can. J. Biochem.* **42**, 1803.

Nowacki, E. (1960) *Genet. Polon.* **1**, 119.

Nowacki, E. (1961) *Wiad. Bot.* **5**, 134.

Nowacki, E. (1964) *Genet. Polon.* **5**, 189.

Nowacki, E. (1968) *Monografie Biochemiczne,* Vol. 16, Polskie Towarzystwo Biochemiczne PWN, Warszawa; *Genet. Polon.* **5** (1964) 88–222.

Nowacki, E. (1969) *Z. Pflanzenzucht.* **61**, 232.

Nowacki, E., and Byerrum, R. U. (1962) *Life Sci.* **5**, 157.

Nowacki, E., and Drzewiecka-Roznowicz, E. (1961) *Genet. Polon.* **2**, 35.

Nowacki, E., and Nowacka, D. (1965) *Wiad. Bot.* **9**, 207.

Nowacki, E., and Prus-Glowacki, W. (1972) *Wiad. Bot.* **16**, 19–33.

Nowacki, E., and Przybylska, J. (1961) *Bull. Acad. Pol. Sci. Ser. Sci. Biol.* **9**, 279.

Nowacki, E., and Waller, G. R. (1973) *Flora (Jena)* **162**, 108–117.

Nowacki, E., and Waller, G. R. (1975a) *Phytochemistry* **14**, 155–159.

Nowacki, E., and Waller, G. R. (1975b) *Phytochemistry* **14**, 161–164.

Nowacki, E., and Waller, G. R. (1975c) *Phytochemistry* **14**, 165–171.

Pfeifer, S. (1962) *Pharmazie* **17**, 536.

Pfeifer, S., and Heydenreich, K. (1961) *Naturwissenschaften* **48**, 222.

Pfeifer, S., and Mann, I. (1966) in *Abh. Dtsch. Akad. Wiss. Kl., Chem. Geol. Biol., Berlin; 3. Internationales Symposium Biochemie und Physiologie der Alkaloide*, Halle (Saale) Germany, June, 1965, 315–317.

Phillipson, J. D., Sariyar, G., and Baytop, T. (1973) *Phytochemistry* **12**, 2431.

Piattelli, M., and Minale, L. (1964) *Phytochemistry* **3**, 547.

Piattelli, M., Minale, L., and Nicolaus, R. A. (1965) *Phytochemistry* **4**, 817.

Piechowski, M., and Nowacki, E. (1959) *Bull. Acad. Polon. Sci. Ser. Sci. Biol.* **7**, 1965.

Price, J. R. (1963) "The distribution of alkaloids in the Rutacea" in *Chemical Plant Taxonomy* (T. Swain, ed.), Academic Press, New York, p. 429.

Reifer, I., Wiewiórowski, M., Nizilek, S., Stawicka, D., and Bratek, M. D. (1962) *Bull. Acad. Pol. Sci. Ser. Sci. Biol.* **10**, 161–165.

Reuter, G. (1957) *Flora (Jena)* **145**, 326.

Robinson, R. (1917) *J. Chem. Soc.* **111**, 876–899.

Robinson, R. (1955) *The Structural Relations of Natural Products,* Clarendon Press, Oxford.

Robinson, T. (1965) *Phytochemistry* **4**, 67.

Robinson, T., and Cepurneek, C. (1965) *Phytochemistry* **4**, 75.

Romeo, J. T., and Bell, E. A. (1974) *Lloydia* **37**, 543.

Santávy, F. (1962) *Collect. Czech. Chem. Commun.* **27**, 1717–1725.

Santávy, F. (1963) in *Abh. Dtsch. Akad. Wiss. Kl., Chem. Geol. Biol., Berlin; 2. Internationales Symposium Biochemie und Physiologie der Alkaloid*, Halle (Saale), Germany, June, 1962, pp. 235–239.

Santávy, F. (1966) in *Abh. Dtsch. Akad. Wiss. Kl., Chem Geol. Biol., Berlin; 3. Internationales Symposium Biochemie und Physiologie der Alkaloide*, Halle (Saale), Germany, June, 1965, p. 43.

Šantávy, F., Kaul, J. L., Hruban, L., Dolejs, L., Hanus, V., Blaha, K., and Cross, A. D. (1965) *Collect. Czech. Chem. Commun.* **30**, 335, 3479.

Šantávy, F., Maturova, M., and Hruban, L. (1966) *Chem. Commun.* **36**, 144.

Sárkány, S., Dános, B., and Sárkány-Kiss, I. (1959a) *Ann. Univ. Sci. Budapest, Sec. Biol.* **2**, 211.

Sárkány, S., Sárkány-Kiss, I., Dános, B., and Farkas-Riedel, L. (1959b) *Acta Bot. Acad. Sci. Hung.* **5**, 97.

Sastry, S. D., and Waller, G. R. (1972) *Phytochemistry* **11**, 2241–2245.

Schöpf, C. (1949) *Angew. Chem.* **61**, 31.

Schuette, H. R. (1960) *Arch. Pharm.* **293**, 1006.

Schuette, H. R. (1961) *Atompraxis* **7**, 91.

Schuette, H. R. (1965) *Beit. Biochem. Physiol. Naturst. Festschr.* **1965**, 435.

Schuette, H. R., Bohlmann, F., and Reusche, W. (1961) *Arch. Pharm.* **294**, 610.

Scott, A. I., Reichardt, P. B., Slayton, M. B., and Sweeney, J. G. (1973) *Rec. Adv. Phytochem.* **6**, 117–146.

Shalaby, A. F., and Steinegger, E. (1964) *Pharm. Acta Helv.* **39**, 752.

Sharapov, N. I. (1962) *Zakonomernosti khimizma rastenii,* Izd. Akad. Nauk. U.S.S.R., Leningrad Otd., Leningrad.

Skursky, L., and Novotny, M. (1966) in *Abh. Dtsch. Akad. Wiss. Kl., Chem. Geol. Biol., Berlin; 3. Internationales Symposium Biochemie und Physiologie der Alkaloide,* Halle (Saale), Germany, June, 1965, p. 577.

Slavík, J. (1955) *Collect. Czech. Chem. Commun.* **20**, 198–202.

Slavík, J. (1965) *Collect. Czech. Chem. Commun.* **30**, 2864–2869.

Slavík, J. and Slavikova, L. (1956) *Collect. Czech. Chem. Commun.* **21**, 211.

Slavík, J., and Slavikova, L. (1963) *Collect. Czech. Chem. Commun.* **28**, 1728–1732.

Slavík, J., Dolejś, L., Vokéc, K., and Hanuś, V. (1965) *Collect. Czech. Chem. Commun.* **30**, 2864.

Sokolov, V. S. (1952) *Alkaloidhältige Pflanzen der USSR,* Verlag Akad. Wiss., Moscow.

Spencer, I. D. (1966) *Lloydia* **29**, 71–89.

Steinegger, E., and Herdt, E. (1968) *Pharm. Acta Helv.* **43**, 331.

Steinegger, E., and Moser, C. (1967) *Pharm. Acta Helv.* **42**, 177.

Steinegger, E., and Weber, P. (1968) *Helv. Chim. Acta* **51**, 206.

Steinegger, E., and Wicky, K. (1965) *Pharm. Acta Helv.* **40**, 610.

Steinegger, E., Bernasconi, R., and Ottaviano, G. (1963) *Pharm. Acta Helv.* **38**, 371.

Steinegger, E., Moser, C., and Weber, P. (1968) *Phytochemistry* **7**, 161.

Stermitz, F. R. (1968) *Rec. Adv. Phytochem.* **7**, 161.

Stermitz, F. R., and Coomes, R. M. (1969) *Phytochemistry* **8**, 611.

Stermitz, F. R., Nicodem, D. E., Wei, C. C., and McMurtrey, K. D. (1969) *Phytochemistry* **8**, 615.

Stermitz, F. R., Kim, D. K., and Larson, K. A. (1973a) *Phytochemistry* **12**, 1355–1357.

Stermitz, F. R., Ito, R. J., Workman, S. M., and Klein, W. M. (1973b) *Phytochemistry* **12**, 381–382.

Stermitz, F. R., Stermitz, J. R., Zanoni, T. A., and Gillespie, J. P. (1974) *Phytochemistry* **13**, 1151–1153.

Strasburger, E., Noll, F., Schenck, H., and Schimper, A. F. W. (1960) *Botanikapodrecznik dla szkól wyzszych,* 26th ed., PWRiL, Warszawa.

Swain, T. (1966) *Comparative Phytochemistry,* Academic Press, New York and London.

Szafer, W., and Kostyniuk, M. (1963) *Zarys Paleobotaniki,* 2nd ed., PWN, Warszawa.

Takhtajan, A. (1959) *Die Evolution der Angiospermen,* G. Fischer, Jena.

Taylor, E. H., and Shough, H. R. (1967) *Lloydia,* **30**, 197.

Tétényi, P. (1973) *Chem. Bot. Classif., Proc. Nobel Symp.* 25th, 1973, 67.

Tétényi, P., and Lorincz, G. (1965) *Kiserl. Kozlemeny., C. Kertesz.* **57**, 75–93.

Tétényi, P., Lorincz, C., and Szabo, E. (1961) *Pharmazie* **16**, 426.

Tétényi, P., and Vágújfalvi, D. (1967) *Herba Hung.* **6**, 123.

Tétényi, P., and Vágújfalvi, D. (1968) *Plant Med. Phytother.* **3**, 97.

Troll, H.-J. (1965) *Züchter* **35**, 233.

Tschiersch, V. B. (1959) *Flora (Jena)* **147**, 405.

Tschiersch, V. B., and Hanelt, P. (1967) *Flora (Jena)* Abt. A **157**, 389.

Turner, B. L. (1967) *Pure Appl. Chem.* **14**, 189.

Vágújfalvi, D. (1963) *Herba Hung.* **2**, 5.

Vágújfalvi, D., and Tétényi, P. (1967) *Herba Hung.* **6**, 221.

Vaquette, J., Pousset, J. L., Paris, R. R., and Cavé, A. (1974) *Phytochemistry* **13**, 1257–1259.

Waller, G. R. (1969) *Prog. Chem. Fats Other Lipids* **10**, 151.

Waller, G. R. (1975) unpublished.

Waller, G. R., Yang, K. S., Gholson, R. K., Hadwiger, L. A., and Chaykin, S. (1966) *J. Biol. Chem.* **241**, 4411.

Warren, F. L. (1966) in *Abh. Dtsch. Akad. Wiss. Kl., Chem. Geol. Biol., Berlin; 3. Internationales Symposium Biochemie und Physiologie der Alkaloide,* Halle (Saale), Germany, June, 1975, p. 571.

Wenkert, E. (1959) *Experientia* **15**, 65.

Wenkert, E. (1968) *Acc. Chem. Res.* **1**, 78.

Wettstein, R. (1895) "Solanaceae" in *Die natürlichen Pflanzenfamilien* (A. Engler and K. Prantl, eds.), Englemann, Leipzig, Vol. 4, p. 3, 4–38.

Wettstein, R. (1935) *Handbuch der systematischen Botanik*, 4th ed., F. Deutike, Leipzig und Wien.

White, E. P. (1943) *New Zealand J. Sci. Tech.* **25**, 93.

White, E. P. (1946) *New Zealand J. Sci. Tech.* **27**, 474.

White, E. P. (1954) *New Zealand J. Sci. Tech.* **35**, 452.

Wicky, K., and Steinegger, E. (1965) *Pharm. Acta Helv.* **40**, 658.

Winterstein, E., and Trier, G. (1910) *Die Alkaloide*, Bornträger, Berlin.

Wocker, G. (1953) *Die Chemie der natürlichen Alkaloide*, 3 parts, F. Enke, Stuttgart.

Zachow, F. (1967) *Züchter* **37**, 35–42.

Zukowski, P. (1951) *Botanika*, PWRiL, Warszawa.

References for Chapter 2, Genetic Control of Alkaloid Production

Allard, R. W. (1960) *Principles of Plant Breeding*, J. Wiley and Sons, New York, p. 18.

Barton, D. H. R., Kirby, G. W., Steglich, W., and Thomas, G. M. (1965) *Proc. Chem. Soc. (London)*, 1965, 3990.

Battersby, A. R., Binks, R., Francis, R. J., McCaldin, D. J., and Ramuz, H. (1964) *J. Chem. Soc.*, 1964, 3600–3610.

Battersby, A. R., Brown, R. T., Clements, J. H., and Inverbach, G. C. (1965) *Chem. Commun.*, 1965, 230.

Beadle, G. W., and Tatum, E. L. (1941) *Proc. Natl. Acad. Sci.* **27**, 499.

Becker, G. (1956) *Tagungsber. Dtsch. Akad. Landwirtschaftswissen. Berlin*, **27**, 71–98.

Blaim, K., and Berbéc, J. (1968) *Genet. Polon.* **8**, 10–15.

Böhm, H. (1965) *Planta Med.* **13**, 234–240.

Böhm, H. (1969) "Genetik der Alkaloidmerkmale" in *Biosynthese der Alkaloide* (Mothes, K., and Schütte, H. R., eds.) VEB Deutscher Verlag der Wissenschaften, Berlin, pp. 21–39.

Burk, L. G., and Jeffrey, R. N. (1958) *Tobacco Sci. 2*, 139–141.

Buzzati-Traverso, A. A. (1960) "Paper chromatography in relation to genetics and taxonomy" in *New Approaches in Cell Biology* (P. M. B. Walker, ed.), Academic Press, New York, pp. 95–123.

Dawson, R. F., Christman, D. R., D'Adamo, A., Solt, M. L., and Wolf, A. D. (1960) *J. Am. Chem. Soc.* **82**, 2628.

Dewey, L., Byerrum, R., and Ball, C. (1955) *Biochim. Biophys. Acta* **18**, 141.

Dános, B. (1965) *Pharmazie* **20**, 727–730.

Fedotov, V. S. (1932) *Bull. Appl. Bot. Gen. Plant Breeding (Leningrad)* Suppl., **54**, 29–42.

Feigl, F. (1966) *Spot Tests in Organic Analysis* (R. E. Oesper, tr.), Elsevier, New York.

Games, D. E., Jackson, A. H., Kahn, N. A., and Millington, D. S. (1974) *Lloydia* **37**, 581–588.

Gladstones, J. S. (1974) *Tech. Bull. West. Aust. Dept. Agri.* **26**.

Goodspeed, T. H. (1959) *The Genus Nicotiana*, Chronica Botanica Company, Waltham, Mass.

Griffith, R. B., Valleau, W. D., and Stokes, G. W. (1955) *Science* **121**, 343–344.

Hackbarth, J. (1961) *Z. Pflanzenzüchtung* **45**, 334–344.

Hackbarth, J., and Sengbusch, R. V. (1939) *Züchter* **6**, 249–255.

Hackbarth, J., and Troll, H.-J. (1941) *Züchter* **13**, 63–65.

Hackbarth, J., and Troll, H.-J. (1955) *Z. Pflanzenzücht, 34,* 409–420.

Hagberg, A. (1950) *Hereditas* **36**, 228–230.

Harborne, J. B. (1963) *Phytochemistry* **2**, 85.

Harding, J., and Mankinen, C. (1968) personal communication.

Hargreaves, R. T., Johnson, R. D., Millington, D. S., Mondal, M. H., Beavers, W., Becker, L., Young, C., and Rinehart, K. L., Jr. (1974) *Lloydia* **37**, 569–580.

Ivanov, N. N. (1931) *Bull. Appl. Bot. Gen. Plant Breeding (Leningrad)* **25**, 344–347.

Ivanov, N. N., and Lavrova, M. N. (1931) *Bull. Appl. Bot. Gen. Plant Breeding (Leningrad)* **25**, 291.

Ivanov, N. N., and Smirnova, M. I. (1932) *Bull Appl. Bot. Gen. Plant Breeding (Leningrad),* Suppl., 54.

Jeffrey, R. N. (1959) *Tobacco Sci.* **3**, 89–93.

Jeffrey, R. N. and Tso, T. C. (1964) *Plant Physiol.* **39**, 48.

Kawatani, T., and Asahina, H. (1959) *Japan. J. Genet.* **34**, 353–362.

Kazimierski, T., and Nowacki, E. (1961a) *Flora (Jena)* **151**, 202–209.

Kazimierski, T., and Nowacki, E. (1961b) *Genet. Polon.* **2**, 93–96.

Kazimierski, T., and Nowacki, E. (1964) *Biul. Hod. Rosl. Nasienn* **3–4**, 1.

Koelle, G. (1961) *Züchter* **31**, 346–351.

Koelle, G. (1965a) *Züchter* **35**, 222–228.

Koelle, G. (1965b) *Z. Pflanzenzücht.* **54**, 61–69.

Koelle, G. (1966a) *Z. Pflanzenzücht.* **55**, 375–382.

Koelle, G. (1966b) *Abh. Dtsch. Akad. Wiss. Kl., Chem. Geol. Biol. Berlin; 3. Internationales Symposium Biochemie und Physiologie der Alkaloide,* Halle (Salle), Germany, June, 1965, pp. 179–180.

Koelle, G. (1975) *Z. Pflanzenzücht.* **75**, 71–79.

Kraft, D. (1953) *Pharmazie* **8**, 170–173.

Leete, E., and Medekel, M. R. (1972) *Phytochemistry* **11**, 2751–2756.

Manske, R. H. F., and Holmes, H. L., eds., *The Alkaloids,* Vols. I–XV, 1950–1975, Academic Press, New York.

Mikolajczyk, J. (1961) *Genet. Polon.* **2**, 19–92.

Mikolajczyk, J. and Nowacki, E. (1961) *Genet. Polon.* **2**, 55–64.

Millington, D. S., Steinman, D. H., and Rinehart, K. L., Jr. (1974) *J. Am. Chem. Soc.* **96**, 1909–1917.

Mizusaka, S., Tanabe, Y., Noguchi, M., and Tamaki, E. (1972) *Phytochemistry* **11**, 2757–2762.

Moldenhauer, K. (1961) "Spezielle Probleme des Arzneigewürzpflanzenbau" Wiss. Zeitsch. Univ. Leipzig.

Mothes, K. (1960) "Alkaloids in the plant" in *The Alkaloids,* Vol. VI (R. H. F. Manske and H. L. Holmes, eds.) Academic Press, New York, p. 1.

Mothes, K., Romeike, A., and Schroeter, H. B. (1955)*Naturwissenschaften* **42**, 214.

Neumann, D., and Schroeter, H. B. (1966) *Abh. Dtsch. Akad. Wiss. Kl., Chem. Geol. Biol. Berlin; 3. Internationales Symposium Biochemie und Physiologie der Alkaloide,* Halle (Salle), Germany, June, 1965, pp. 161–165.

Nowacki, E. (1958) *Bull. Acad. Polon. Sci. Ser. Sci. Biol.* **6**, 11–13.

Nowacki, E. (1959) *Postepy Nauk. Roln.* **5 (59)**, 15–20.

Nowacki, E. (1963a) *Rocz. Nauk Roln. Ser. A* **88**, 135–141.

Nowacki, E. (1963b) *Genet. Polon.* **4**, 161–202.

Nowacki, E. (1964a) *Genet. Polon.* **5**, 189–222.

Nowacki, E. (1964b) *Postepy Nauk Roln.* **4 (88)**, 19–34.

Nowacki, E. (1966) *Postepy Nauk Roln.* **2 (98)**, 17–34.

Nowacki, E. (1968) *Biosynteza Alkaloidiw Chinolizydynowych,* PWN, Warszawa.

Nowacki, E., and Dunn, D. B. (1964) *Genet. Polon.* **5**, 47–56.

Nowacki, E., and Waller, G. R. (1975) *Phytochemistry* **19**, 161–164.

Nowacki, E., Bragdo, M., Duda, A., and Kazimierski, T. (1961) *Flora (Jena)* **151**, 120.

Nowacki, E., Byerrum, R. U., Kazimierski, T., and Nowacka, D., (1966) *Abh. Dtsch. Akad. Wiss. Kl., Chem. Geol. Biol. Berlin; 3. Internationales Symposium Biochemie und Physiologie der Alkaloide,* Halle (Salle), Germany, June, 1965, pp. 205–211.

Peters, J. A., (1959) *Classic Papers in Genetics,* Prentice Hall, Englewood Cliffs, N.J.

Piechowski, M., and Nowacki, E. (1959) *Bull. Acad. Polon. Sci. Ser. Sci. Biol.* **7**, 165–168.

Porsche, W. (1964) *Züchter* **34**, 251–256.

Pufahl, K., and Schreiber, K. (1961) *Experientia* **17**, 302.

Pufahl, K., and Schreiber, K. (1963) *Züchter* **33**, 287–290.

Reuter, G. (1962) *Planta Med.* **10**, 226–231.

Romeike, A. (1958) *Planta Med.* **6**, 426–427.

Romeike, A. (1961) *Kulturpflanze* **9**, 171–180.

Romeike, A. (1962) *Kulturpflanze* **10**, 140–148.

Schreiber, K., Hammer, U., Ithal, E., Ripperger, H., Rudolph, W., and Weissenborn, A., (1961) "Über das Alkaloid-Vorkommen verschiedener Solanum-Arten" in *Chemie und Biochemie der Solanum-Alkaloide. Tagungsberichte Nr. 27,* Deutsche Akademie der Landwirtschaftswissenschaften zu Berlin, pp. 47–73.

Schreiber, K., Aurich, O., and Pufahl, K. (1962) *Arch. Pharm.* **295**, 271–275.

Schwarze, P. (1963) *Züchter* **33**, 275–281.

Schwarze, P., and Hackbarth, J. (1957) *Züchter* **27**, 332–341.

Schröck, O. (1941) *Züchter* **13**, 115–117.

Schroeter, H. B. (1958) *Proceedings of the 2nd International Science Tob. Congress, Brussels,* 426; *Biol. Abstr.* **35**, 35736 (1960).

Schuette, H. R. (1960) *Arch. Pharm.* **293**, 1006.

Scott-Moncrieff, R. (1936) *J. Genet.* **32**, 117.

Seehofer, E. (1957) *Züchter* **27**, 244–245.

Sengbusch, R. (1930) *Züchter* **2**, 1–2.

Sengbusch, R. (1931) *Züchter* **3**, 93–109.

Sengbusch, R. (1938) *Züchter* **10**, 91–95.

Sengbusch, R. (1942) *Landwirtsch. Jahrb.* **91**, 719–762.

Smith, H. H., and Abashian, D. V. (1963) *Am. J. Bot.* **50**, 435–447.

Smith, H. H., and Smith, C. R. (1942) *J. Agr. Res.* **65**, 347–359.

Srb, A., and Owen, R. (1952) *General Genetics,* W. H. Freeman, San Francisco.

Stermitz, F. R., and Rapoport, H. (1961) *J. Am. Chem. Soc.* **83**, 4045.

Sweeley, C. C., Young, N. D., Holland, J. F., and Gates, S. C. (1974) *J. Chromat.* **99**, 507–517.

Tatum, E. L., and Bonner, D. M. (1957) *J. Biol. Chem.* **151**, 349.

Tétényi, P., Lörincz, C., and Szabo, E. (1961) *Pharmazie* **16**, 426.

Troll, H.-J., (1958) *Z. Pflanzenzücht.* **39**, 35–46.

Tswett, M. S. (1903) *Trudy Warzawskogo objzczestwo Otd. Biologii* **14**, 20.

Valleau, W. D. (1949) *J. Agr. Res.* **78**, 171–181.

Waller, G. R. (ed.) (1972) *Biochemical Application of Mass Spectrometry,* Wiley-Interscience, New York.

Wegner, E. (1956) *Tabakforschung* **2**, 39–40.

Weybrew, J. A., and Mann, T. J. (1963) *Tob. Sci.* **7**, 28–36, published in 1963, *Tob. Int.* **156**, 30–38.

Weybrew, J. A., Mann, T. J., and Moore, E. L. (1960) *Tob. Sci.* **4**, 190–193, published in 1960, *Tob. Int. (N.Y.)* **151**, 24–27.

References for Chapter 3, Environmental Influences on Alkaloid Production

Andreeva, N. M. (1952) *Sb. Trud. Belorus. Inst. Zemled. Bull.* **5**.

Appel, H. H., and Streeter, P. (1970a) *Rev. Latinoamer. Quim.* **1**, 63; *Chem. Abstr.* **74**, 72815v (1971).

Appel, H. H., and St. eeter, P. (1970b) *Scientia (Valparaiso)* **36**, 105.

Auda, H., Juneja, P., Eisenbraun, E. J., Waller, G. R., Kays, W. R., and Appel, H. H. (1967) *J. Am. Chem. Soc.* **89**, 2476.

Auda, H., Waller, G. R., and Eisenbraun, E. J. (1967) *J. Biol. Chem.* **242**, 4157; (1968) **243**, May 25 Errata.

Avramova, B. (1955) *Farmaciya (Bulg.)* **4**.

Bachmanowa, S., and Byszewski, W. (1959) *Hodowld. Rosl. Aklim. Nasieun.* **3**, 475.

Bennett, W. D. (1963) *New Zealand J. Agr. Res.* **6**, 310–313.

Boit, H. G., and Ehmke, H. (1956) *Chem. Ber.* **89**, 2093.

Bush, L. P., and Buckner, R. C. (1973) *Crop Sci. Soc. Am. Spec. Publ.* No. **4**, 99–112.

Byszewski, W. (1959) *Hodowla. Rosl. Aklim. Nasienn.* **3**, 439.

Byszewski, W. (1960) *Hodowla. Rosl. Aklim. Nasienn,* **4**, 69.

Dafert, O., and Siegmund, O. (1932) *Heil, Gewürz-Pflanz.* **14**, 98.

Duchnowska, A., and Pawelczyk, E. (1960) *Biol. Inst. Rosl. Lecz.* **6**, 31.

El-Hamidi, A., Saleh, A. M., and Hamdi, H. (1966) *Abh. Dtsch. Akad. Wiss. Kl., Chem. Geol. Biol., Berlin; 3, Internationales Symposium Biochemie und Physiologie der Alkaloide,* June, 1965, pp. 567–570.

Ermakov, A. I., Prizemina, E. P., Sharapov, N. I., and Shifrina, Kh. B. (1935) *Tr. Prikl. Bot. Genet. Sel.* **10**.

Frost, G. M., Hale, R. L., Waller, G. R., Zalkow, L. H., and Girotra, N. N. (1967) *Chem. Ind. (London)* **1967**, 320.

Gentry, C. E., Chapman, R. A., Henson, L., and Buckner, R. C. (1969) *Agron. J.* **61**, 313 – 316.

Il'inskaya, T. N., and Yosifova, M. G. (1956) *Bull. Narcotics* **8**, 38.

James, W. O. (1950) "Alkaloids in the plant" in *The Alkaloids* (R. F. Manske and H. L. Holmes, eds.) Vol. I, Academic Press, New York, p. 71.

Johnson, R. D. (1972), Ph.D. Dissertation, Oklahoma State University, Stillwater, p. 122.

Keeler, R. F. (1975) *Lloydia* **38**, 56–86.

Kleinschmidt, G., and Mothes, K. (1958) *Pharmazie* **13**, 357.

Koch, K., and Mangel, K. (1972) *Z. Pflanzenernähr. Bodenk.* **131**, 148–154.

Kuzmenko, A. A., and Tikhvinska, V. D. (1935) *K Fizyologi, Nikotino Obrazovanya u Tabaka,* Botanik, Izdat., Kiev.

Malarski, H., and Sypniewski, J. (1923) *Pamietnik PINGW, Pulawy* **4**, 302.

Marten, G. C., Simons, A. B., and Frelich, J. R. (1974) *Agron. J.* **66**, 363.

Matveyev, N. D. (1959) *Osnovy Sortovodno-Semennovo Dela po Lekarstvom Kulturam,* Selkhozgiz, Moscow.

McNair, J. B. (1942) *Lloydia* **5**, 208.

Mika, E. S. (1962) *Lloydia* **25**, 291.

Mironenko, A. V. (1965) *Fiziologya I Biokhimiya Lupina,* Nauka i Tekhnika, Minsk.

Mironenko, A. V. (1975) *Biokhimiya Lupina,* Nauka i Tekhnika, Minsk.

Mothes, K. (1928) *Planta* **5**, 563.

Nowacki, E. (1958) *Rocz. Nauk Roln. Ser. A* **79**, 505.

Nowacki, E. (1969) *Problemy* **25**, 264.

Nowacki, E., and Waller, G. R. (1972) *Abh. Dtsch. Akad. Wiss. Kl., Chem. Geol. Biol. Berlin;*

4. *Internationales Symposium Biochemie und Physiologie der Alkaloide,* June, 1969, p. 314.

Nowacki, E., and Weznikas, Th. (1975) *Pamiet. Pulawski* **64,** 5–23 and 25–44.

Nowacki, E., Kazimierski, T., Golankiewicz, K., Dezor-Mazur, M., and Boczoń, W. (1973) *Acta Agrobot.* **26,** 123–138.

Nowacki, E., Jurzysta, M., and Gorski, P. (1975) *Bull. Acad. Pol. Sci. Ser. Sci. Biol.* **23,** 219–225.

Nowacki, E., Jurzysta, M., Gorski, P., Nowacka, D., and Waller, G. R. (1976) *Biochem. Physiol. Pflanz.* **169,** 231.

Nowotnówna, A. (1928) *Pamietnik PINGW Pulawy* **9,** 5.

Oram, R. N., and Williams, J. D. (1967) *Nature* **213,** 946.

Proskurina, N. F. (1955) *Zh. Obshch. Khim.* **25,** 839.

Sabalitschka, Th., and Jungermann, C. (1925) *Biochem. Z.* **164,** 279.

Sandberg, F., (1961) *Pakistan J. Sci. Ind. Res.,* **4,** 280–283.

Schmid, H. (1948) *Ber. Schweiz. Bot. Ges.* **58,** 5.

Sharapov, N. I. (Szarapow) (1956) *Chemizm Roslin A. Klimat,* PWRil, Warszawa.

Sharapov, N. I. (1961) *Zakonomernosti khimizma Rastenii, Izv. Akad. Nauk SSSR,* Leningrad Otd., Leningrad. C. A. (1962) **51,** 10238.

Sinclair, C. (1967) *Nature* **213,** 214–215.

Smirnova-Ikonikova, M. I. (1938) *Biokhimiya Kulturnikh Rasteni,* Selkhozgiz, Moscow.

Smith, T. A. (1965) *Phytochemistry* **4,** 599.

Sokolov, V. S. (1952) *Alkaloidhältige Pflanzen der USSR,* Verlag Akad. Wiss., Moscow.

Sokolov, V. S. (1959) *Symp. Soc. Exp. Biol.* **13,** 230.

Stillings, E. W., and Laurie, A. (1943) *Proc. Am. Soc. Hort. Sci.* **42,** 590.

Stuczyński, E. (1969) *Pamiet. Pulaswki* **36,** 69–116.

Stuczyński, E., Stuczyńska, J., and Skalacki, S. (1970) *Pamiet. Pulawski* **39,** 103–128.

Stuczyński, E., Stuczyńska, J., Jakubowski, S., and Jasińska, B. (1971) *Pamiet. Pulawski* **44,** 119–144.

Sukhorukov, K. T., and Borodulina, N. A. (1932) *Trudy Vses. Inst. Lekarst. I. Aromat. Rast.* Vol. I.

Suzuki, M., and McLeod, L. B. (1970) *Can. J. Plant Sci.* **50,** 445–450.

Tétényi, P., Lörincz, C., and Szabo, E. (1961) *Pharmazie* **16,** 426.

Timofieyuk, K. M. (1929) *Nauchno-Agron. Zh.* **11.**

Tso, T. C. (1972) *Physiology and Biochemistry of Tobacco Plants,* Dowden, Hutchinson, and Ross, Stroudsberg, Pa., pp. 42–81.

Tso, T. C., Kasperbauer, M. J., and Sorokin, T. P. (1970) *Plant Physiol.* **45,** 330–337.

Tso, T. C., Sorokin, T. P., and Engelhaupt, M. E. (1973) *Plant Physiol.* **51,** 805–806.

Unger, W. (1912) *Apoth. Ztg.* **27,** 763.

Vágújfalvi, D. (1963a) *Herba Hungarica* **2,** 5.

Vágújfalvi, D. (1963b) *Botanikai. Koziemenyek* **50,** 42.

Vágújfalvi, D. (1964) *Herba Hungarica* **3,** 165.

Vogel, Z., and Weber, E. (1922) *Z. Pflanzenernähr. Dueng.* **1.**

Voseurusa, J. (1960) *Pharmazie* **15,** 552.

Wallenbrock, J. C. J. (1940) *Recl. Trav. Bot. Néerl.* **37,** 78.

Waller, G. R., and Lawrence, R. H., Jr. (1975) in 23rd Annual Conference American Society for Mass Spectrometry, Houston, Texas, May 25–30, abstract no. Q-12, p. 99.

Waller, G. R., Tang, May S-I., Scott, M. R., Goldberg, F. J., Mayes, J. S., and Auda, H. (1965) *Plant Physiol.* **40,** 803.

Weeks, W. W. (1970) "Physiology of alkaloids in germinating seed of *Nicotiana tabacum,*" Ph.D. Thesis, University of Kentucky, Lexington.

Weeks, W. W., and Bush, L. P. (1974) *Plant Physiol.* **53,** 73.

Weeks, W. W., Davis, D. L., and Bush, L. P. (1969) *J. Chromatog.* **43**, 506.
Weevers, Th., and von Oort, H. D. (1929) *Proc. K.Ned. Akad. Wet.* **32**, 364.
Weybrew, J. A., Long, R. C., Dunn, R. A. and Woltz, W. G. (1974) in *Mechanisms of Regulation of Plant Growth* (R. L. Bieleski, A. R. Ferguson, and M. M. Creswell, eds.), Bulletin 12, The Royal Society of New Zealand, Wellington, pp. 843–847.
Woodhead, S., and Swain, T. (1974) *Phytochemistry* **13**, 953–956.
Yates, S. G., and Tokey, H. L. (1965) *Aust. J. Chem.* **18**, 53.
Yoshida, D. (1973) *Bull. Hatano Tobacco Exp. Sta.* **73**, 239–244.
Yunusov, S. Y., and Akramov, S. T. (1955) *Zh. Obshch. Khim.* **25**, 1813.
Yunusov, S. Y., and Akramov, S. T. (1960) *Zh. Obshch. Khim.* **30**, 3132.
Zolotnicka, S. A. (1949) *Trudy Bot. Inst. Akad. Nauk Arm. SSR.* **2**, 2.

References for Chapter 4, Sites of Alkaloid Formation

Benveniste, P., Hirth, L., and Ourisson, G., (1966) *Phytochemistry* **5**, 31–44.
Birecka, H., Szklarek, D., and Mazan, A. (1960) *Bull. Acad. Polon. Sci. Ser. Sci. Biol.* **8**, 167–173.
Chojecki, Sz. (1949) *Prace Tytoniowe,* PWN, Warszawa, Vol. 1, pp. 49–87.
Cranmer, M. F., (1965) "Biochemical and biosystematic investigation of lupine alkaloids of baptisia," Ph.D. thesis, University of Texas, Austin.
Cromwell, B. T. (1943) *Am. J. Bot.* **31**, 351.
Dawson, R. F. (1941) *Science* **94**, 396.
Dawson, R. F. (1942) *Am. J. Bot.* **29**, 66, 813.
Dawson, R. F. (1944) *Am. J. Bot.* **31**, 351.
Dawson, R. F. (1948) *Adv. Enzym.* **8**, 203.
Dawson, R. F. (1951) *J. Am. Chem. Soc.* **73**, 4218.
Dawson, R. F., and Osdene, T. S. (1972) *Rec. Adv. Phytochem.* **5**, 317.
Dewey, L. J., Byerrum, R. U., and Ball, C. D. (1955) *Biochim. Biophys. Acta* **18**, 141.
Diaper, D. G. M., Kirkwood, A., and Marion, L. (1951) *Can. J. Chem.* **29**, 964.
Elze, H., and Teuscher, E. (1967) *Flora (Jena) Abt. A* **158**, 127–132.
Evans, W. C., and Partridge, M. W. (1953) *J. Pharm. Pharmacol.* **5**, 772.
Furuya, K., Kojima, H., and Syono, K. (1966) *Chem. Pharm. Bull.* **14**, 1189–1190.
Furuya, K., Kojima, H., and Syono, K. (1971) *Phytochemistry* **10**, 1529–1532.
Glowacki, W. (1975) *Genet. Polon.* **16**, 47.
Guseva, A. R., and Paseshnichenko, W. A. (1958) *Biokhimia* **23**, 412.
Hadwiger, L. A., and Waller, G. R. (1964) *Plant Physiol.* **39**, 244–247.
Haga, P. R. van (1956) *Biochim. Biophys. Acta* **19**, 562.
Hagberg, A. (1950) *Hereditas* **36**, 228–230.
Heine, K. (1942) *Planta* **33**, 185.
Heinze, T. M. (1975) "The formation of nicotine and a study of morphology in callus derived from Maryland-872 tobacco," M.S. Thesis, Oklahoma State University, Stillwater.
Heinze, T. M., and Waller, G. R. (1975) unpublished results.
Hemberg, T., and Fluck, H. (1953) *Pharm. Acta Helv.* **28**, 74–85.
Hills, K. L., Trautner, E. M., and Rodwel, C. N. (1946) *Aust. J. Sci.* **9**, 24.
Ill'in, G. S. (1948) *Biokhimia* **13**, 193; *Chem. Abstr.* **42**, 7841 (1948).

Ill'in, G. S. (1955) *Dokl. Akad. Nauk SSSR* **105**, 777; *Chem. Abstr.* **50**, 7960 (1956).

James, W. O. (1953) *J. Pharm.* **5**, 809.

James, W. O., and Butt, V. S. (1957) *Abh. Dtsch. Akad. Wiss. Kl., Chem. Geol. Biol. Berlin; 1. Internationales Symposium Biochemie und Physiologie der Alkaloide,* Halle (Salle), Germany, June, 1959, p. 182.

Kazimierski, T., and Nowacki, E. (1960) *Bull. Acad. Polon. Sci. Ser. Sci. Biol.* **8**, 587–589.

Kisaki, T., and Tamaki, E. (1966) *Phytochemistry* **5**, 293–300.

Kleinschmidt, G., and Mothes, K. (1959) *Z. Naturforsch.* **14B**, 52–56.

Kleinschmidt, G., and Mothes, K. (1960) *Arch. Pharm.* **293**, 948–953.

Kuzdowicz, A. (1955) *Acta Soc. Bot. Polon.* **24**, 549–566.

Lashuk, G. I. (1948) *Dokl. Akad. Nauk SSSR* **60**, 1357; *Chem. Abstr.* **42**, 7374 (1948).

Lee, H. J., and Waller, G. R. (1972) *Phytochemistry* **11**, 965–973.

Leete, E. (1955) *Chem. Ind. (London)* **1955**, *537*.

Leete, E. (1957) *Chem. Ind. (London)* **1957**, *1270*.

Leete, E., Kirwood, S., and Marion, L. (1952) *Can. J. Chem.* **30**, 749.

Leete, E., Marion, L., and Spenser, I. D. (1954) *Can. J. Chem.* **32**, 1116.

Maciejewska-Potapczykowa, W., and Nowacki, E. (1959) *Acta Soc. Bot. Polon.* **28**, 83.

Mashkovtsev, M. F., and Sirotenko, A. A. (1956) *Sb. Rab. Vses. Inst. Tabaka I Makhorki,* **149**; *Chem. Abstr.* **50**, 7962 (1956); *Fiziol. Rastenii* **3**, 79.

Massicot, J., and Marion, L. (1957) *Can. J. Chem.* **35**, 1.

Meyer, A., and Schmidt, E. (1910) *Flora (Jena)* **100**, 317.

Mironenko, A. V. (1965) *Fizyologia i Biokhimia Lupina,* Nauka i Tekhnika, Minsk. USSR.

Mironenko, A. V., and Rikhovska, M. A. (1958) *Dokl. Akad. Nauk Belorus. SSR* **2**, 348; *Chem. Abstr.* **54**, 13293 (1960).

Mironenko, A. V., and Spiridonova, G. I. (1959) *Dokl. Akad. Nauk Belorus. SSR* **3**, 311; *Chem. Abstr.* **54**, 15558 (1960).

Mironenko, A. V., Godneva, M. T., and Masiko, A. A. (1959) *Dokl. Akad. Nauk Belorus. SSR* **3**, 171; *Chem. Abstr.* **53**, 20309 (1959).

Moens, B. (1882) *Dekinocultur in Azii.* Batavia.

Moerloose, P. de (1954) *Pharm. Weekb.* **89**, 541–547.

Mothes, K. (1955) *Ann. Rev. Plant Physiol.* **16**, 393–432.

Mothes, K. (1969) *Experientia* **25**, 225.

Mothes, K., and Engelbrecht, L. (1956a) *Flora (Jena)* **143**, 428–472.

Mothes, K., and Engelbrecht, L. (1956b) in *Biochemie Kulturpflanze* (K. Mothes and R. Mansfield, eds.), Akademie-Verlag, Berlin, Vol. 1, pp. 258–259.

Mothes, K., and Kretschmer, D. (1946) *Naturwissenschaften* **35**, 26.

Mothes, K., Engelbrecht, L., Tschoepe, K. H., and Trefftz-Hutschereuter, G. (1957) *(Jena)* **144**, 518.

Nalborczyk, T. (1968) *Bull. Acad. Pol. Sci. Ser. Biol.* **16**, 317.

Neumann, D., and Mueller, E. (1967) *Flora (Jena) Abt. A* **158**, 479–491.

Neumann, D., and Tschoepe, K. H. (1966) *Flora (Jena) Abt. A* **156**, 521–542.

Nowacki, E. (1958) *Roczn. Nauk Roln.* **79A**, 505–525; *Chem. Abstr.* **53**, 11541 (1959).

Nowacki, E., and Nowacka, D. (1966) *Flora (Jena) Abt. A* **156**, 457–463.

Nowacki, E., and Waller, G. R. (1972) *Abh. Dtsch. Akad. Wiss. Kl., Chem. Geol. Biol., Berlin; 4. Internationales Symposium Biochemie and Physiologie der Alkaloide,* Halle (Salle), Germany, June, 1969, pp. 187–195.

Nowacki, E. K., and Waller, G. R. (1973) *Z. Pflanzenphysiol.* **69**, 228–241.

Nowacki, E. and Waller, G. R. (1975) *Transactions of the 2nd Conference of Translocation and Accumulation of Nutrients in Plant Organisms,* pp. 113–124.

Nowacki, E., Weznikas, Th., and Nowacka, D. (1967) *Flora (Jena) Abt. A* **158**, 461–467.

Nowotńowa, A. (1928) *Pamietniki PINGW* **9**, 5–15, Pulawy.

Pal, B. P., and Hath, B. V. (1944) *Proc. Ind. Acad. Sci.* **320**, 79.

Peters, L., Schwanitz, F., and Sengbusch, R. (1956) in *Biochemie Kulturpflanze* (K. Mothes and R. Mansfield, eds.), Akademie-Verlag, Berlin, Vol. 1, pp. 247–257.

Pöhm, M. (1955) *Monatsh. Chem.* **86**, 875.

Pöhm. M. (1957) *Monatsh. Chem.* **88**, 597.

Pöhm, M., (1966) *Abh. Dtsch. Akad. Wiss. Kl., Chem. Geol. Biol., Berlin; 3. Internationales Symposium Biochemie und Physiologie der Alkaloide,* Halle (Salle), Germany, June, 1965, p. 251.

Prokoshev, S. M., Petrochenko, E. I., Ill'in, G. S., Baranova, W. Z., and Lebedeva, N. A. (1952) *Dokl. Akad. Nauk SSSR* **83**, 881; *Chem. Abstr.* **46**, 8201 (1952).

Rabitzsch, G. (1958) *Planta Med.* **6**, 103.

Reifer, I., and Kleczkowska, D. (1957) *Acta Biochem. Polon.* **4**, 2.

Romeike, A. (1953) *Pharmazie* **9**, 688–723.

Romeike, A. (1956) *Flora (Jena)* **143**, 67–86.

Romeike, A. (1959) *Flora (Jena)* **148**, 306–320.

Romeike, A. (1964) *Flora (Jena)* **154**, 163–173.

Sabalitschka, T., and Jungermann, C. (1925) *Biochem. Z.* **164**, 279.

Schiedt, U., Boeck-Behrens, G., and Delluva, A. M. (1962) *Hoppe-Seyl. Z. Physiol. Chem.* **330**, 46.

Schmidt, H. (1948) *Ber. Schweiz. Bot. Ges.* **58**, 5.

Schroeter, H. B. (1955) *Pharmazie* **10**, 141–157.

Schuette, H. R. (1961) *Atompraxis* **7**, 91.

Schuette, H. R., and Nowacki, E. (1959) *Naturwissenschaften* **46**, 493.

Shibata, S., and Imaseki, I. (1953) *Pharm. Bull. (Tokyo)* **1**, 285.

Shibata, S., and Imaseki, I. (1956) *Pharm. Bull. (Tokyo)* **3**, 277.

Shiio, I., and Ohta, S. (1973) *Agric. Biol. Chem.* **37**, 1857–1864.

Shmuk, A. A., Smirnova, A., and Ill'in, G. S., (1941a) *Dokl. Akad. Nauk SSSR* **32**, 5.

Shmuk, A., Smirnova, A., and Ill'in, G. S. (1941b) *C. R. Acad. Sci. URSS* **32**, 365; *Chem. Abstr.* **37**, 2777 (1943).

Skursky, L., Burleson, D., and Waller, G. R. (1969) *J. Biol. Chem.* **244**, 3238–3242.

Smirnova, M. I., and Moshkov, B. S. (1940) *Vest. Sots. Rastenievod.* **1940**, 68–77; *Chem. Abstr.* **35**, 4414 (1941).

Solt, M. (1957a) *Plant Physiol.* **32**, 484.

Solt, M. (1957b) *Plant Physiol.* **32**, 480.

Solt, M., and Dawson, R. F. (1958) *Plant Physiol.* **33**, 375–381.

Speake, T., McCloskey, P., and Smith, W. K. (1964) *Nature* **201**, 614–615.

Steward, F. C. (1964) *Science* **143**, 22.

Strasburger, E. (1885) *Ber. Deut. Bot. Ges.* **3**, 34.

Strasburger, E. (1906) *Ber. Deut. Bot. Ges.* **24**, 599.

Tabata, M., Yamamoto, H., and Hiraoka, N. (1968) *Jpn. J. Genet.* **43**, 319–322.

Tabata, M., Yamamoto, H., Hiraoka, N., Marumoto, Y., and Konoshima, M. (1971) *Phytochemistry* **10**, 723–729.

Tomaszewski, Z. (1957) *Hod i Aklimatyzezja Roslin* **1**, 3; *Chem. Abstr.* **53**, 20291 (1959).

Tschiersch, B. (1962) *Pharmazie* **17**, 621–623.

Wada, E., Kisaki, T., and Ihida, M. (1959) *Arch. Biochem. Biophys.* **80**, 258–267.

Waller, G. R., and Nakazawa, K. (1963) *Plant Physiol.* **38**, 318–322.

Waller, G. R., and Skursky, L. (1972) *Plant Physiol.* **50**, 622–626.

Warren-Wilson, P. M. (1952) *New Phytol.* **51**, 301–316.

White, H. A., and Spencer, M. (1964), *Can. J. Bot.* **42**, 1481.

Yang, K. S., and Waller, G. R. (1965) *Phytochemistry* **4**, 881–889.

References for Chapter 5, The Role of Alkaloids in Plants

Aniol, A., Kazimierski, T., and Nowacki, E. (1968) *Z. Pflanzenzücht* **59**, 317–326.

Aniol, A., Kazimierski, T., Nowacki, E., and Waller, G. R. (1972) *Abh. Dtsch. Akad. Wiss. Kl., Chem. Geol. Biol., Berlin; 4. Internationales Symposium Biochemie und Physiologie der Alkaloide,* Halle (Saale) Germany, June, 1969, pp. 481–492.

Arnold, G. W., and Hill, J. I. (1971) "Chemical factors affecting selection of food plants by ruminants" in *Phytochemical Ecology* (J. B. Harborne, ed.), Academic Press, New York, p. 92.

Basler, E. (1975) personal communication.

Blaszczak, W., and Statkun, T. (1958) *Roczn. Nauk Roln. Ser. A* **79**, 593–634.

Boling, J. A., Bush, L. P., and Buckner, R. C. (1973) in *Proceedings of the Fescue Toxicity Conference,* Northern Regional Research Laboratory, Agricultural Research Service, USDA, Peoria, Illinois, pp. 91–97.

Brower, L. P., Ryerson, W. N., Coppinger, L. L., and Glazier, S. C. (1968) *Science* **161**, 1349.

Bu'Lock, J. D. (1961) *Adv. Microbiol.* **3**, 293.

Bu'Lock, J. D. (1965) *The Biosynthesis of Natural Products,* McGraw-Hill Book Co., New York, pp. 9–11.

Cronquist, A. (1968) *The Evolution and Classification of Flowering Plants,* Houghton Mifflin Co., Boston.

Dawson, R. F., and Osdene, T. S. (1972) *Rec. Adv. Phytochem.* **5**, 317.

Downey, J. C., and Dunn, D. B. (1964) *Ecology* **45**, 172.

Duffus, C. M., and Duffus, J. H. (1969) *Experientia* **25**, 581.

Duke, J. A. (1975) *Rec. Adv. Phytochem.* **9**, 83–117.

Eigsti, O. J., and Dustin, P. (1955) *Colchicine: in Agriculture, Medicine, Biology, and Chemistry,* Iowa State College Press, Ames.

Ehrlich, P. R., and Raven, P. H. (1964) *Evolution* **18**, 586.

Ehrlich, P. R., and Raven, P. H. (1967) *Sci. Am.* **216**, no. 6 (June), 104.

Emlen, S. T. (1971) *Science* **173**, 462–463.

Errera, L. (1887) *Brussels Ac. Bull.* **13**, 272 (cited in *Royal Society Catalogue of Scientific Papers,* Vol. 14).

Farnell, D. R., Tutrell, M. C., Watson, V. H., Poe, W. E., and Coats, R. E. (1975) *J. Environ. Qual.* **4**, 120–122.

Fowden. L. (1974) *Rec. Adv. Phytochem.* **8**, 95.

Fraenkel, G. S. (1959) *Science* **219**, 1466.

Fraenkel, G. S. (1969) *Entom. Exp. Appl.* **12**, 473–476.

Geissman, T. A., and Crout, D. H. G. (1969) *Organic Chemistry of Secondary Plant Metabolism,* Freeman, Cooper and Co., San Francisco.

Gentry, C. E., Chapman, R. A., Henson, L., and Buckner, R. C. (1969) *Agron. J.* **61**, 313–316.

Greathouse, G. A. (1938) *Phytopathology* **28**, 592–593.

Greathouse, G. A. (1939) *Plant Physiol.* **14**, 377–380.

Greathouse, G. A., and Rigler, N. E. (1940a) *Am. J. Bot.* **27**, 99–107.

Greathouse, G. A., and Rigler, N. E. (1940b) *Phytopathology* **30**, 475–485.

Grove, M. D., Tookey, H. L., and Yates, S. G. (1973) in *Proceedings of the Fescue Toxicity Conference,* Northern Regional Research Laboratory, Agricultural Research Service, USDA, Peoria, Illinois, pp. 124–130.

Hadwiger, L. A., and Swochaw, M. E. (1971) *Plant Physiol.* **47**, 346.

Hagman, J. I., Marten, G. C., and Hovin, A. W. (1975) *Crop Sci.* **15**, 41–44.

Hassall, K. A. (1969) *World Crop Protection,* Vol. 2, *Pesticides,* Chemical Rubber Co., Cleveland, and Heffe Books, Ltd., London, pp. 128–138.

Hoffman, D., Brunnemann, K. D., Gori, G. B., and Wynden, E. L. (1975) *Rec. Adv. Phytochem.* **9**, 63–81.

Hogetsu, T., Shibaoka, H., and Shimokoniyama, M. (1974) *Plant Cell Physiol.* **15**, 265.

Hovin, A. W., and Marten, G. C. (1975) *Crop Sci.* **15**, 705–707.

Hsiao, T. H., and Fraenkel, G. S. (1968) *Ann. Entom. Soc. Am.* **61**, 485–493.

James, W. O. (1950) "Alkaloids in the plant" in *The Alkaloids* (R. H. F. Manske and H. L. Holmes, eds.), Vol. 1, Academic Press, New York, p. 16.

Kazimierski, T., and Nowacki, E. (1971) *Genet. Polon.* **12**, 347.

Keeler, R. F. (1969) *J. Agr. Food Chem.* **17**, 473.

Keeler, R. F. (1975) *Lloydia* **38**, 56–86.

Kendall, W. A., and Sherwood, R. T. (1975) *Agron. J.* **67**, 667.

Kern, H. (1952) *Phytopath. Z.* **19**, 351–382.

Krzymanska, J. (1967) *Biul. Inst. Ochir. Rosl.* **36**, 237–47.

Kuc, J. (1975) *Rec. Adv. Phytochem.* **9**, 139–150.

Kupchan, S. M., and By, A. W. (1968) "Steroid alkaloids—The veratrum group" in *The Alkaloids* (R. H. F. Manske and H. L. Holmes, eds.), Vol. 10, Academic Press, New York, p. 143.

Kupchan, S. M. (1975) *Rec. Adv. Phytochem.* **9**, 167–188.

Lamprecht, H. (1964a) *Agr. Hort. Gen.* **22**, 56–148.

Lamprecht, H. (1964b) *Agr. Hort. Gen.* **22**, 1–55.

Lang, K., and Keuer, H. (1957) *Biochem. Z.* **329**, 277–282.

Lawrence, R. L., Jr., and Waller, G. R. (1973a) *Fed. Proc.* **32**, 521.

Lawrence, R. L., Jr., and Waller, G. R. (1973b) Ninth International Congress of Biochemistry, Stockholm, Sweden, 1-7 Abstract No. 9b 30, p. 403.

Lawrence, R. L., Jr., and Waller, G. R. (1975) unpublished results.

Liebig, J. von (1840) *Die Organische Chemie und ihre Anwendung auf Agricultur und Physiologie,* Brunswick (English trans. by L. Playfair, 1940).

Liebig, J. von (1841) *Ann. Chim. Phys.* **39**, 129.

Luckner, M. (1972) *Secondary Metabolism in Plants and Animals* (T. N. Vasudevan, tr.), Academic Press, New York.

Mabry, T. J., and Difeo, D. R. (1973) "The role of the secondary plant chemistry in the evolution of the Mediterranean scrub vegetation" in *The Mediterranean-Type Ecosystems* (F. di Castri and H. A. Mooney, eds.), Springer-Verlag, New York, p. 121.

Mabry, T. J., and Dreiding, A. S. (1968) *Rec. Adv. Phytochem.* **1**, 145–160.

Mabry, T. J., Kimler, K., and Chang, C. (1972) *Rec. Adv. Phytochem.* **5**, 106–134.

Mackiewicz, H. (1958) *Rocz. Nauk Roln. Ser. A* 79, 103.

Mahler, H. R., and Baylor, M. B. (1967) *Proc. Natl. Acad. Sci.* **58**, 256.

Marten, G. C., Barnes, R. F., Simons, A. B., and Wooding, F. J. (1973) *Agron. J.* **65**, 199–201.

Massingill, J. L., Jr., and Hodgkins, J. E. (1967) *Phytochemistry* **6**, 977.

Mattocks, A. R. (1968a) *J. Chem. Soc. C* **1968**, 235.

Mattocks, A. R. (1968b) *Nature* **217**, 723.

Maysurian, N. A. (1957) *Dokl. Akad. Nauk Arm. SSR* **22**, 2.

McKee, R. K. (1955) *Ann. Appl. Biol.* **43**, 147–148.

McKee, R. K. (1961) *Tagungsber Dtsch. Akad. Landwirtschaftswiss. Berlin* 27, 277–289.

Mitchell, J. C. (1975) *Rec. Adv. Phytochem.* **9**, 119–138.

Mitscher, L. A. (1975) *Rec. Adv. Phytochem.* **9**, 243–282.

Mothes, K. (1953) *Ann. Rev. Plant Physiol.* **6**, 393–432.

Mothes, K. (1960) "Alkaloids in the plant" in *The Alkaloids,* Vol. 6 (R. H. F. Manske and H. L. Holmes, eds.), Academic Press, N.Y., p. 1.

Mothes, K. (1966) *Naturwissenschaften* **53**, 317.

Mothes, K. (1969) *Experientia* **25**, 225.

Mothes, K. (1975) *Rec. Adv. Phytochem.* **10**, 385–401.
Mothes, K., and Romeike, A. (1954) *Flora (Jena) Abt. A* **142**, 109.
Mothes, K., and Schuette, H. R. (1969) *Biosynthese der Alkaloide,* VEB Deutscher Verlag der Wissenschaften, Berlin, pp. 1–39.
Nishikawa, H., and Shiio, I. (1969) *J. Biochem. (Tokyo)* **65**, 523.
Nowacki, E. (1958) *Roczn. Nauk Roln. Ser. A* **79**, 33–42.
Nowacki, E. (1961) *Genet. Polon.* **2**, 103–196.
Nowacki, E., Nowacka, D., and Rudnicka, A. (1975) *Biul. Hod. Rosl.* **5**, 3–5.
Nowacki, E. (1974) *Wszechswiat* **1974**, 169–173.
Nowacki, E., and Kazimierski, T. (1971) *Z. Pflanzenzücht.* **66**, 249–259.
Nowacki, E., and Waller, G. R. (1972) *Abh. Dtsch. Akad. Wiss. Kl., Chem. Geol. Biol. Berlin;* *4. Internationales Symposium Biochemie und Physiologie der Alkaloide,* Halle (Salle) Germany, June, 1969, p. 314.
Nowacki, E., and Waller, G. R. (1973) *Bull. Acad. Polon. Sci., Ser. Sci. Biol.* **21**, 459–463.
Nowacki, E., and Wezyk, S. (1960) *Roczn. Nauk Roln. Ser. B* **75**, 385–399.
Nowacki, E., Junzysta, P., Gorski, D., Nowacka, D., and Waller, G. R. (1976) *Biochem. Physiol. Pflanz.* **169**, 231.
Nowotny-Mieczynska, A., and Zientkiewicz, I. (1955) *Roczn. Nauk Roln. Ser A* **72**, 1.
Olmsted, J. B., and Borisy, G. G. (1973) *Ann. Rev. Biochem.* **42**, 521–522.
Olney, H. O. (1968) *Plant Physiol.* **43**, 293.
Overland, H. (1966) *Am. J. Bot.* **53**, 423.
Piattelli, M., and Minale, L. (1964) *Phytochemistry* **3**, 547.
Pöhm, M. (1966) *Abh. Dtsch. Akad. Wiss. Kl., Chem. Geol. Biol. Berlin; 3. Internationales Symposium Biochemie und Physiologie der Alkaloide,* Halle (Salle) Germany, June 1969, p. 251.
Ramshorn, K. (1955) *Flora (Jena)* **143**, 601.
Raven, P. H. (1973) "The evolution of Mediterranean floras" in *The Mediterranean-Type Ecosystems* (F. di Castri and H. A. Mooney, eds.) Springer-Verlag, New York, p. 213.
Ressler, C. (1975) *Rec. Adv. Phytochem.* **9**, 151–166.
Reznik, H. (1955) *Z. Bot.* **43**, 499.
Robbins, J. P., Wilkinson, S. R., and Burdich, D. (1973) in *Proceedings of the Fescue Toxicity Conference,* Northern Regional Research Laboratory, Agricultural Research Service, USDA, Peoria, Illinois, pp. 98–107.
Robinson, T. (1969) personal communication.
Robinson, T. (1974) *Science* **184**, 430–435.
Sastry, S. D., and Waller, G. R. (1971) *Phytochemistry* **10**, 1961.
Savile, D. B. O. (1954) *Science* **120**, 583.
Schiedt, U., Boech-Behrens, G., and Delluva, A. M. (1962) *Hoppe-Seyl. Z. Physiol. Chem.* **330**, 46.
Schultes, R. E. (1975) *Rec. Adv. Phytochem.* **9**, 1–28.
Sengbusch, R. (1934) *Naturwissenschaften* **22**, 278.
Sherline, P., Leung, J. T., and Kipnis, D. M. (1975) *J. Biol. Chem.* **250**, 5481–5486.
Shibaoka, H. (1974) *Plant Cell Physiol.* **15**, 251–263.
Simons, A. B., and Marten, G. C. (1971) *Agron. J.* **63**, 915.
Skursky, L., and Waller, G. R. (1972) *Abh. Dtsch. Akad. Wiss. Kl., Chem. Geol. Biol. Berlin;* *4. Internationales Symposium Biochemie und Physiologie der Alkaloide,* Halle (Salle) Germany, June, 1969, pp. 181–186.
Smith, B. D. (1966) *Nature* **212**, 213.
Taubenhaus, J. I., and Ezekiel, W. N. (1936) *Texas Agr. Exp. Sta. Bull.* **527**, 1–52.
Tsukamoto, H., Oguri, K., Watabe, T., and Yoshimura, H. (1964) *J. Biochem. (Tokyo)* **55**, 394–400.

Virtanen, A. (1958) *Angew. Chem.* **70**, 544–552.

Virtanen, A., Hietala, P. K., and Wahlroos, O. (1957) *Arch. Biochim. Biophys.* **69**, 486–500.

Wall, M. E. (1975) *Rec. Adv. Phytochem.* **9**, 29.

Waller, G. R., and Nakazawa, K. (1963) *Plant Physiol.* **38**, 318.

Waller, G. R., and Burstrom, H. (1969) *Nature* **222**, 576.

Wegorek, W., and Czaplicki, E. (1966a) *Nachrichtenbl. Dtsch. Pflanzenschutzdienst (Berlin)* **20**, 22–25.

Wegorek, W. and Jasienska - Obrebska, E. (1964) *Biul. Inst. Ochr. Rosl.* **27**, 17–26.

Weinburg, E. D. (1971) *Perspect. Biol.* **14**, 565–577.

Wood, R. K. S. (1967) *Physiological Plant Pathology,* Blackwell Scientific Publications, Oxford, pp. 421–423.

Yamamoto, I. (1965) *Adv. Pest Control Res.* **6**, 231–260.

Zachow, F. (1965) *Züchter* **37**, 35.

Zalewski, K., Blaszczak, W., and Glaser, T. (1959) *Pr. Kom. Mat. Przyr. Poznan Tow. Przyj. Nauk* **10**, 1–7.

References for Chapter 6, Metabolic (Catabolic) Modifications of Alkaloids by Plants

Aldercreutz, H., Tikkamen, M. J., and Humerman, D. H. (1974) *J. Steroid Biochem.* **5**, 211.

Altschul, S. von Reis, *Bot. Mus. Leaflets* (Harvard) (1970) **22 (10)**, 337–343.

Altschul, S. von Reis (1973) *Drugs & Foods from Little Known Plants,* Harvard Univ. Press, Cambridge, Mass.

Alworth, W. L., and Rapoport, H. (1965) *Arch. Biochem. Biophys.* **112**, 45.

Alworth, W. L., Lieberman, L., and Ruckstahl, J. A. (1969) *Phytochemistry* **8**, 1427.

Ames, B. (1975) personal communication.

Antoun, M. D., and Roberts, M. F. (1975a) *Planta Med.* **28**, 6.

Antoun, M. D., and Roberts, M. F. (1975b) *Phytochemistry* **14**, 907.

Antoun, M. D., and Roberts, M. F. (1975c) *Phytochemistry* **14**, 1275.

Arcamone, F., Chain, E. B., Ferretti, A., Minghetti, A., Pennella, P., and Tonolo, A. (1962) *Biochim. Biophys. Acta* **57**, 174.

Axelrod, J., and Reichenthal, J. J. (1953) *J. Pharm. Exp. Ther.* **107**, 519.

Barton, D. H. R., Kirby, G. W., Steglich, W., Thomas, G. M., Battersby, A. R., Dabson, T. A., and Ramsey, H. (1965) *J. Chem. Soc.* **1965**, 2423.

Battersby, A. R., Foulkes, D. M., and Binks, R. (1965) *J. Chem. Soc.* **1965**, 3323.

Baumann, T. W., and Wanner, H. (1972) *Planta* **108**, 11–19.

Beaudin-Dufour, D., and Muller, L. E. (1971) *Turrialba (Costa Rica)* **21(4)**, 387–392.

Bertrand, G. (1905) *C. R. Acad. Sci.* **141**, 209.

Birecka, H., and Zebrowska, J. (1960) *Bull. Acad. Polon. Sci. Ser. Sci. Biol.* **8**, 339; *Chem. Abstr.* **55**, 697h.

Boecklin, G. E. (1975) *Tea and Coffee Trade J.* **147**, 32–33.

Brady, L. R., and Tyler, V. E., Jr. (1958) *Plant Physiol.* **33**, 334.

Bu'Lock, J. D. (1965) *The Biosynthesis of Natural Products,* McGraw-Hill Book Co., New York, pp. 81–82.

Burg, A. W. (1976) *Drug Metab. Rev.* **4**, 199–228.

Burlingame, A. L., Cox, R. E., and Derrick, P. (1974) *Anal. Chem.* **46**, 248.

Chandler, J. L. R., and Gholson, R. K. (1972) *Phytochemistry* **11**, 239–242.

Chevalier, A. (1937) *Rev. Bot. Appl. Agric. Tropicale* **17**, 821.

Clautriau, G. (1892) *Bull. Soc. Belge. Microsc.* **18**, 80 cited in James (1950).

Coscia, C. J. (1976) personal communication.

Crafts, A. S. (1961) *Translocation in Plants,* Holt, Rinehart, and Winston, New York.

Cromwell, B. T. (1933) *Biochem. J.* **27**, 860.

Daddona, P. E., and Hutchinson, C. R. (1976) *Phytochemistry* **15**, 941–945.

Dawber, R. T. (1976) *Z. Ernährungwiss.* **15**, 52–53.

Dawber, T., Kannel, W., and Gordon, T. (1974) *New England J. Med.* **291**, 871–874.

Dawson, R. F. (1948) *Adv. Enzymol.* **8**, 203.

Dawson, R. F. (1951) *J. Am. Chem. Soc.,* **73**, 4218.

Dawson, R. F. (1952) *Am. J. Bot.,* **39**, 250.

Dawson, R. F., and Osdene, T. J. (1972) *Rec. Adv. Phytochem.* **5**, 317–338.

Dawson, R. F., Christman, D. R., D'Adamo, A., Solt, M. L., and Wolf, A. D. (1960) *J. Am. Chem. Soc.* **82**, 2628.

Dempewolff, R. F. (1975) *Sci. Dig.* **77**, 31–36.

Dickenson, P. B., and Fairbairn, J. W. (1975) *Ann. Bot. (London)* **39**, 707–712.

Dietrich, S. M. C., and Martin, R. O. (1969) *Biochemistry* **8**, 4163.

Digenis, G. A. (1969) *J. Pharm. Sci.* **58**, 39.

Digenis, G. A., Faraj, B. A., and Abou-Charr, C. I. (1966) *Biochem. J.* **101**, 27c.

Ebbighouser, O. R., Mowat, J. M., Stearns, H., and Vestergaard, P. (1975) *Biochem. Mass Spectrom.* **1**, 305.

Eberspaecher, F., Uesseler, H., and Lingens, F. (1970) *Hoppe-Seyl. Z. Physiol. Chem.* **351**, 1465.

Egri, L. (1957) *Fachliche Mitt. Oesterr. Tabakreg.* **19** (CA **51**, 18490) (cited by Stepka and Dewey, 1963)

El-Hamidi, A., and Wanner, H. (1964) *Planta* **61**, 90–96.

Fairbairn, J. W. (1965) in *Beiträge zur Biochemie und Physiologie von Naturstoffen: Festschrift: Kurt Mothes zum 65 Geburtstag,* Jena, VEB Gustav Fischer Verlag, p. 113.

Fairbairn, J. W., and Ali. A. A. E. R. (1968a) *Phytochemistry* **7**, 1593.

Fairbairn, J. W., and Ali, A. A. E. R. (1968b) *Phytochemistry* **7**, 1599.

Fairbairn, J. W., and Challen, S. B. (1959) *Biochem. J.* **72**, 556.

Fairbairn, J. W., and El-Masry, S. (1967) *Phytochemistry* **6**, 499.

Fairbairn, J. W., and El-Masry, S. (1968) *Phytochemistry* **7**, 181.

Fairbairn, J. W., and Paterson, A. (1966) *Nature* **210**, 1163.

Fairbairn, J. W., and Suwal, P. N. (1961) *Phytochemistry* **1**, 38.

Fairbairn, J. W., and Wassel, G. (1964) *Phytochemistry* **3**, 253.

Fairbairn, J. W., and Wassel, G. (1965) *Phytochemistry* **3**, 583.

Fairbairn, J. W., and Wassel, G. M. (1967) *J. Chem. U. A. R.* **10**, 275.

Fairbairn, J. W., Paterson, A., and Wassel, G. (1964) *Phytochemistry* **3**, 577.

Fairbairn, J. W., Djote, M., and Paterson, A. (1968a) *Phytochemistry* **7**, 2111.

Fairbairn, J. W., Palmer, J. M., and Paterson, A. (1968b) *Phytochemistry* **7**, 2117.

Fairbairn, J. W., Hakim, F., and Dickenson, P. B. (1973) *J. Pharm. Pharmac.* **25**, 113P suppl.

Fairbairn, J. W., Hakim, F., and El Kheir, Y. (1974) *Phytochemistry* **13**, 1133–1139.

Fairbairn, J. W., and Steele, M. (1977) personal communication.

Floss, H. G., Robbers, J. E., and Heinstein, K. F. (1974) *Rec. Adv. Phytochem.* **8**, 141–178.

Fodor, G. (1976) *Spec. Period. Rep.* (Chem. Soc. of London) **6**, 65.

Frank, A. W., and Marion, L. (1956) *Can. J. Chem.* **34**, 1641.

Friedman, G. D., Siegelaub, A. B., and Seltzer, C. C. (1974) *New Engl. J. Med.* **290**, 469–473.

Frost, G. M., Yang, K. S., and Waller, G. R. (1967) *J. Biol. Chem.* **242**, 887.

Gentry, C. E. (1968) Interrelationship of *Rhizoctonia solani* Kuehn, Environment, and Genotype on the Alkaloid Content of Tall Fescue, *Festuca arundinacea* Schrebs, Ph.D. Thesis, Univ. of Kentucky.

Gholson, R. K. (1966) *Nature* **212**, 933.

Gholson, R. K., and Kori, J. (1964) *J. Biol. Chem.* **239**, 2399.

Gholson, R. K., Ueda, I., Ogasawara, N., and Henderson, L. M. (1964) *J. Biol. Chem.* **239**, 1208.

Gholson, R. K., Sakakibara, S. and Wicks, F. D. (1976) American Society for Microbiology 76th Annual Meeting, Atlantic City, N.J., May 2–7, Session 72.

Griffith, G. D., and Griffith, T. (1964) *Plant Physiol.* **39**, 970.

Griffith, G. D., Griffith, T., and Byerrum, R. (1960) *J. Biol. Chem.* **235**, 3536.

Gross, D., Lehmann, H., and Schuette, H. R. (1970) *Z. Pflanzenphysiol.* **63**, 1.

Gruetzmann, K. D. and Schroeter, H. B. (1966) in *Abh. Dtsch. Akad. Wiss., Kl., Chem. Geol. Biol. Berlin; 3. Internationales Symposium Biochemie und Physiologie der Alkaloide,* Halle (Saale) Germany, June, 1965, p. 347.

Heftmann, E., and Schwimmer, S. (1972) *Phytochemistry* **11**, 2783.

Heinstein, P. F., Lee, S. L., and Floss, H. G. (1971) *Biochem. Biophys. Res. Commun.* **44**, 1244.

Hennekins, C. H., Drolette, M. E., Jessee, M. J., Daves, J. E., and Hutchinson, G. B. (1976) *New Engl. J. Med.* **294**, 633–636.

Herndlhofer, E. (1933), *Boletim de Agricultura (Brasil)* **34**, 163–251.

Hiles, R. A., and Byerrum, R. U. (1969) *Phytochemistry* **8**, 1927.

Ill'in, G. S. (1965) *Pharmazie* **20**, 724.

Ill'in, G. S. (1966) in *Abh. Dtsch. Akad. Wiss., Kl., Chem. Geol. Biol., Berlin; 3. Internationales Symposium Biochemie und Physiologie der Alkaloide,* June, 1965, Halle (Saale) Germany, p. 347.

Ill'in, G. S., and Lovkova, M. Ja. (1972) in *Abh. Dtsch. Akad. Wiss., Kl., Chem. Geol. Biol., Berlin; 4. Internationales Symposium Biochemie und Physiologie der Alkaloide,* June, 1965, Halle (Saale) Germany, pp. 207–208.

Ill'in, G. S., and Lovkova, M. Ja. (1966) *Biokhimyia* **31**, 174.

Ill'in, G. S., and Serebrovskaya, K. B. (1958) *Dokl. Akad. Nauk SSSR* **118**, 139; *Chem. Abstr.* **52**, 8293h.

Ill'in, G. S., Lovkova, M. Ya., and Klimentjeva, N. I. (1968) *Izv. Akad. Nauk. SSSR Ser. Biol.* **1968**, 871.

Imsande, J. (1961) *J. Biol. Chem.* **236**, 1494.

James, W. O. (1950) "Alkaloids in the plant" in *The Alkaloids* (R. H. F. Manske and H. L. Holmes, eds.), Vol. I, Academic Press, New York, p. 15.

James, W. O. (1953) *Endeavour* **12**, 76.

Jick, H., Miettinen, O. S., Neff, R. K., Shapiro, S., Heinonen, O. P., and Slone, P. (1973) *New Engl. J. Med.* **289**, 63–67.

Jindra, A., Cihak, A., and Kovacs, P. (1964) *Collect. Czech. Chem. Commun.* **29**, 1059.

Johnson, R. D., and Waller, G. R. (1974) *Phytochemistry* **13**, 1493–1500.

Joshi, J. G., and Handler, P. (1960) *J. Biol. Chem.* **235**, 2981.

Joshi, J. G., and Handler, P. (1962) *J. Biol. Chem.* **237**, 929, 3185.

Kahn, Y., and Blum, J. J. (1968) *J. Biol. Chem.* **243**, 1448.

Kalberer, P. (1964) *Ber. Schweiz. Bot. Ges.* **74**, 62.

Kalberer, P. (1965) *Nature* **205**, 597.

Kannel, W. B., and Dawber, T. R. (1973) *New Engl. J. Med.* **289**, 100–101.

Kaplan, N. O. (1961) in *Metabolic Pathways* (D. M. Greenberg, ed.), Academic Press, New York, Vol. 2, p. 627.

Keller, J., Liersch, M., and Grunicke, D. (1971) *Eur. J. Biochem.* **22**, 263.

Keller, H., Wanner, H., and Baumann, T. W. (1972) *Planta* **108**, 339–350.

Kirby, G. W., Mossey, S. R., and Steinreich, D. (1972) *J. Chem. Soc., Perkin Trans. I,* 1642.

Kisaki, T., and Tamaki, E. (1961a) *Arch. Biochem. Biophys.* **92**, 351.

Kisaki, T., and Tamaki, E. (1961b) *Arch. Biochem. Biophys.* **94**, 252.

Kisaki, T., and Tamaki, E. (1964a) *Nippon Nogei Kagaku Kaishi* **38**, 549; *Chem. Abstr.* **63**, 7059b.

Kisaki, T., and Tamaki, E. (1964b) *Nippon Nogei Kagaku Kaishi* **38**, 392; *Chem. Abstr.* **63**, 3325h.

Kisaki, T., and Tamaki, E. (1964c) *Agr. Biol. Chem.* **28**, 492.

Kisaki, T., and Tamaki, E. (1966) *Phytochemistry* **11**, 965–973.

Klatsky, A. L., Friedman, G. D., and Siegelaub, A. B., (1973) *J. Am. Med. Assoc.* **226**, 540.

Lee, H. J., and Waller, G. R. (1972a) *Phytochemistry* **11**, 965–973.

Lee, H. J., and Waller, G. R. (1972b) *Phytochemistry* **11**, 2233–2240.

Lee, M. G., and Millard, B. J. (1975) *Biochem. Mass Spectrom.* **2**, 78.

Leete, E., (1968) *Tetrahedron Lett.* **1968**, 4433.

Leete, E. (1969) *Adv. Enzymol.* **32**, 373.

Leete, E. (1973) in First Philip Morris Science Symposium Proceedings, p. 92.

Leete, E. (1974) *Phytochemistry* **14**, 471–474.

Leete, E., and Bell, V. M. (1959) *J. Am. Chem. Soc.* **81**, 4358.

Leete, E., and Chedekel, M. R. (1974) *Phytochemistry* **14**, 1853–1859.

Lingens, F. (1972) in *Abh. Dtsch. Akad. Wiss. Kl., Chem. Geol. Biol., Berlin; 4. Internationales Symposium Biochemie und Physiologie der Alkaloide,* June, 1969, Halle (Saale) Germany, pp. 65–73.

Lingens, F., Goebel, W., and Uesseler, H. (1967) *Eur. J. Biochem.* **2**, 442.

Looser, E., Baumann, T. W., and Wanner, H. (1974) *Phytochemistry* **13**, 2515–2518.

Lovkova, M. Ya. (1964) *Acta Biol. Acad. Sci. Hung.* **14**, 273.

Lovkova, M. Ya., Ill'in, G. S., and Minozhedinova, N. S. (1973) *Prikl. Biokhim. Mikrobiol.* **9**, 595–598; *Chem. Abstr.* **79**, 133282.

Luckner, M. (1972) *Secondary Metabolism in Plants and Animals* (T. N. Vasudevan, tr.), Academic Press, New York, p. 212.

Madyastha, K. M., Guarnaccia, R., Baxter, C., and Coscia, C. J. (1973) *J. Biol. Chem.* **248**, 2497.

Madyastha, K. M., Meehan, T. D., and Coscia, C. J. (1976) *Biochemistry* **15**, 1097.

Madyastha, K. M., Ridgway, J. E., Dwyer, J. G., and Coscia, C. J. (1977) *Abs. 1st Int. Congress on Cell Biology, Sept. 5–10,* Boston, Mass., published in *J. Cell. Biol.* **316**, 272.

Mann, D. F., and Byerrum, R. V. (1974a) *J. Biol. Chem.* **249**, 6817–6823.

Mann, D. F., and Byerrum, R. V. (1974b) *Plant Physiol.* **53**, 603–609.

McCann, J., Choi, E., Yamasaki, S. D., and Ames, B. N. (1975) *Proc. Natl. Acad. Sci.* **72**, 5135–5190 and references cited.

McFarlane, J., Madyastha, K. M., and Coscia, C. J. (1975) *Biochem. Biophys. Res. Commun.* **66**, 1263.

Meehan, T. D., and Coscia, C. J. (1973) *Biochem. Biophys. Res. Commun.* **53**, 1043.

Melo, M., Carvalho, A., and Monaco, L. C. (1975) in Congresso da Sociedade Brasileiro paro o Progresso da Ciência, 27°, Reunião, Secão 281.

Merker, J., and Pyriki, C. (1966) in *Abh. Dtsch. Akad. Wiss. Kl., Chem. Geol. Biol., Berlin; 3. Internationales Symposium Biochemie und Physiologie der Alkaloide,* June, 1965, Halle (Saale) Germany, p. 173.

Miller, K. J., Jolles, C., and Rapoport, H. (1973) *Phytochemistry* **12**, 597.

Mizusaki, S., Kisaki, T., and Tamaki, E. (1968) *Plant Physiol.* **43**, 93.

Mizusaki, S., Tanabe, Y., Noguchi, M., and Tamaki, E. (1971) *Plant Cell Physiol.* **12**, 633.

Mizusaki, S., Tanabe, Y., Noguchi, M., and Tamaki, E. (1972) *Phytochemistry* **11**, 2757.

Mizusaki, S., Tanabe, Y., Noguchi, M., and Tamaki, E. (1973) *Plant Cell Physiol.* **14**, 103–110.

Mothes, K. (1928) *Planta* **5**, 563.

Mothes, K. (1965) *Naturwissenschaften* **52**, 571.

Mothes, K. (1966) in *Abh. Dtsch. Akad. Wiss. Kl., Chem. Geol. Biol., Berlin; 3. Internationales Symposium Biochemie und Physiologie der Alkaloide,* June, 1965, Halle (Saale) Germany, p. 27.

Mothes, K. (1975) *Rec. Adv. Phytochem.* **10**, 385–405.

Mothes, K., and Schuette, H. R. (1969) *Biosynthese der Alkaloide,* VEB Deutscher Verlag der Wissenschaften, Berlin.

Mothes, K., Engelbrecht, L., Tschoepe, K. H., and Hutschenreuter-Trefftz, G. (1957) *Flora (Jena)* **144**, 518.

Mothes, K.. Richter, I., Stolle, D., and Groeger, D. (1965) *Naturwissenschaften* **52**, 431.

Mudzhiri, K. S. (1955) *Sbornik Trudov Tbilisi Nauch.-Issledovatel. Khim. Farm. Inst.* **7**, 19–38 (*Chem. Abstr.* **54**, 13286).

Neumann, D., and Tschoepe, K. H. (1966) *Flora (Jena) Abt. A* **156**, 521.

Newell, J., and Waller, G. R., *Proc. Okla. Acad. Sci.* **45**, 143 (1965).

Nowacki, E. K., and Waller, G. R. (1973) *Z. Pflanzenphysiol.* **69**, 228–291.

Nowacki, E. K., and Waller, G. R. (1975a) *Phytochemistry* **14**, 161–164.

Nowacki, E. K., and Waller, G. R. (1975b) *Phytochemistry* **14**, 165–171.

O'Donovan, D., and Leete, E. (1963) *J. Am. Chem. Soc.* **85**, 461.

Ogutuga, D. B. A., and Northcote, D. H. (1970) *Biochem. J.* **117**, 715–720.

Petrack, B., Greengard, P., Craston, A., and Kalinsky, H. J. (1963) *Biochem. Biophys. Res. Commun.* **13**, 472.

Phillipson, D. J., Handa, S. S., and El-Dabbas, S. W. (1976) *Phytochemistry* **15**, 1297–1301.

Pöhm, M. (1966) in *Abh. Dtsch. Akad. Wiss. Kl., Chem. Geol. Biol., Berlin; 3. Internationales Symposium Biochemie und Physiologie der Alkaloide,* June, 1965, Halle (Salle) Germany. p. 251.

Preiss, J., and Handler, P. (1958) *J. Biol. Chem.* **233**, 493.

Ramstad, E., and Agurell, S. (1964) *Ann. Rev. Plant Physiol.* **15**, 143.

Rapoport, H. (1966) *Abh. Dtsch. Akad. Wiss. Kl.. Chem. Geol. Biol.. Berlin; 3. Internationales Symposium Biochemie und Physiologie der Alkaloides,* June, 1965, Halle (Saale) Germany, p. 111.

Ridgway, J. E., and Coscia, C. J. (1976) personal communication.

Robbers, J. E., and Floss, H. G. (1970) *J. Pharm. Sci.* **59**, 702.

Roberts, M. F. (1971) *Phytochemistry* **10**, 3021.

Roberts, M. F. (1974) *Phytochemistry* **13**, 119.

Roberts, M. F. (1975) *Phytochemistry* **14**, 2393.

Robinson, T. (1971) personal communication.

Robinson, T. (1974) *Science* **184**, 430–435.

Roddick, J. G. (1976) *Phytochemistry* **15**, 475.

Romeike, A. (1962) *Naturwissenschaften* **18**, 426.

Romeike, A. (1964) *Flora (Jena)* **154**, 163.

Ryrie, I. J., and Scott, K. J. (1969) *Biochem. J.* **115**, 679.

Sadykov, A. S., Aslanov, Kh. A., and Kushmuradov, Yu. K. (1975) *Quinolizidine Alkaloids,* Science Publishing House, Moscow.

Sander, H. (1956) *Planta* **47**, 374.

Sanders, H. J. (1969) *Chem. Eng. News* **47** (**21**) (May 19) 54–68.

Sárkány, S. (1962) *Magy. Tud. Akad. Biol.* **5**, 93–107.

Sárkány, S., Sárkány-Kiss, J., and Varzar-Petri, G. (1966) in *Abh. Dtsch. Akad. Wiss. Kl., Chem. Geol. Biol., Berlin; 3. Internationales Symposium Biochemie und Physiologie der Alkaloide,* June, 1965, Halle (Saale) Germany, pp. 355–361.

Sárkány, S., and Dános, B. (1958) *Acta Bot. Acad. Sci. Hung.* **3**, 293; *Chem. Abstr.* **52**, 1030b.

Sastry, S. D. (1972) in *Biochemical Applications of Mass Spectrometry* (George R. Waller, ed.), Wiley-Interscience, New York, pp. 665–709.

Sastry, S. D., and Waller, G. R. (1972) *Phytochemistry* **11**, 2241–2745.

Schroeter, H. B. (1958) in *2nd International Science Tobacco Congress* (Brussels), p. 426; *Biol. Abstr.* (1960) **35**, 35736.

Schroeter, H. B. (1966) in *Abh. Dtsch. Akad. Wiss. Kl., Chem. Geol. Biol., Berlin; 3. Internationales Symposium Biochemie und Physiologie der Alkaloide,* June, 1965, Halle (Saale) Germany, p. 157.

Schuette, H. R. (1969) in *Biochemie der Alkaloide* (K. Mothes and H. R. Schuette, eds.), VEB Deutscher Verlag der Wissenschaften, Berlin, p'. 240.

Schultes, R. E. (1975) *Rec. Adv. Phytochem.* **9**, 1.

Schultes, R. E. (1976) *Philip Morris Science Symposium, 2nd*, pp. 133–171.

Schumauder, H. P., and Groeger, D. (1973) *Biochem. Physiol. Pflanz.* **164**, 41.

Scott, A. I. (1970) *Acc. Chem. Res.* **3**, 151.

Scott, A. I. (1974) *Science* **184**, 760–764.

Scott, A. I. (1976) *Philip Morris Science Symposium, 2nd*, pp. 101–132.

Scott, A. I., Reichardt, P. B., Slayton, M. B., and Sweeny, J. G. (1962) *Bio-organic Chem.* **1**, 157.

Scott, A. I., Reichardt, P. B., Slayton, M. B., and Sweeny, J. G. (1973) *Rec. Adv. Phytochem.* **6**, 117.

Skursky, L., and Waller, G. R., (1972) in *Abh. Dtsch. Akad. Wiss. Kl., Chem. Geol. Biol., Berlin; 4. Internationales Symposium Biochemie und Physiologie der Alkaloide*, June, 1969, Halle (Saale) Germany, pp. 181–186.

Skursky, L.. Burleson, D.. and Waller, G. R. (1969) *J. Biol. Chem.* **244**, 3238.

Solt, M. L., Dawson, R. F., and Christman, D. R. (1960) *Plant Physiol.* **35**, 886.

Spencer, R. L., and Preiss, J. (1967) *J. Biol. Chem.* **242**, 385.

Stagg, F. V., and Millin, D. J. (1975) *J. Sci. Food Agric.* **26**, 1439–1459.

Steinegger, E., and Bernasconi, R. (1964) *Pharm. Acta Helv.* **39**, 480.

Stepka, W., and Dewey, L. J. (1961) *Plant Physiol.* **36**, 592.

Stepka, W., and Dewey, L. J. (1963) *Plant Physiol.* **38**, 283.

Stermitz, F. R., and Rapoport, H. (1961) *J. Am. Chem. Soc.* **83**, 4045.

Suzuki, T. and Takahashi, E. (1976) *Phytochemistry* **15**, 1235–1239.

Sylvain, P. G. (1976) personal communication.

Trigg, C. W. (1922) "The chemistry of the coffee bean" in *All About Coffee* (W. H. Ukers, ed.), The Tea and Coffee Trade Journal Co., p. 155.

Tso, T. C., and Jeffrey, R. N. (1957) *Plant Physiol.* **32**, 86.

Tso, T. C., and Jeffrey, R. N. (1959) *Arch. Biochem. Biophys.* **80**, 46.

Tso, T. C., and Jeffrey, R. N. (1961) *Arch. Biochem. Biophys.* **92**, 253.

United States Department of Health, Education and Welfare, National Heart Institute (1966) *The Framingham Heart Study: Habits and Coronary Heart Disease* (PHS Publication No. 1515), US GPO, Washington, D. C.

Waller, G. R. (1969) *Progr. Chem. Fats Other Lipids* **10**, 151–238.

Waller, G. R., and Lee, J. L-C. (1969) *Plant Physiol.* **44**, 522.

Waller, G. R., and Skursky, L. (1972) *Plant Physiol.* **50**, 622.

Waller, G. R., Tang, M.S-I., Scott, M. R., Goldberg, F. J., Mayes, J. S., and Auda, H. (1965) *Plant Physiol.* **40**, 803.

Waller, G. R., Yang, K. S., Gholson, R. K., Hadwiger, L. A., and Chaykin, S. (1966) *J. Biol. Chem.* **241**, 4411–4418.

Waller, G. R., Yang, K. S., Sastry, S. D., Frost, G. F., and Johnson, R. D. (1975) unpublished results.

Wanner, H. (1963) in *Abh. Dtsch. Akad. Wiss. Kl., Chem. Geol. Biol., Berlin; 2. Internationales Symposium Biochemie und Physiologie der Alkaloide*, June, 1959, Halle (Saale) Germany, pp. 65–79.

Wanner, H., and Kalberer, P. (1966) in *Abh. Dtsch. Akad. Wiss. Kl., Chem. Geol. Biol. Berlin; 2. Internationales Symposium Biochemie und Physiologie der Alkaloide*, June, 1965, Halle (Saale) Germany, p. 607.

Wanner, H., Pešáková, N., Baumann, T. W., Charubala, R., Guggisberg, A., Hesse, M., and Schmid, H. (1975) *Phytochemistry* **14**, 747–750.

Weevers, Th. (1932) *Recl. Trav. Bot. Neerl.* **30**, 336.

Wenkert, E. (1954) *Experientia* **10**, 346.

Weybrew, J. A. (1975) personal communication.

Weybrew, J. A., Mann, T. J., and Moore, E. L. (1960) *Tob. Int.`NY* **151**, 15–24.

Weybrew, J. A., and Mann, T. J. (1963) *Tob. Int. NY.* **156**, 8–30.

Weygand, F., and Floss, H. G. (1963) *Angew. Chem. Int. Ed. (Engl.)* **2**, 243.

Willuhn, G. (1966) in *Abh. Dtsch. Akad. Wiss. Kl., Chem. Geol. Biol., Berlin; 3. Internationales Symposium Biochemie und Physiologie der Alkaloide,* June, 1965, Halle (Saale) Germany, pp. 97–103.

Yang, K. S., Gholson, R. K., and Waller, G. R. (1965) *J. Am. Chem. Soc.* **87**, 4184.

Yoshida, D. (1962) *Plant Cell Physiol.* **3**, 391.

Yoshida, D. (1971) *Hatano Tabako Shikinjo Jokoko* 69, 63–68: *Chem. Abstr.* (1973) *79,* 2873.

Zielke, H. R., Byerrum, R. U., O'Neal, R. M., Burns, L. C., and Koeppe, R. E. (1968) *J. Biol. Chem.* **243**, 4757.

Zsadon, B., and Kaposi, P. (1970) *Tetrahedron Lett.* **1970,** 4615.

Zsadon, B., and Kaposi, P. (1972) *Acta Chim. Acad. Sci. Hung.* **71**, 115.

Zsadon, B., and Oha, K. H. (1971) *Acta Chim. Acad. Sci. Hung.* **69**, 87.

Zsadon, B., Oha, K. H., and Tétényi, P. (1975) *Acta Chim. Acad. Sci. Hung.* **84**, 71.

Author Index

Subject Index

The letter "s" following a page number indicates that the structural formula is shown.